Vegetable Cultivation and Health Benefits

Vegetable Cultivation and Health Benefits

Logan Evans
Julia Elliot

Kruger Brentt
Publishers
2026

Kruger Brentt Publishers UK. LTD.
Company Number 9728962

Regd. Office: 68 St Margarets Road, Edgware, Middlesex HA8 9UU

ISBN: 978-1-78715-630-2

For information on all our publications visit our website at http://krugerbrentt.com/

PREFACE

Vegetable Cultivation and Health Benefits offers a comprehensive exploration of both the science of growing vegetables and the nutritional advantages they provide for human health. Vegetables are a vital component of a balanced diet, supplying essential vitamins, minerals, fiber, and bioactive compounds that support wellness and prevent chronic diseases. This book is designed for students, researchers, farmers, nutritionists, and health professionals, providing a holistic view of vegetable production, cultivation techniques, and their role in promoting health.

The book begins with an introduction to vegetable cultivation, including soil preparation, climate considerations, seed selection, planting techniques, irrigation, fertilization, pest management, and post-harvest handling. Emphasis is placed on sustainable and organic practices that enhance yield, quality, and environmental stewardship.

A major focus is placed on the nutritional composition and health-promoting properties of various vegetables. Topics include vitamins, minerals, antioxidants, phytochemicals, and their roles in disease prevention, immune support, and overall well-being. The book also examines ways to maximize nutritional content through cultivation practices, post-harvest handling, and food preparation.

Case studies, practical examples, and illustrative tables highlight successful vegetable cultivation practices and their impact on health outcomes. Emerging trends, such as functional vegetables, biofortification, and urban gardening, are also discussed to provide a forward-looking perspective on vegetable production and consumption.

We hope that Vegetable Cultivation and Health Benefits will serve as a valuable resource, combining scientific cultivation methods with insights into nutrition, enabling readers to produce and consume vegetables in ways that maximize both yield and health benefits.

Logan Evans

Julia Elliot

CONTENTS

Chapter 1 NUTRITIONAL COMPOSITION OF VEGETABLES

1. INTRODUCTION: AN OVERVIEW

The term vegetable originates from the Medieval Latin vegetabilis, meaning "growing" or "flourishing" (pertaining to a plant), which evolved from a Late Latin definition of "to be enlivening" or "quickening." The contemporary meaning of "vegetable" denotes a "plant cultivated for consumption." [1]

Vegetables are defined as the portions of plants that provide sustenance for humans and other animals. It is commonly seen as consumable plant material, encompassing flowers, fruits, stems, leaves, roots, and seeds.

Hunter-gatherers foraged for wild vegetables. This progressed to the cultivation of vegetables in various regions globally between 10,000 BC and 7000 BC, subsequent to the emergence of a novel agricultural lifestyle. Primarily, locally grown plants would have been cultivated. Over time, numerous exotic crops from other places were introduced to the local market. Currently, the cultivation of the majority of vegetables globally is significantly reliant on climatic conditions, and crops may be grown to safeguard the environment in less suitable regions. The extent of vegetable production fluctuates based on its intended purpose; subsistence farmers cater to their families' food requirements, while others serve agro-allied enterprises.

Furthermore, vegetables are consumable both raw and cooked, playing a crucial part in human nutrition due to their low fat and carbohydrate content, while being rich in vitamins, minerals, and dietary fiber. Nutrition experts recommended that individuals increase their intake of fruits and vegetables, suggesting five or more servings daily. Vegetables are generally abundant in minerals, particularly calcium and iron, as well as vitamins A and C. Furthermore, nearly all varieties of vegetables serve as sufficient supplies of antioxidants and dietary fiber. [2].

1.1 Categories of vegetables

There are various descriptions on categories of vegetables namely:

a. Fruits of some other plants fall into the vegetable group. Such as cucumber, okra, pepper, tomatoes, sweet corn, eggplant, squash.

b. Leafy vegetables such as brussels sprouts, waterleaf, cabbage, spinach, pumpkin leaves, celery, jute plant, cassava leaves are common examples in this group.
c. Root vegetables: these are plants whose underground parts are used as vegetables. Examples include beets, carrot, radish, sweet potato, turnips.
d. Seed vegetables: the seed vegetables are actually the legumes and pods such as beans and peas.
e. Stem vegetables: these are plants where the shoot parts are considered.

Moreso, the **Table 1** below indicate some examples of vegetable, description, and nutritional value.

Table 1. Some examples of vegetable, description, and nutritional value.

Cultivars type	Description	Nutritional value
Abelmoschus esculentus (okra, ladies' fingers)	An edible green seed pod of the mallow family.	About 90% water in the raw stage. Also contains thiamine, magnesium, folate, iron, copper, niacin, phosphorus, dietary fiber, vitamins C and K.
Allium ampeloprasum L. (leek, elephant garlic)	Leeks are a tasty vegetable that adds a lot of flavor when used in cooking. They are long, with greenish-white stalks reminiscent of green onions, another member of the allium family.	Leek contains carbohydrate, fiber, sugar, fat, protein, vitamin K1, vitamin A, manganese, vitamin C, and folate.
Allium cepa (onion, spring, scallion, shallot)	These are fleshy, hollow cylindrical biennial bulb.	Contains carbohydrate, fiber and some amount of vitamins like thiamine, vitamin C, B6, niacin, riboflavin, and minerals like Ca, Fe, Zn, K, Mn.
Allium sativum (garlic)	A perennial flowering bulb.	Carbohydrate, fiber, protein, manganese vitamin B6, vitamin C, selenium, and calcium.
Apium graveolens (celery)	A biennial plant with pinnate leaves. Consumption assist in weight lose.	Has about 95% water, a very good source of vitamin K with Zn, Na, K, P, Mg Fe, and Cu.
Asparagus officinalis (asparagus)	A tall scaly leaves plant with stout stems and feathery foliage. It maybe green, white, or yellow in coloration.	Carbohydrate, fiber, sugar, fat, protein, vitamin K1, A, B1, folate, iron.

Beta vulgaris (beetroot, sea beet, Swiss chard, sugar beet)	Nitrate helps to lower blood pressure.	Contains a lot of nitrates, carbohydrate, fiber, sugar, fat, protein, folate, manganese, potassium, vitamin C, magnesium.

Brassica oleracea (kale)	A member of the cruciferous family of vegetables, and it shares some similarities to cabbage and broccoli. Kale has a shout for being one of the healthiest vegetables.	It provides an enormous amount of vitamin K1, beta-carotene, and vitamin C. Also consist of carbohydrate, fiber, fat, protein, manganese, and calcium.
Brassica oleracea var. *capitata* (cabbage)	This is a leafy green, red-purple, or pale green vegetable crop.	This has about 96% water, low carbohydrate, fiber, negligible fat, protein. Rich in vitamins K1, C, and B_6, folate, manganese.
Brassica oleracea var. *gemmifera* (Brussels Sprouts)	Resembles miniature cabbage. The edible part is the bud and the leaves.	Contains about 85% water with low carbohydrate, fiber, sugar, fat, and protein with vitamins K, C, and B_6, manganese, folate.
Brassica oleracea var. *italica* (broccoli)	From the cabbage family. The flowering head, leaves, and stalk serve as a vegetable. Particularly helpful in preventing cancer.	Contains lots of calories. A rich source of vitamins C and K. Also contains carbohydrate, fiber, sugar, fat, protein, folate, manganese.
Brassica rapa var. *rapa* (turnip, Chinese cabbage, napa cabbage, bokchoy)	Root vegetable cultivated for its fleshy taproot.	Little quantity of carbohydrates and protein. Contains 93% water and vitamins K, A, and C. The boiled leaves contain substantial lutein.
Capsicum annuum (pepper, bell pepper, sweet pepper)	This belongs to the nightingale family and is cultivated for the thick fruit.	Rich in vitamins C, carbohydrate, fiber, sugar, fat, protein, vitamin B6, folate, potassium, manganese.
Celosia argentea var. *argenta* (Lagos spinach, quail grass, soko, celosia or feather cockscomb)	A broadleaf annual of the Amaranth family.	This is a great source of vitamins A and C, also good in protein, iron, and calcium.
Cichorium endivia endive (chicory)	A leafy green vegetable that has a bitter, yet slightly sweet taste. It belongs to the same family of vegetables as radicchio and curly endive.	High in calories, carbohydrates, fiber. Good source of protein, vitamin K, vitamin A, folate, manganese, vitamin C.
Corchorus olitorius (Jew's mallow, bush okra, nalta jute, jute plant)	The jute plant is used as fiber while the leaves and young fruits serve as a vegetable.	A rich source of vitamin A and C, fiber, zinc, and thiamine.

Cucumis sativus (cucumber)	A creeping vine that bears cylindrical fruits used as vegetables.	Contains 95% of water with carbohydrates and proteins, also minerals such as Ca, Fe, Mg, Mn, K, P, Na, and Zn, with loads of vitamins.
Cucurbita pepo (pumpkin. squash, marrow, zucchini)	The plump nutritious orange vegetable.	Rich in vitamin C, potassium, fiber. Super rich in provitamin A beta-carotene and vitamin A.
Cynara cardunculus var. scolymus (artichoke)	A kind of thistle or prickly flower plant with a cone shape. The edible part is the flower bud before the flower blooms.	Rich source of fiber and reasonably high in vitamins C and K1. Also contains carbohydrate, sugar, fat, protein, folate, magnesium, iron and manganese.
Daucus carota subsp. sativus (carrot)	Root vegetables of typical orange color. It's a biennial plant.	Carbohydrate, fiber, sugar, fat, protein, vitamin A, K1, C, and B6, with diverse minerals.
Dioscorea spp. (yam)	Tuberous starchy root. There are a variety of forms.	Dietary values include K, Mn, vitamins B6 and C, thiamine, fiber.
Eruca vesicaria ssp. sativa (arugula)	An annual rosette basal leafy plant that grows close to the ground.	Rich source of fibers and phytochemicals but low in calories, sugar, carbohydrates, and fat. Also contains a lot of vitamins such as vitamins A, C K, folate, potassium, calcium,
Gnetum africanum (eru, African jointfir)	A perennial wild vine. Belonging to the gymnosperm.	A good source of protein as well as essential and non-essential amino acids.
Ipomoea batatas (sweet potato)	Herbaceous vine with alternate heart-shaped leaves.	Contains a substantial amount of carbohydrate with negligible fat and rich content of vitamins A, Bs, and C, Mn, and K.
Lactuca sativa (lettuce, celtuce)	An annual leafy vegetable of the Aster family.	Contain substantial amount of carbohydrate, fiber, protein, vitamin K1, vitamin A, manganese, vitamin C, and folate.
Manihot esculenta (cassava leaves)	A vegetable with long tapering root.	Contains vitamins B6 and C, Fe, Ca, Mg. The leaves are rich in leucine. While the root is high in dietary fiber carbohydrate and sugar.
Mushroom	These are the flea-shy spore-producing fruiting structure of edible fungi.	Contains much water, carbohydrate, and proteins. Also rich in B vitamins and minerals like Zn, P, and K.
Ocimum grassitimum (clove basil)	An aromatic perennial herb.	Constituents include eugenol, Mg, and essential oil.

Pachyrhizus erosus (jicama)	Jicama is a kind of bulb/ root vegetable which looks a little bit like an onion, with a yellow exterior and a white inner. In terms of taste and texture, it is sweet and crunchy.	This contains carbohydrates, fiber, sugar, protein, vitamin C, potassium, magnesium, manganese, and a negligible amount of fat. Also, consist of a large amount of water.
Pastinaca sativa (parsnip)	This is a biennial plant with a rosette with roughly hairy leaves that have a pungent odor when crushed.	Contains a high amount of carbohydrate, dietary fiber, potassium. Also vitamin C, B6, iron, and magnesium.
Phaseolus vulgaris (green bean, French bean runner bean, haricot bean, lima bean)	These are unripe young fruits that are usually in enclosed pods from different cultivars of beans.	A rich source of folic acid fibers and vitamins like A, C, and K. Also a good source of Fe, Mg, P, Ca, and folate.
Piper guineense (uziza leaves, Ashanti pepper leaves, Benin pepper leaves)	These are vines useful in culinary. The leaves have a peppery taste and the leaves are green when fresh.	Contains a high level of vitamins C and E. Also contains protein, dietary fiber, and essential oils.
Pisum sativum (pea, snap pea, snow pea, split pea)	A small spherical seed pod-shaped vegetable.	Has a reasonable amount of carbohydrates, protein, and fiber. Also vitamins A, C, K and thiamine.
Raphanus sativus (radish, daikon, seed pod varieties)	An annual or biennial swollen taproot. The abundant anthocyanin gives the characteristic range of skin colors.	High calories, with a low amount of vitamin C, carbohydrate, and protein.
Solanum lycopersicum (tomatoes)	Edible vines with berry.	A major source of antioxidants is lycopene. Also contains 95% water, protein, carbohydrate, fiber, K vitamins B and E.
Solanum melongena (eggplant)	Belongs to the nightshade family and is cultivated for its purple spongy absorbent fruits.	Fiber, fat, protein, Mg, K, Cu, folate, vitamin C and vitamin B6.
Solanum tuberosum (potato)	A herbaceous perennial tubers.	Contains phosphorus, calcium, magnesium, zinc. Also a rich source of vitamin B6, vitamin C, and fiber.

Solanum tuberosum (potato)	A herbaceous perennial tubers.	Contains phosphorus, calcium, magnesium, zinc. Also a rich source of vitamin B6, vitamin C, and fiber.
Spinacia oleracea (spinach)	An annual leafy vegetable. The leaves are simple, oval to triangular, alternate with broad leaves on the lower part of the stem thinning at the apex.	Rich source of vitamin A, vitamin C, vitamin K, Mg, Mn, folate. Also a good source of B vitamins, vitamin E, potassium, calcium and dietary fiber.
Talinum fruticosum (waterleaf, Florida spinach, Lagos bologi, sweetheart, potherb, fame flower)	An erect plant with small, broad, and fleshy leaves.	A very good source of vitamins A and C. Also high in Fe and Ca.
Vernonia amygdalina (bitter leaf)	A shrub and member of the daisy family with elliptical leaves and rough bark. Grows up to 20 cm long.	High protein content, crude fiber. Also contains minerals like NA, Fe, Mg, Zn, Ca, K.
Vicia faba (broad beans)	Member of the pea and bean family. An erect annual plant where the seeds in the pod serve as a vegetable.	Carbohydrate, proteins, and fat. High in dietary mineral-like Mg, Fe, Mn, P, folate, and B vitamins.

Sources: [3, 4].

2. COMMON VEGETABLES AND PARTS USED AS FOOD

Most vegetables as we all know are different across various continents. **Table 2** shows common vegetables, parts used, cultivars, scientific names, and origin.

Table 2. Showing common vegetables.

Part used	Cultivars	Scientific name	Origin
Bulbs	Garlic	*Allium sativum*	Asia
Bulbs, leaves	Onion, spring, scallion, shallot	*Allium cepa*	Asia and Africa
Fruits	Pepper, bell pepper, sweet pepper	*Capsicum annuum*	North and South America, Africa
Fruits	Cucumber	*Cucumis sativus*	Southern Asia
Fruits	Tomatoes	*Solanum lycopersicum*	South America
Fruits	Eggplant	*Solanum melongena*	South and East Asia

Fruits, flowers	Pumpkin. Squash, marrow, zucchini,	*Cucurbita spp.*	Mesoamerica
Flower, bud	Artichoke	*Cynara cardunculus var. scolymus*	Mediterranean area and North Africa
Leaf sheath	Leek, elephant garlic	*Allium ampeloprasum*	Europe and Middle East
Leaves	Clove basil	*Ocimum gratissimum*	Africa and Southern Asia
Leaves	Arugula	*Eruca vesicaria ssp. sativa*	Mediterranean and Middle-East
Leaves	Spinach	*Spinacia oleracea*	Central and Southwest Asia
Leaves	Lagos spinach (also known as quail grass, soko, celosia or feather cockscomb)	*Celosia argentea var. argentea*	Central and West Africa
Leaves	Fluted gourd, fluted pumpkin, 'Ugwu'	*Telfairia occidentalis*	West Africa
Leaves	Kale	*Brassica oleracea*	Eastern Mediterranean and Asia
Leaves	Bitter leaf	*Vernonia amygdalina*	Tropical Africa
Leaves	Waterleaf, Florida spinach, Lagos bologi, sweetheart, potherb, fame flower	*Talinum fruticosum*	Central and South America, Caribbean, West Africa
Leaves	Uziza leaves, Ashanti pepper leaves, Benin pepper leaves	*Piper guineense*	West Africa
Leaves	Eru, wild spinach	*Gnetum africanum*	Tropical Africa
Leaves, axillary buds, stems, flower heads	Cabbage, Brussel sprouts, cauliflower, broccoli, kale, kohlrabi, collard greens	Brassica oleracea	Europe
Leaves, seed	Jute plant, Jew's mallow, buch okra, nalta jute, jute mallow	Corchorus olitorius	Tropical Asia and Africa
Leaves, stem, seed oil	Lettuce, celtuce	Lactuca sativa	Egypt
Pods, seeds	Green bean, French bean runner bean, haricot bean, lima bean	*Phaseolus vulgaris, Phaseolus coccineus, Phaseolus lunatus*	Central and South America

Pods, seeds	Broad beans	*Vicia faba*	Mediterranean and Middle East
Pods, seeds, sprout	Pea, snap pea, snow pea, split pea	*Pisum sativum*	Mediterranean and Middle East
Root	Parsnip	*Pastinaca sativa*	Eurasia
Root, leaves	Beetroot, sea beet, Swiss chard, sugar beet	*Beta vulgaris*	Europe and Near East
Root, leaves	Turnip, Chinese cabbage, napa cabbage, bokchoy	*Brassica rapa*	Asia
Root, leaves, stems	Carrot	*Daucus carota*	Persia
Seed pods, seed oil, sprouting, roots, leaves	Radish, daikon, seed pod varieties	*Raphanus sativus*	South Eastern Asia
Seed, pod	Okra, ladies' fingers	*Abelmoschus esculentus*	West Africa, South Asia, and Ethiopia
Tubers	Yam	*Dioscorea spp.*	Tropical Africa
Tubers	Cassava	*Manuhot esculenta*	South America
Tubers	Potato	*Solanum tuberosum*	South America
Tubers, leaves, shoots	Sweet potato	*Ipomoea batatas*	Central and South America

Sources: [4, 5].

2.1 Health benefit of vegetables

Some of the under listed health benefits of vegetables were mentioned by [6] and they include:

- Vegetables act as suppliers of dietary fibers and are important sources of essential vitamins, minerals, and trace elements.
- They also serve as antioxidants and loads of vitamins such as A, C, and E. A diet laden with vegetables leads to a reduction in the terminal incidence of cancer, stroke, cardiovascular disease, and other chronic ailments.
- The fiber contents of vegetables can assist in keeping hunger in check because it fills the stomach.
- Vegetables have a positive effect on the blood sugar of consumers.
- Consumption of vegetables also assists in the control of high cholesterol levels and blood pressure in the body.
- It aids to reduce inflammation and aid digestion.
- Generally, vegetables have a positive effect on the blood sugar of consumers.

3.DESCRIPTIONONTHERAPEUTICBENEFITSOFPHYTOCHEMICALS IN VEGETABLES

The words "Phytochemicals" often called secondary metabolites are non- nutritive chemical compounds produced by plants via several chemical pathways. Recent studies have demonstrated that a large number of phytochemicals can be beneficial to the function of human cells. Several studies also indicate the effects of phytochemical-rich foods on health. While phytochemicals can help to improve health.

Also, phytochemical substances are found in plant-based foods such as fruits, vegetables, whole grains, nuts, seeds, and legumes. They give plants their color, flavor, and aroma. There are thousands of different phytochemicals in which scientists are discovering the different roles they play in human health. Much of the current evidence on the benefits of phytochemicals came from observing people who eat mainly plant- based diets. These people have been shown to have significantly lower rates of certain types of cancers and heart disease [7, 8]. Phytochemicals may have the potential to:

- Aid the function of the immune system
- Protect cells and DNA from damage that may lead to cancer
- Improve the health conditions
- Reduce inflammation
- Slow the growth rate of some cancer cells
- Help regulate hormones

You will find some examples of specific phytochemicals and their potential benefits, along with some of the foods in which they are found in **Table 3.**

Each plant food has many different phytochemicals; for instance, there are more than 100 phytochemicals in carrots alone. You may not find a single food item with all the essential nutrients. There is a need to include a little portion of every food in your diet for great health benefits. Different phytochemicals have different functions in the body, and many of them complement one another. Evidence shows that taking phytochemicals in supplement form may not provide the same benefits as eating whole plant foods, because phytochemicals in supplements may not be as easily absorbed by the body as those from food sources. So the best way to have a variety of phytochemicals and other essential nutrients in the diet is to eat a rainbow of plant-based foods. The fruits and vegetables with deeper and brighter colors or with stronger flavors are often the best sources of phytochemicals. Larger concentrations of phytochemicals are also often found in the skins or peels of fruits and vegetables [8].

Table 3. Description of some vegetables and their nutritional values. Some extract.

Phytochemical	Foods	Potential benefit
Carotenoids (beta carotene, lycopene)	Cooked tomatoes, orange squash,carrots, sweet potatoes, and green plants, such as broccoli.	May inhibit cancer cell growth, reduce risk of cardiovascular disease, and boost immunity.
Flavonoids	Berries, apples, citrus fruits, soybeans, coffee, tea, walnuts, whole grains.	May fight inflammation, decrease damage to DNA, and reduce tumor growth.
Anthocyanins	Berries	It may help lower blood pressure.
Isothiocyanates (sulforaphane)	Cruciferous vegetables, such as broccoli, cauliflower, and kale.	May protect against cancer and cardiovascular disease.
Lutein and zeaxanthin	Dark, leafy greens, such as spinach and chard.	May promote eye health.
Sources: [7, 8].		

3.1 Preparing vegetables for healthy living

Most essential nutrients in vegetables can easily be washed and bleached off thorough preparatory procedures. Preparing food to preserve the nutrients for maximum effect ensures healthy eating. Soluble nutrients such as vitamins B and C can easily be lost during preparation [9, 10]. Following the required preservative, methods to keep nutrients in food intact enable us to gain access to the nutrients embedded in the vegetables. Useful tips to follow in preparing vegetables include:

1. Handling is essential for keeping vegetables fresh and healthy for consumption. Therefore, wash your hands before handling or preparing the vegetables for consumption or commercial purposes.
2. It is very important to wash vegetables before consuming them.
3. Use of healthy cooking methods like broiling, grilling, steaming, and roasting. As much as it is possible deep frying should be avoided because it leads to the destruction of the nutrients and addition of fats to the vegetable.
4. Ensure you cook vegetables at a safe temperature using a food thermometer. To avoid contaminants, it is essential to refrigerate within 2 h of purchase or harvest. You may as prepare.
5. Vegetables should be cooked in little water for a short period to prevent the soluble nutrients from being washed off.
6. Avoid dicing or cutting into small chunks. This can lead to the wash off of the nutrients such as the B and C vitamins.
7. Avoid using salt as flavor instead use a variety of spices.
8. Avoid packaged or processed vegetables because these are likely to contain salt, sugar, and fats.

4.TECHNIQUESFORPROCESSINGANDPRESERVINGVEGETABLES

The production, availability, and consumption of fresh vegetables vary significantly among cultures and regions due to the distinct growing and harvesting seasons of various plants. The primary purpose of processing is to extend shelf life by inhibiting microbial spoilage and spontaneous physiological degradation of plant cells. [9, 11].

According to [10] processing and preserving are essential keys for foods product in order to avoid spoilage of such items. Some methods of processing and preserving vegetables can be:

- Freezing
- Canning
- Blanching
- Fermenting
- Packaging
- Dehydration
- Irradiation

REFERENCES

1. Vegetables—Origin and meaning of vegetable by Online Etymology Dictionary [Internet]. 2021. Available from: www.etymonline.com [Accessed: August 21, 2021]
2. Teresa AL, Harry SP. Italian horticultural and culinary records of summer squash (*Cucurbita pepo* Cucurbitaceae) and emergence of the zucchini in 19th-century Milan. Annals of Botany. 2016;**118**:53-69. DOI: 10.1093/ aob/ mcw080
3. Harper D. Vegie. Online Etymology Dictionary. Available from: www. etymonline.com [Accessed: August 21, 2021]
4. Wikipedia. Vegetable. Available from: https://en.wikipedia.org/wiki/ Vegetable [Accessed: October 01, 2021]
5. Joseph M. 56 different types of vegetables (and their nutrition profiles). Avai able from: nutritionadvance.com/ healthy-foods/types-of-vegetables [Ac cessed: August 15, 2021]
6. Pem D, Jeewon R. Fruit and Vegetable Intake: Benefits and Progress of Nutrition Education Interventions— Narrative Review Article. Iranian Journal of Public Health. 2015;**44**(10): 1309-1321
7. González-Vallinas M, González-Castejón M, Rodríguez-Casado A, Ramírez de Molina A. Dietary phytochemicals in cancer prevention and therapy: A complementary approach with promising perspectives. Nutrition Reviews. 2013;**71**:585-599. DOI: 10.1111/nure.12051

8. Probst YC, Guan VX, Kent K. Dietary phytochemical intake from foods and health outcomes: A systematic review protocol and preliminary scoping. BMJ Open. 2017;7:e013337. DOI: 10.1136/ bmjopen-2016-013337
9. Sinha NK, Hui YH, Evranuz EÖ, Siddiq M, Ahmed J. Handbook of Vegetables and Vegetable Processing. New Jersey: Wiley-Blackwell; 2010. p. 788. ISBN: 978-0-813-81541-1
10. Vainio H, Bianchini F. Fruits and vegetables. In: IARC Handbooks of Cancer Prevention. Lyon: IARC; 2003. p. 9. ISBN 978-92-832-3008-3
11. Joseph JJ. Vegetable Processing. Britannica. Available from: https://www.britannica.com/technology/vegetable-processing/Fermentation-and-pickling. [Accessed: August 13, 2021]

Chapter 2 VEGETABLES ABUNDANT IN ANTIOXIDANTS: EFFECTS ON HUMAN HEALTH

1. INTRODUCTION: AN OVERVIEW

Currently, the prevailing trend in food intake within our society emphasizes "natural and healthy" options, incorporating vegetables and fruits into the diet. "Western" diets are characterized by a high caloric intake and elevated levels of animal protein, accompanied by a low consumption of vegetables. This type of diet combined with physical inactivity leads to the onset of diseases such as diabetes, obesity, and cancer [1]. The pursuit of vegetables increasingly attracts attention due to their bioactive substances with medical properties, which can promote a balanced diet, enhance the body's defenses, and aid in illness prevention. [2].

Vegetables are a broad category of plants that fulfill many roles in medicine, food, and the beverage industry. Their leaves, stems, seeds, and flowers are beneficial for nourishment, flavoring, and coloring. They are plants of significant relevance in our diet. They are protective foods that significantly contribute to health maintenance, bodily construction and repair, and illness prevention due to their abundant vital nutrients. They contain a wealth of antioxidants, including carotenoids, ascorbic acid, flavonoids, folic acid, and minerals such as calcium, iron, and phosphorus, along with phenolic compounds, proanthocyanidins, vitamins, and saponins. These phytochemicals demonstrate many biological effects, including antioxidant, anti-inflammatory, antibacterial, and anti-cancer properties. The antioxidant activity of phenolic compounds is well-established, attributable to their chain-breaking and free radical scavenging capabilities, which eliminate free radical intermediates and provide protection against the generation of reactive oxygen species [3]. Excessive generation of reactive oxygen species (ROS) can induce oxidative damage to biological macromolecules, including nucleic acids and proteins. [4].

In rural areas, vegetables form an integral component of the food and nutrition of the local population as they are traditionally reckoned for their medicinal, therapeutic and nutritional values since time immemorial. They are either consumed as raw or cooked as traditional delicacies and the sales from the surplus of these vegetables serve as an additional income to many families. Moreover, to this set of population, vegetable consumption gives variety to their food and add flavours to the diet. It is rich in various nutritive elements and can make up for the dietary shortfalls of vitamins and minerals necessary for the human diet. Generally, malnutrition and food shortages are prevalent among the rural population. Thus, the cultivation of vegetables will contribute to increased food production, balanced nutrition, food and health security and poverty eradication for them.

Study revealed that before the nineteenth century, there has been a particular interest in vegetables and herbs because of the beneficial effects of phytochemicals present in them [5]. They were being used for therapeutic purposes until synthetic drugs were developed. Their consumption has drawn great attention due to the discovery of the fact that, their regular intake has an alliance with declined rates of heart diseases, cancers, diabetes, and other degenerative diseases. Protection that vegetable offer has been ascribed to the presence of antioxidative compounds such as α- tocopherol, ascorbic acid [6] and phytochemicals like carotenoids, flavonoids, lycopene, phenolics and β-carotene [7–9]. These chemicals function as free radical scavengers, hydrogen-donating agents, and metal ion chelators, with the capacity to limit the production of free radicals and reactive oxygen species (ROS) [10].

This review highlights the bioactive chemicals found in vegetables, their biological significance, and their involvement in disease prevention and health promotion.

2. ANTIOXIDANT POTENTIAL OF VEGETABLES

Antioxidants are defined as substances which when present at low concentration compared to those of an oxidizable substrate [11], significantly delay or prevent the oxidation of that substrate. Some of the mechanisms of action of antioxidants involve prevention of lipid peroxidation, oxidative damage to membranes, glycation of proteins and inactivation of enzymes caused by free radicals. Oxidative stress can arise due to the generation of ROS including free radicals and non-free radicals. Evidences have shown that they have roles in the development of several diseased conditions such as lipid peroxidation, protein oxidation, DNA damage and cellular degeneration [12, 13].

Normally, during cellular metabolism, free radicals and other reactive oxygen species are continuously released in the body. They can also be produced from sources such as drugs, food, exhausts and other pollution from the environment. Organisms are endued with endogenous and exogenous antioxidant defence systems against free radical generation. However, when free radical produced in the body overwhelms the antioxidant system, oxidative stress ensues. This has implications in the aetiology of several pathological conditions [14]. This is the reason for the special attention being given to the use of antioxidants especially of natural origin.

Antioxidant phytochemicals have been recognised for the role they play in the prevention and management of chronic diseases [15]. Phytochemicals are proven to have antioxidant capacities in humans. Consumption of vegetables with high contents of these compounds is liable to raise the antioxidant capacity of the body system. For instance, serum total antioxidant capacity was found to be elevated significantly following the consumption of spinach in elderly women [16]. Further, a study has also supported a significant increase in antioxidant capacity caused by the daily consumption of 10 servings of fruit and vegetables for a period of 15 days [17].

Given the growing prospect observed in these phytochemicals, there is a need to identify and quantify them, elucidate their mechanisms of action and assess their potential health benefits. This information could serve as a basis for intervention strategies.

The phytochemical content of some vegetables like Broccoli, Brussels sprout, green cabbage have been revealed [18]. α- and β-carotene are richly present in broccoli (1 and 779 mg/100 FW), carrot (4.6 and 8.8 mg/100 FW), tomato (112 and 393 mg/100 FW), pea (19 and 485 mg/100 FW), and sweet pepper (59 mg/100 FW). However, some factors such as level of growth, handling during post harvesting, and processing could contribute to the significant variation that sometimes occur in structure and function of phytochemicals from vegetable to vegetable [19]. Vegetables such as beans, broccoli, cabbage, cauliflower, cress, pea, spinach, spring onion, and sweet peppers are reported to be rich in ascorbic acid [20]. Asparagus, Brussels sprout, cabbage, carrot, cauliflower, kale, lettuce, spinach, sweet potato, and turnip are abundant in vitamin E [18] while red pepper is high in vitamin C content (144 mg/100 g) [21].

Furthermore, the structure can influence the antioxidant capability of a phytochemical in vegetables. For instance, the antioxidant properties of flavonoids depend on their C-ring structure. Flavonoids like rutin and luteolin with full substituted C-ring and an ether bonded with three oxygen display superior free radical scavenging capacity and higher reaction rate when compared to flavonoids that lack one or more C ring structural elements. Also, individual phenolic units in vegetables have been found to show better antioxidant activity than total phenolics. Generally, the high scavenging property of vegetables with phenolic content may be as a result of hydroxyl groups existing in the phenolic compounds' chemical structure that can provide the needed component for radical scavenging activity (**Figure 1**). The hydrogen atoms of the adjacent hydroxyl groups, situated in various positions of the rings A, B and C, the double bonds of the benzene ring, and the double bond of the oxo functional group (-C=O) of some flavonoids, offer them their high antioxidant activity. These structural features are shown in the benzene rings of flavonoids as shown in **Figure 2.** The antioxidant ability centre of phenolic acids is in the phenolic hydroxyl group, such that the positioning of phenolic hydroxyls is directly related to their antioxidant activity [22].

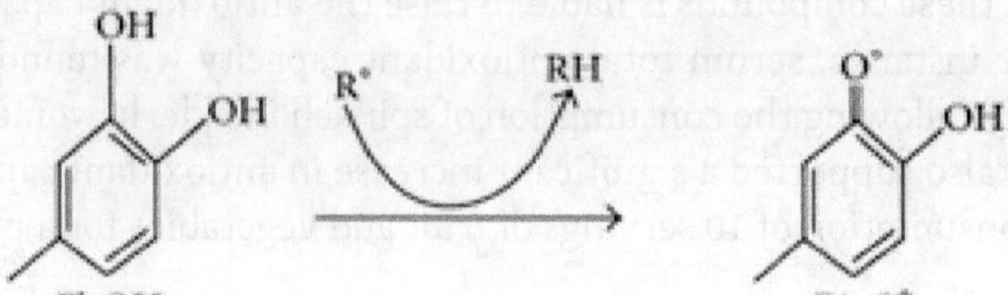

Figure 1. Free radical scavenging of flavonoids where R is the radical and Fl-OH is the flavonoid.

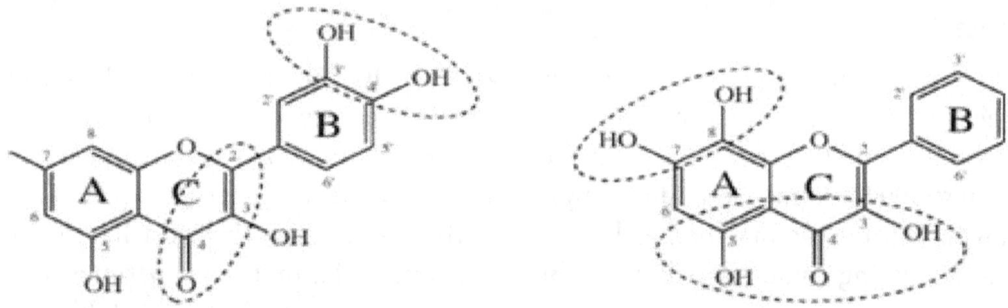

Figure 2. Structural characteristics of flavonoids with high antioxidant activity; presence of hydroxyl groups and double bonds (circled) in the benzene rings.

The impartation of colours is also an important factor in the antioxidative activity of phytochemicals in vegetables (**Table 1**). It is said to be directly related to the pigment content such as carotene, phytoene, chlorophyll, lycopene and anthocyanin, and their relative quantities at different maturity stages. It is an important trait that largely reflects quality, type of phytochemical as well as antioxidant activity [30]. Jaganath and Crozier [31] inferred that colour difference as observed in vegetables and fruits is an indication of accumulation of phytochemicals such as flavonoids and carotenoids. Red pigment conferred on red bell peppers is linked to a large content of lycopene, a member of the carotenoid family, localised in the prostate gland. Lycopene is a powerful antioxidant that has a connection with lessened risk of some cancers, especially prostate cancer [23], and protection against heart attacks. The yellow or orange colouration noticed in vegetables such as spinach, sweet potatoes and carrots represents produce rich in both α and β-carotene. They are also members of the carotenoid family where β-carotene can be converted to vitamin A in the body, a nutrient that plays a crucial role in vision. The mechanism involves the cleavage of β-carotene into two molecules of vitamin A which is then converted to II-cis-retinal and thereafter combines with opsin to form a protein called rhodopsin. When light hits the rods, metarhodopsin is produced [24]. Beta-carotenes in vegetables are also involved in improving immune function, skin and bone health as well as prevention of cancer [25].

Table 1. Impartation of colours in vegetables and their health benefits.

Vegetable colour	Pigment/ Phytochemical content	Health benefits	References
Red (in red bell peppers)	Lycopene anthocyanin	Reduced risk of some cancers, especially prostate cancer delays many diseases associated with ageing	[23]
Yellow or orange (in spinach, sweet potatoes and carrots)	α and β-carotene, flavonoids	plays a crucial role in vision, improves immune function, improves skin and bone health, prevents cancer	[24, 25]
Green (as in spinach, parsley)	Lutein	Maintains good vision	[26]
Blue/purple (as in eggplant)	Anthocyanin and phenolics	Reduced risk of ageing, cancers and heart disease	[27, 28]
White, tan or brown (as in onion, garlic, potato)	Thiosulphides,	Lowers blood pressure, regulate cholesterol levels and lower the risk of some types of cancers	[29]

3. MECHANISMS OF ANTIOXIDATIVE ACTION OF VEGETABLE PHYTOCHEMICALS

Generally, the protective effect of vegetables has been credited to their antioxidant components. Antioxidants prolong the onset of free radical generation due to their capacity to supply hydrogen atoms or chelate metals implicated in ROS formation [18]. The mode of operation by which antioxidants negate the influence of free radicals involves various mechanisms among which is the termination of the free radicals [32] and post-modification of resultant bioactive compounds during metabolism [33].

Cells can respond to the effect of antioxidant phytochemicals by interacting with receptors and enzymes involved in signal transduction, or through modification of gene expressions that may affect the redox status of the cell and subsequent induction of series of redox-dependent reactions [34]. Reference [34] also presented an evolving evidence that phytochemicals like flavonoids may participate in the modulation of intracellular signalling cascades. Intracellular signalling pathways serve as major avenues of connection between the plasma membrane and regulatory targets in various intracellular compartments [35]. This signalling process also leads to the activation of protein kinases by phosphorylation, and then affects the activity of transcription factors that regulates gene expression [36]. Ruiz et al. [37] showed that the signalling cascades enable the cells to regulate processes such as growth, proliferation, and apoptosis. Phytochemicals can modulate effects in cells through selective actions on different components of the signalling cascades.

Oneofthemechanismsutilisedbycarotenoidsisinhibitionoftheoxidationinitiatedby singlet oxygen. Flavonoids possess antioxidant, anti-inflammatory, antimutagenic and anti-carcinogenic properties. Apigenin is a flavone found in parsley and celery. It prevents inflammation by hampering the production of proinflammatory cytokines and cyclooxygenase 2 (COX-2) expression through the inhibition of nuclear factor-κB (NF-κB) (**Figure 3**), phosphoinositide-3-kinase (PI3K/Akt) and activating transcription factor-3/cyclic adenosine monophosphate (ATF/cyclic AMP) responsive element signalling pathways [38]. Kaempferol, a flavonol present in broccoli suppresses the inflammatory activities of inducible nitric oxide synthase (iNOS) and COX-2 by blocking signal transducer and activator of transcription 1 (STAT-1), NF-κB, and activator protein 1 (AP-1) signalling pathways as observed in activated macrophages and human endothelial cells [39]. Quercetin, known to be present in green leafy vegetables, onions and broccoli, exerts its potent antioxidant and anti-inflammatory activities by inhibiting the expression of pro-inflammatory cytokines [40] and suppressing (tumour necrosis factor) TNF-induced NF-κB (**Figure 3**) [41]. Furthermore, it can also regulate lipid profile thereby reducing glycaemia through the inhibition of 11β-hydroxysteroid dehydrogenase type 1 [42].

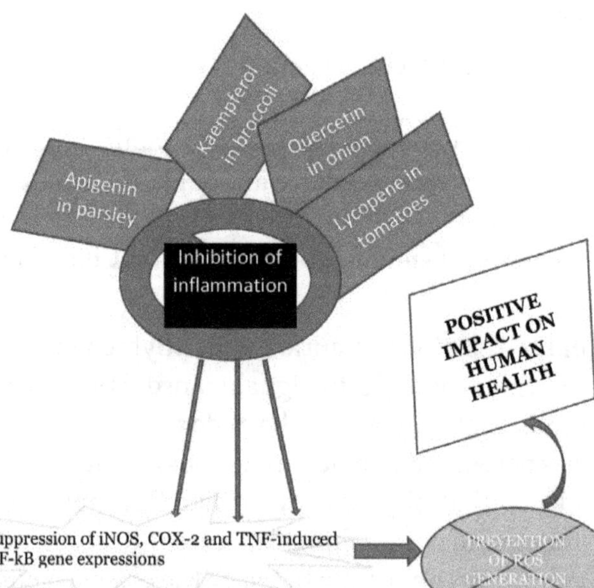

Figure 3. Mechanism of inhibition of inflammation and oxidative stress by

phytochemicals in vegetables.

Lycopene, a carotenoid present in tomatoes, diminishes inflammatory response by reducing the gene expressions of iNOS and COX-2 [43] as well as IL-12 production. This is achieved by obstructing mitogen-activated protein kinase (MAPK) signalling and the activation of NF-κB [44]. Moreover, β-carotene in green-coloured leafy vegetables prevents the genetic expressions of LPS-induced iNOS, COX-2, and TNF-α by reducing phosphorylation and degradation of I-kappa B-related protein (IκBR) and nuclear translocation of NF-κB in macrophages [45]. Lutein, known for the ability to repress LPS- and hydrogen peroxide-induced pro-inflammatory gene expression by diminishing the activities of PI3K and NF-κB inducing kinase (NIK) and phosphorylation of Akt [46].

4. CLASSIFICATION OF VEGETABLES

There are thousands of species of plants used as vegetables globally.

Classification of these species can be done by taking into considerations some common features such as the part of the plant used for nutrition and the specific nutritional value. A summary of the health benefits of vegetables described below and their antioxidants content are displayed in **Table 2**.

4.1 Leafy vegetables

4.1.1 Lettuce

Lettuce (Lactuca sativa L.) is from the Asteraceae (Compositae) family. It is a vegetable that is extensively cultivated globally, commonly consumed fresh and as one of the salad ingredients owing to its health-promoting effects [68]. It comes in different textures, colours, leaf shapes and in a wide variety of head formations. According to Mou [69], it is classified into six major types which are butterhead, Cos or Romaine, Crisphead, Leaf or Cutting, Stalk or Stem, and Latin. lettuce is known to be a great source of flavonoids and vitamin B9. [70, 71] reported in their studies that, its vitamin B9 and flavonoid contents were higher compared to spinach, which is another popular green leafy vegetable that is widely consumed.

Table 2. Summary of the vegetables and their health functions.

Vegetables	Antioxidants Content	Health Functions	References
Lettuce	Vitamins C and E, carotene, lutein	reduction in the occurrence of cancers like lung, prostate, and colon, heart disease and vision impairment, reduction in total cholesterol and LDL levels	[47, 48]

Watercress	Vitamins C and E, quercetin, lutein, β-carotene, anthocyanins	prevention of various liver diseases, reduction in susceptibility to cancer of the colon and oxidative stress, prevention against type 2 diabetes and cardiovascular diseases	[49, 50]
Vernonia Amygdalina (Bitter Leaf)	Luteolin	Treatment of malaria, fever and cough, management of postpartum haemorrhage, Reduction of plasma glucose and triglyceride/cholesterol levels, inhibition of human breast cancer cells' growth.	[51–53]
Telfairia Occidentalis (Fluted Pumpkin)	ascorbic acid	Treatment of anaemia, malaria convulsion and gastrointestinal disorders, lowering of blood glucose level, elevation of haematological parameters, improvement of sperm quality, management of infertility	[54–56]
Corchorus olitorius (Jute Mallow)	Vitamins C and E, Carotenoids, quercetin	Treatment of folic acid, iron shortage and anaemia, blood purification	[57, 58]
Onion	kaempferol, quercetin	Regulation of fasting blood glucose levels, improvement of glucose tolerance, support of growth of beneficial bacteria, decreased incidence of stomach and intestine cancers	[59–61]
Carrot	Vitamins C and E, carotenoid	Inhibition of mutagenesis, lowering of blood glucose level, function as immune enhancer, anticarcinogen and antioxidant	[62]
Tomato	Lycopene, carotene, tocopherols, lutein, ascorbic acid	Reduction in the development of stomach and rectal cancers, reduction in vulnerability to lipid peroxidation, reduction in prostate cancer risk and an increased apoptotic cell death in carcinomas, elevation of antioxidant defence system	[63–65]
Eggplant	lycopene, lutein, α-carotene, myricetin and kaempferol	control of the level of high blood cholesterol inhibition of type 2 diabetes and hypertension	[66, 67]

Variation with regards to quality/quantity of vitamins and phytochemicals depends on some pre-harvest factors such as agricultural practices, maturity, genotype, maturity and environmental circumstances [72].

However, genetic composition has a great influence on the determination of synthesis and bioaccumulation of carotenoids, chlorophylls, vitamin E, phenolic compounds, vitamin C and antioxidant molecules [73]. Vitamins are vital micronutrients present in lettuce and have been implicated in the reduction of certain diseases such as cardiovascular and degenerative diseases [74]. Variations in the amount of vitamins found in lettuce may depend on leaf type, colouration and butterhead. Romaine lettuces have been particularly found to be good sources of folate [70], with green lettuce having the highest vitamin C concentration [75].

Also, the content of carotenoids such as carotene and lutein and colours across the 52 genotypes of lettuce was discovered to vary as assessed by [76] and categorised in the following order: Romaine and green leaf > red leaf > butterhead > crisphead. Report given about these two carotenoids that they were remarkably and positively correlated with chlorophyll a and b as well as with total chlorophyll content was found to be quite contrary to the findings of [77] who argued that the content of these carotenoids may not correlate with leaf green pigmentation, since the contents of carotenoids seemed to be lower in green compared to red-pigmented lettuce. This contradictory opinion may suggest that the content in carotenoids may not be consistently and outrightly related to leaf pigmentation [78]. Nevertheless, the frequent consumption of carotenoids-rich lettuce has been linked with a reduction in the occurrence of chronic diseases such as certain types of cancer (lung, prostate, and colon), heart disease and vision impairment [47].

Further, many researchers also reported variation in the content of the secondary metabolites present in lettuce with respect to the genotypes and leaf colours [79]. According to [80], red lettuce contains just a single anthocyanin, called cyanidin-3-O-(6″-malonyl-_-glucopyranoside) which is further converted to two cyanidin derivatives named as (cyanidin-3-O-(6″-malonyl-_-glucopyranoside methyl ester) and cyanidin-3-O--glucopyranoside). These cyanidin derivatives possess antioxidant activities against lipid peroxidation and cyclooxygenase activity.

The significance of leaf pigmentation was further accentuated by [78] in their study where leaf pigmentation was found to correlate with the concentration of phenolic compounds such as flavonoids, and anthocyanins. For instance, the total phenolic content in red butterhead, red leaf and red romaine lettuces was higher than the green counterparts [81, 82]. The red colour of lettuce has been associated with a high level of total phenolics, popular for imparting a higher antioxidant activity than vitamins C and E [83]. The health benefits of red-pigmented lettuce have been highlighted in an in vivo study done by Lee et al. [48] on mice fed with a high-fat diet and supplemented with red lettuce. The results indicated that total cholesterol and low-density lipoprotein (LDL) were reduced, thus underscoring the prospects of red-pigmented lettuce consumption against cardiovascular disease.

Similarly, a study carried out by [84] demonstrated that rats fed with red oak-leaf lettuce reduced appreciably LDL and cholesterol levels. Also, the new cultivar B-2 of red-pigmented lettuce, characterised by a high concentration of flavones, anthocyanins and phenolic acids has been reported by [80] to contribute to decline in diseases caused by oxidative stress, leading to anti-tumour activities against some cancer cell lines.

Based on the above considerations, clinical studies have validated the inherent benefits of frequent consumption of fresh lettuce, in particular the red-pigmented varieties.

4.1.2 Brassica leafy vegetables

This category of vegetable was formerly referred to as cruciferous vegetables and it includes a broad range of species with promising health-benefitting properties. These species include kale, pack Choi, mizuna, watercress, wild and salad rocket DC and Eruca vesicaria [L.] Cav. Database has shown that, about 12% of the vegetables grown globally belong to the Brassicaceae family. They are rich sources of phytochemicals [49], vitamins C, E and K, carotenoids, and phenolic compounds.

Genetic factor has been considered as a factor that influences and modulates the biosynthesis and accumulation of phytochemicals in Brassica leafy vegetables [49]. Variations in the amount of phytochemicals have been observed in a comparative study of antioxidant molecules involving four Brassica leafy vegetables [85]. Watercress showed the highest polyphenol and vitamin C content, while salad and wild rocket showed high concentrations of kaempferol and quercetin derivatives. Mizuna displayed remarkable concentrations of isorhamnetin and sinapic acid [85]. The potential value of salad Brassica leafy vegetables as dietary sources of antioxidants has been highlighted. Its positive effects against type 2 diabetes and cardiovascular diseases have been affirmed.

Brassica leafy vegetables, in particular kale, are considered as a valuable source of carotenoids such as lutein and β-carotene as well as chlorophyll a and b. During the analysis to determine the concentrations of carotenoid of 33 kale cultivars, zeaxanthin was the most abundant carotenoid in 21 cultivars. American and hybrid cultivars were shown to have high concentrations of zeaxanthin, while German landraces, German commercial varieties, Italian, and red-coloured kale varieties exhibited high concentrations of chlorophyll a and b [86].

Coloured Brassica leafy vegetables like violet kale or pack Choi containing anthocyanins have taken the attention of nutritionists and horticulturists. The anthocyanins content could serve as a marker for differentiating between varieties/ cultivars. The phytochemical concentrations in pack Choi are dependent on the particular colour. Aiyeloja and Bello [87] observed that red pack Choi produced higher concentrations of total flavonoids, total phenolic compounds, glucosinolates, carotenoids and anthocyanins than its green counterpart. Regular intake of leafy vegetables containing anthocyanin could contribute to prevention of various liver diseases, reduction in susceptibility to cancer of the colon and oxidative stress [50].

4.1.3 Bitter leaf (VA)

Vernonia amygdalina from *Asteraceae* family is popularly known as the bitter leaf in English. Its petiolate leaves of are about 6 mm in diameter and elliptic in shape. The green leaves have a characteristic bitter taste [88]. They are well distributed in tropical Africa and Asia. The leaves of VA serve as condiments in soup after being washed or boiled to remove the bitter taste. In folk medicine. It has a long history of being used in the treatment of malaria fever and cough [51].

In traditional medicine, many practitioners make use of the different parts of the plants for the treatment of antihelminth, antimalaria [89]. Many others use the aqueous extract got from the leaves as a tonic, an appetiser and for wound healing [90]. Traditional birth attendants from Malawia and Uganda find it useful in aiding removal of the placenta after birth, post-pertum uterine contraction, induction of breast milk production and management of postpartum haemorrhage.

The local use of VA in various parts of Africa for treatment of several ailments and general well-being has been backed up scientifically. Antidiabetic potential of the aqueous extract of VA in streptozotocin-induced diabetic rats has been reported [52]. The finding showed that VA was capable of diminishing plasma glucose, levels of triglyceride/LDL-cholesterol and malondialdehyde (an index of lipid peroxidation). This can occur via scavenging of the reactive oxygen species or by promoting the synthesis of antioxidant enzymes [91] which can subsequently lead to reduction in oxidative stress.

Reference [88] described the antidiabetic effect of VA when combined with another vegetable named *Gongronema latifolium*, on the pancreatic β – cells of rats induced with streptozotocin. The animals administered with the extracts were observed to gain body weight as compared to weight loss experienced in the diabetic group. Further, blood glucose level significantly declined after 28 days of treatment with the combined extracts. Regeneration of islets cells was believed to be the explanation as this would induce a rise in insulin production and secretion [92]. Active ingredients such as flavonoids are believed to be present in VA [44] which may be responsible for their potentials in altering pancreatic damage initiated by streptozotocin or alloxan in experimental animals. In addition, the bitter principle of VA may also be responsible for insulin production, stimulation and release of pancreatic islets from the beta-cells [93]. Likewise, tannin, flavonoids glycosides and phytosterols of this plant could also inhibit the action of alpha-glucosidase inhibitor which may have contributed to the hypoglycemic effect being exhibited by this plant.

VA is increasingly becoming a powerful and strong challenger for cancer management as coumarins, flavonoids, sesquiterpene lactones have been implicated as the active principles in VA that may be responsible for its anticancer activity [94]. Aqueous extracts of VA was found to exhibit a cytostatic action on cultured human breast tumour cells (MCF-7) growth *in vitro* implying its tumour stabilisation and protective effects *in vivo* [94]. Its potential effects in inhibiting DNA synthesis even at physiological concentrations have been demonstrated in cancer cells [95]. Its hexane, chloroform, butanol and ethylacetate fractions were found to be capable of inhibiting the growth of human breast cancer cells even at concentrations as low as 0.1 mg/ml to 1 mg/ml, with an inhibition rate as high as 98% [53].

Other findings have established the usefulness of VA and its biopeptides (derived from the aqueous extracts of its leaves) against cancer via apoptotic mitogen-activated protein kinases and signal transduction pathways [96].

4.1.4 Fluted pumpkin (*Telfairia occidentalis*)

T. occidentalis (TO) commonly called fluted pumpkin is from the family: Cucurbitaceae. It is a popular vegetable that occurs in the forest zone of West and Central Africa. It is a perennial vine, growing to 10 m or more in length with its stems having branching tendrils and the leaves divided into 3–5 leaflets. The leaf is widely consumed due to its diverse benefits. The young succulent shoots and leaves are consumed as vegetables in the eastern part of Nigeria. Its herbal preparation has been applied in treating anaemia, malaria convulsion, gastrointestinal disorders [54]. Also, in addition to its nutritional value, this vegetable has agricultural and industrial importance [97].

Some scientific researchers have discovered its free radical scavenging and antioxidant properties. The leaves are rich in ascorbic acid and phenols [97]. Utilisation of the leaves in folk medicine in the treatment of some diseases in which the involvement of generation of free radicals have been implicated could be as a result of the antioxidative and radical scavenging ability [98].

Kim et al. [55] reported that aqueous extract of TO leaves lowers blood glucose level and elevates haematological parameters in rats. The chemical composition of TO shows vitamin A and C as part of its constituents which are well-known antioxidants and capable of scavenging free radicals [99]. They are well-established haemopoietic factors that have a direct impact on blood production in the bone marrow. Amino acids are also derived from TO and could also be useful in production of the globin component of the haemoglobin, contributing to elevation in haemoglobin concentration.

The leaf extract has also been documented to have the ability to improve sperm parameters which can assist in improving sperm quality [56]. Some of its active ingredients possess spermatogenic activities. Therefore, the leaves may be very applicable in the treatment and management of infertility especially those linked with a reduction in sperm performance.

The anti-anaemic potentials of the aqueous extract of leaves against phenyl hydrazine-induced anaemia in rabbits have been investigated [100]. The finding revealed that the leaves are notably rich in iron and play a major role in curing anaemia.

4.1.5 Basil (Ocimum)

Ocimum basilicum and *Ocimum gratissimum* are known for the management of different diseases in Africa. They belong to the Lamiaceae family. The leaves can be petiolate or sessile and most times toothed at the margin. The presence of volatile oil which contain up to 75% of thymol, gives them a characteristic pleasant aroma. The leaf of *O. gratissimum* or even the whole plant is a well-known remedy for diarrhoea and other diseases [101].

In folklore medicine, *O. basilicum* (basil) is a medicinal plant used for various ailments, such as cough, diarrhoea, headaches, constipation and kidney malfunction. Its oil commonly referred to as basil oil contains camphor with antibacterial properties. The vapour of the boiling leaves can be inhaled by people with catarrh and colds while the leaves may be rubbed between the palms and sniffed for treatment of a cold.

4.1.6 Jute mallow (Corchorus olitorius)

C. olitorius (Linn) is a leafy vegetable belonging to the family Tiliaceae. It is commonly called jute mallow in English and "ewedu" in south western Nigeria. The plant is characterised by the viscosity of its leaves which usually forms a thick viscous soup after being cooked and can be added to stew or soup. The leaves are rich sources of vitamin and minerals.

In folklore medicine, the leaf extract is employed in the treatment of pain, fever, gonorrhoea and tumour. It is rich in minerals, vitamins B1, B2, C and E, carotenoids, [57]. In some parts of Nigeria, the leaves' decoction is used to treat shortage of folic acid and iron, as well as anaemia. The leaves are also used as a blood purifier [58] while the cold leaf extract infusion is consumed to restore appetite and strength and the leaves are used for treating fever, tumours, gonorrhoea and piles [102].

The hepatoprotective effect of the ethanolic extracts against carbontetrachloride-induced hepatotoxicity in rats has been studied [103]. The extracts produced a significant hepatoprotective effect by reducing the levels of liver function enzymes and lipid peroxidation. Some of the phenolic antioxidants in the leaves are 3, 5-dicaffeoylquinic acid, quercetin 3-galactoside, phenolic [5-caffeoylquinic acid (chlorogenic acid), quercetin 3-(6-malonylglucoside), quercetin 3-glucoside, and quercetin 3-(6-malonylgalactoside).

4.1.7 Amaranth globe (*Gongronema latifolium*)

Gongronema latifolium commonly known as the amaranth globe is from Asclepiadaceae family. It is a tropical rainforest plant in Nigeria, used as spice and vegetable in traditional folk medicine [104]. It has a sharp-bitter, sweet taste. Useful in making sauces, preparation of salads and soup. In West Africa, it is widely used for nutritional and medicinal purposes. The aerial parts can be prepared as an infusion to treat malaria, intestinal worms, cough and dysentery. It can be taken as a tonic to address the loss of appetite. The decoction made from its stem with lime juice is taken to treat stomach-ache, diabetes and high blood pressure. Senegal and Ghana believed that the leaves when rubbed on the joints of small children could assist them to walk. The latex is applied to teeth affected by caries, used in weight loss in lactating women and for general health management. A decoction of the roots, combined with other plant species, is utilised in the treatment of sickle cell anaemia. The leaves, when macerated in alcohol is used to treat bilharzia and hepatitis [105].

Screening of Gongronema latifolium vegetable revealed the presence of phytochemicals such as alkaloids, tannins, glycosides, polyphenols, saponins and flavonoids [106]. Its antidiabetic properties have been revealed in streptozotocininduced diabetic rats during the oral administration of its aqueous and ethanolic extracts [104]. Its antibacterial activity and ability to maintain healthy blood glucose level has been documented.

Gongronema latifolium is a vegetable with pool of antioxidants with capability to prevent and treat many diseases.

4.2 Root, bulb and tuber vegetables

4.2.1 Alliums

Alliums vegetables belong to the ***Alliaceae family*** and they include garlic, onion, leek, chive, welsh onion, among others. Alliums are very rich in thiosulphides, which have an association with the reduction of various chronic diseases. Variations in the total thiosulphide content among alliums, even when grown under identical conditions have been reported [107]. The report revealed that, the total thiosulphide contents in green onion leaves, chive, and onion bulbs were found to be 0.2, 0.72, and 1.02 g/kg fresh weight, respectively. Even, the type of thiosulphides in these vegetables were found to also vary. For instance, onion bulbs contain 34% methiin, 5% ethiin, 6% propiin, 5% alliin, and 49% isoalliin [107], while garlic cloves contained about 92% alliin, 8% methiin, and trace amounts of ethiin, propiin, and isoalliin [108]. Flavonoids such as anthocyanins and flavonols like quercetin and kaempferol are found to be present in red onions and yellow fresh cultivars respectively.

Onion and garlic are rich and abundant sources of calcium, potassium and manganese providing close to 10% of their daily requirements in humans. Onion and garlic can acquire selenium if they are cultivated in soils rich in selenium, in the form of selenocysteine and seleno-proteins. This led to the proposal by [109] that selenium-enriched garlic and onion could provide a safe efficient delivery system of selenium into the body for cancer prevention [110]. Onions are also good source of chromium relevant in diabetes prevention. The mechanism appears to be through the potentiation of insulin receptor kinases [29]. A lot of clinical studies have demonstrated the ability of chromium to regulate fasting blood glucose levels, which can lead to improved glucose tolerance, decreased insulin levels and improved lipid profile levels in diabetic patients.

Onions are also rich source of dietary fibres like inulin with varying degrees of health benefits [111]. Its prebiotic properties are reflected in its preference fermentation by beneficial bowel bacteria like *Lactobacilli* and *Bifidobacteria*, thereby changing the bacterial microflora of the intestine to make pathogenic, or diseasecausing bacteria less abundant [59]. Fructans are abundant in onions and they are excellent supporters of the growth of beneficial bacteria [60]. In addition, fructans facilitate absorption of calcium and this could serve a useful purpose in the prevention of osteoporosis [112]. Diets high in fructans have also been associated with a decrease in the levels of lipid and blood glucose profiles glucose [113]. Antidiabetic potentials and antihyperglycemic effects of onions have been demonstrated [114].

The therapeutic merit and positive impact of onions, garlic and other *Allium* vegetables have been further confirmed in various epidemiological studies carried out where their consumption has been found to delay the growth of a broad spectrum of cancers. Consumption of onions has been linked to reduced incidence of stomach and intestine cancers [61] and reduction in mortality due to prostate cancer [115].

Routine intake of garlic has been linked with a decline in the occurrence of preneoplastic lesions in individuals infected by *Helicobacter pylori* [116]. It is also involved in reduction in the risk for colorectal and prostate cancers. Presumably, some garlic constituents can inhibit tumour initiation through deactivation and elimination of pro-carcinogens [117]. Some studies presented the ability of onion extracts to inhibit mutation [118], reduce multiplication process of cancer cells [119] and risks for cardiovascular diseases [120], which are effects being attributed to the presence of bioactive molecules such as quercetin **(Table 3)**.

Table 3. List of selected vegetables and their antioxidant-richness.

Vegetable	Phytochemicals present	Antioxidant	References
Broccoli	Flavonoids: anthocyanins, flavanols, flavonols Tetrapenoids: carotenoids, quinones	α carotene, β carotene, α tocopherols,	[121–127]
		Sitosterol, β-sitosterol, sitostanol,	
		campesterol, brassicasterol, stigmasterol,	
		cyanidin, catechin, luteolin, quercetin,	
		procyanidin A1, procyanidin B2	
Brussels sprout	Tetrapenoids: Carotenoids, triterpenoids: tocopherols, tocotrienols, sterols	α carotene, β carotene, α tocopherols, campesterol, β sitosterol	[122, 123, 128, 129]
Eggplant	Phenolic acids: hydrocinnamic acid Flavonoids: anthocyanin, carotenoids	Quercetin, apigenin, rutin, lutein, zeaxanthin, nasunin	[1, 130]
Onion	Triterpenoids: sterols Sulphur compounds: thiosulfinates Flavonoids: anthocyanins, flavonols Lignans	β-sitosterol, allicin, cyanidin, quercetin, kaempferol	[131]
Spinach	Triterpenoids: Phenolic terpenes Tetrapenoids: Carotenoids	Vitamin E, α carotene, β carotene, lycopene	[131]
Cabbage	Glucosinolates: Aromatic glucosinolates, aliphatic glucosinolates Phenolic acids: Hydrocinnamic acids Lignans Tetrapenoids: arotenoids	Glucobrassicin, ferulic acid, chlorogenic acid	[131]

Carrot is an important root vegetable belonging to the family Apiaceae, rich in flavonoids, carotenoids, vitamin C and vitamin E. It is a coloured vegetable with a gold mine of antioxidants such as carotenoids, polyphenols and vitamins which can function as antioxidants, anticarcinogens and immune enhancers. Carotenoids especially the ones present in orange carrots, are potent antioxidants with capacity to neutralise the toxic effect of free radicals, inhibit mutagenesis to decrease the risk of some cancers. The importance of carotenoids in lowering blood glucose level has been identified in a study where high blood glucose levels were observed in some participants with low level of carotenoids. Carotenoid level can decrease in response to severity of glucose intolerance. This suggests the impact that carrot and vitamin A-rich carotenoids could have on diabetics in the management of their condition [62].

4.3 Solanaceous vegetables

4.3.1 Tomato

Tomato is a widely grown vegetable that is globally consumed. It can be consumed fresh or in its processed forms. The phytochemical constituents in tomatoes are carotenoids, lycopene, phytoene, neurosporene, and carotenes [132].

Assessment of lycopene content based on fresh weight shows that tomato (on average) contains about 35 mg/kg of lycopene, the red cultivars have an average of 90 mg/kg of lycopene while the yellow ones have just 5 mg/kg [133].

It has been revealed that tomatoes and tomato-enriched foods are the richest sources of lycopene in the world. Dietary intake of lycopene shows that, about 25 mg is taken each day and 85% is obtained from fresh and processed products of tomatoes [134]. Also, an appreciable amount of α-, β-, γ-, δ-carotene is found in tomatoes [135]. Tomato is also a remarkable source of ascorbic acid, potassium [67], lutein, tocopherols and flavonoids [63].

The cultivar and culture have been proved to have a great influence on the flavonoid content. For instance, cherry tomatoes notably have a flavonoid content that is higher than standard tomato cultivars while the field-grown have higher flavonoid content than the greenhouse-grown [136].

Several research investigations have been ongoing to determine the relationship between dietary intake of tomato/lycopene and the risk of having cancer. Findings on different cancers relative to lycopene and tomato intake showed a great reduction in prostate cancer risk and an increased apoptotic cell death in carcinomas [64]. People subjected to diets rich in tomato and tomato-based products with high lycopene content, were found to unlikely develop stomach and rectal cancers when compared to those who consumed a lower amount of lycopene-rich vegetables [65].

The antioxidant properties of tomatoes have been described. Its daily consumption for an average of 2–4 weeks elevates the antioxidant defence system and reduces susceptibility to lipid peroxidation [137] as oxidative modification of low- density lipoproteins is key to developing atherosclerosis. Comparative studies [138] in healthy individuals and people with type 2 diabetes, showed reduced vulnerability to lipid peroxidation [139] after daily intake of tomatoes or tomato juice.

Possible anti-inflammatory, anti-thrombotic and lipid-lowering effects of tomatoes and their products have also been investigated where an aqueous extract from tomatoes demonstrated antiplatelet activity *in vitro* [140]. In humans, research shows remarkable deductions in *ex vivo* platelet aggregation a few hours after supplementation with tomato extract [140].

4.3.2 Sweet and hot peppers

Peppers are always available in a beautiful array of colours and shapes. They contribute to the flavour and colourful appearances of our dishes. Fresh peppers are excellent sources of vitamins (in form of C, K), carotenoids and flavonoids (quercetin and luteolin) [141]. Vitamins A and C are involved in the prevention of cancer, age-related diseases, reduction of inflammation and they support immune function. Vitamin K improves blood clotting, bone formation, and protects the cells against oxidative damage. Red peppers are rich in lycopene, a phytochemical commonly known for preventing prostate cancer and cancers of the bladder, cervix, and pancreas.

The nutrient content of bell peppers varies with colour as studies have shown that red coloured bell peppers have significantly higher amount of nutrients than the green counterpart. Their role in prevention of blood clot formation and reduction in the risk of heart attacks and strokes could probably be due to vitamin C, capsaicin, and flavonoids content.

Hot peppers are known for their spiciness. The major phytochemicals identified in hot peppers are capsaicinoids. Capsaicin releases about 70% of the pungent flavour in hot pepper, while dihydrocapsaicin constitutes the remaining 30% [142]. The hotness or heat experienced in the taste bud from hot peppers comes from capsaicin. It relays this sensation by acting on pain receptors in the mouth.

Predominantly, capsaicin is located in the white membranes of peppers, imparting its "heat" to seeds as well. It can lower blood cholesterol and triglycerides levels, boost immunity, and reduce the risk of stomach ulcer. Capsaicin also possesses analgesic, anti-bacterial, and antidiabetic properties. Capsaicin is included in many commercial formulations for the treatment of painful diabetic neuropathy, rheumatoid-arthritis, muscle pains, aches in the tooth and gastric ulceration [143].

The levels of vitamins and minerals present in chilli hot peppers are amazingly high. 100 g of chilli hot peppers provides 240% of vitamin C, 39% of vitamin B6 (pyridoxine), 32% of vitamin A, 13% of iron, 14% of copper and 7% of potassium [144]. They are rich in vitamins such as niacin, pyridoxine (vitamin B6), riboflavin and thiamin (vitamin B1) and minerals like manganese, iron, potassium, and magnesium. It is pertinent to note that, potassium is an important component of cell and body fluids, useful in controlling heart rate and blood pressure. Manganese serves as a co-factor for superoxide dismutase.

Hot and sweet peppers contain substances that can stimulate the body's heat production and oxygen utilisation for about 20 minutes after eating. This experience can contribute to losing extra calories and weight loss.

4.3.3 Eggplant

Eggplant is a very common and popular vegetable grown in many countries. It is grown in the subtropics, tropics and Mediterranean areas because of its demand for a long season of warm weather to produce good yields. Eggplant contains phenolic compounds such as caffeic, chlorogenic acid and flavonoids. Chlorogenic acid is a notable phenolic compound found in all eggplant cultivars with potent free radical scavenging activity [145]. Some of the benefits attributed to chlorogenic acid include antimicrobial, antiviral and anticancer activities. In addition to their nutritional potentials, these phenolic acids found in eggplant are accountable for the bitter taste that comes from the flesh when cut. Breeders have already started working on the development of eggplant cultivars that will give and ensure a balance of optimal nutritional value and pleasant taste.

There are other antioxidants such as lycopene, lutein, α-carotene, myricetin and kaempferol present in eggplant [146]. This is also depicted in **Tables 2** and **3**.

Eggplant is a good source of dietary fibre and bone-building manganese and vitamin K that can support digestion and bone-building respectively. It is also an excellent source of molybdenum and potassium, copper, vitamin C, vitamin B6, folate, and niacin [147]. Studies have demonstrated the effectiveness of eggplant in controlling the level of high blood cholesterol [66]. This was shown in a clinical study, where volunteers were fed with eggplant powder and a significant decrease in blood-lipid profile levels was observed [148]. Its relevance in inhibition of invasion of human fibrosarcoma HT-180 cell, type 2 diabetes and hypertension has been revealed [67].

5. CONCISE SUMMARY

Chronic illnesses are the primary causes of mortality. The overproduction of reactive oxygen species has been implicated in the etiology of numerous chronic illnesses. Consequently, antioxidant phytochemicals are regarded as potential agents for the prevention and management of these diseases, attributed to their biological activities and health benefits, including free radical scavenging capabilities, anti-inflammatory effects, anticancer properties, and protective actions against various diseases.

Consistent consumption of a vegetable-rich diet positively influences human health and provides protection against various chronic diseases. This review added to the evidence supporting the biological significance of frequent intake of antioxidant-rich vegetables. The antioxidant properties of vegetables are associated with the presence of phytochemical components within them. Consequently, the antioxidant components may account for the method by which vegetables mitigate disease risk by directly neutralizing free radicals, modifying gene expression, or indirectly engaging in cellular signaling related to redox equilibrium. To obtain the full health benefits associated with a vegetable-rich diet, it is advisable to consume a wide variety of vegetables to guarantee the intake of a distinctive combination of health-promoting phytonutrients.

REFERENCES

1. Septembre-Malaterreb A, Remizeb F, Pouchereta P. Fruits and vegetables, as a source of nutritional compounds and phytochemicals: Changes in bioactive compounds during lactic fermentation. Food Research International. 2018;**104**:86-99. DOI: 10.1016/j. foodres.2017.09.031
2. De Rosas MI, Deis L, Martínez L, Durán M, Malovini E, Cavagnaro JB. Anthocyanins in nutrition: Biochemistry and health benefits. P. Á. Gargiulo, HL Mesones Arroyo eds. Psychiatry and Neuroscience Update. 2019;**12**:143-152. DOI: 10.1007/978-3-319-95360-1
3. Podsędek A. Natural antioxidants and antioxidant activity of *Brassica* vegetables: A review. LWT- Food Science and Technology. 2007;**40**:1-11. DOI: 10.1016/j.lwt.2005.07.023
4. Hawkins CL, Davies MJ. Detection, identification, and quantification of oxidative protein modifications. Journal of Biological Chemistry. 2019;**294**:19683-19708. DOI: 10.1074/ jbc.REV119.006217
5. Wei Z, Shiow YW. Antioxidant activity and phenolic compounds in selected herbs. Journal of Agricultural and Food Chemistry. 2001;**49**:5165- 5170. DOI: 10.1021/jf010697n
6. Agudo A, Cabrera L, Amiano P, Ardanaz E, Barricarte A, Berenguer T, et al. Fruit and vegetable intakes, dietary antioxidant nutrients, and total mortality in Spanish adults: findings from the Spanish cohort of the European Prospective Investigation into Cancer and Nutrition (EPIC-Spain). American Journal of Clinical Nutrition. 2007;**85**(6):1634-1642. DOI: 10.1093/ ajcn/85.6.1634
7. Odukoya OA, Thomas AE, Adepoju-Bello A. Tannic acid equivalent and cytotoxic activity of selected medicinal plants. West African Journal of Pharmacology. 2001;**15**:43-45
8. Atawodi SE. Antioxidant potential of African medicinal plants. African Journal of Biotechnology. 2005; **4**(2):128-133
9. Amin I, Zamaliah MM, Chin WF. Total antioxidant activity and phenolic content in selected vegetables. Food Chemistry. 2004;**87**:581-586. DOI: 10.1016/j.foodchem.2004.01.010
10. Ames BN, Shigenaga MK, Hagen TM. Oxidants, antioxidants, and the degenerative diseases of aging. Proceedings of the National Academy of Sciences. 1993;**90**:7915-7922. DOI: 10.1073/pnas.90.17.7915

11. Halliwell B, Gutteridge JM. Free Radicals in Biology and Medicine. 3rd ed. Oxford: Clarendon Press, Oxford University Press; 1989. pp. 1-25
12. Amin I, Zamaliah MM, Chin WF. Total Antioxidant activity and phenolic content of selected vegetables. Food Chemistry. 2004;**87**:581-586
13. Sahlin E, Savage GP, Lister CE. Investigation of the antioxidant properties of tomatoes after processing. Journal of Food Composition and Analysis. 2004;**17**:635-647. DOI: 10.1016/j.jfca.2003.10.003
14. Ames BN, Shigenaga MK, Hagen TM. Oxidants, antioxidants, and the degenerative diseases of aging. Proceedings of the National Academy of Science. 1993;**90**:7915-7922. DOI: 10.1073/pnas.90.17.7915
15. Sung J, Lee J. Antioxidant and antiproliferative activities of grape seeds from different cultivars. Food Science and Biotechnology. 2010;**19**:321-326. DOI: 10.1007/s10068-010-0046-6
16. Zhang YJ, Gan RY, Li S, Zhou Y, Li AN, Xu DP, et al. Antioxidant phytochemicals for the prevention and treatment of chronic diseases. Molecules. 2015;**20**(12):21138-21156. DOI: 10.3390/molecules201219753
17. Cao G, Booth SL, Sadowski JA, Prior RL. Increases in human plasma antioxidant capacity after consumption of controlled diets high in fruit and vegetables. American Journal of Clinical Nutrition. 1998;**68**:1081-1087. DOI: 10.1093/ajcn/68.5.1081
18. Hounsome N, Hounsome B. Biochemistry of vegetables: Major classes of primary (carbohydrates, amino acids, fatty acids, vitamins and organic acids) and secondary metabolites (terpenoids, phenolics, alkaloids, and sulphur-containing compounds), in vegetables. In: Sinha NK, Hui YH, Evranuz EO, Siddiq M, Ahmed J, editors. Handbook of Vegetables and Vegetable Processing. Iowa, USA: Blackwell Publishing; 2011. pp. 23-58. DOI: 10.1002/9780470958346
19. Tiwari U, Cummins E. Factors influencing levels of phytochemicals in selected fruit and vegetables during pre-and post-harvest food processing operations. Food Research International. 2013;**50**:497-506. DOI: 10.1016/j.foodres.2011.09.007
20. USDA. United States department of agriculture, agricultural research service. In: USDA National Nutrient Database for Standard Reference Release. 2005. Available from: http// www.nal.usda.gov/fnic/foodcomp/Data/2005(18) [Accessed: 18 August 2013]
21. Lugasi A, Biro L, Hovarie J, Sagi KV, Brand S, Barna E. Lycopene Content of Foods and Lycopene Intake in Two Groups of the Hungarian Population. Nutrition Research. 2003;**23**(8):1035-1044. DOI: 10.1016/ S0271-5317(03)00105-2

22. Rodriguez-Bonilla P, Gandia-Herrero F, Matencio A, Garcia-Carmona F, Lopez-Nicolas JM. Comparative study of the antioxidant capacity of four stilbenes using ORAC, ABTS+, and FRAP techniques. Food Analytical Methods. 2017;**10**:2994- 3000. DOI: 10.1007/s12161-017-0871-9
23. Kotíková Z, Lachman J, Hejtmánková A, Hejtmánková K. Determination of antioxidant activity and antioxidant content in tomato varieties and evaluation of mutual interactions between antioxidants. LWT Food Science and Technology. 2011;**44**:1703-1710. DOI: 10.1016/j. Iwt.2011.03.015
24. Delgado-Vargas F, Paredes-Lopez O. Natural Colorants for Food and Nutraceutical Uses. Boca Raton, FL, USA: CRC Press; 2003. ISBN 1587160765
25. Chen P, Zhang W, Wang X, Zhao K, Negi DS, Zhuo L, et al. Lycopene and risk of prostate cancer: A systematic review and meta-analysis. Medicne (Baltimore). 2015;**94**:33. DOI: 10.1097/ MD.0000000000001260
26. Denniston K, Topping JJ, Caret R. Minerals and cellular function. In: General, Organic and Biochemistry. 4th ed. New York, USA: McGraw Hill, Higher Education; 2004. pp. 797-799
27. Produce for Better Health Foundation. "Phytochemical Info Center". Web. 2015
28. Heim KE, Tagliaferro AR, Bobilya DJ. Flavonoid antioxidants: Chemistry metabolism and structure- activity relationships. Journal of Nutritional Biochemistry. 2002;**13**(10):572-584. DOI: 10.1016/ s0955-2863(02)00208-5
29. Mullen W, Boitier A, Stewart AJ, Crozier A. Flavonoid metabolites in human plasma and urine after the consumption of red onions: Analysis by liquid chromatography with photodiode array and full scan tandem mass spectrometric detection. Journal of Chromatography A. 2004;**1058**:163-168. DOI: 10.1016/S0021-9673(04)01476-1
30. Williams RJ, Spencer JPE, Rice-Evans C. Flavonoids: Antioxidants or signalling molecules? Free Radical Biol. Med. 2004;**36**:838-849. DOI: 10.1016/j.freeradbiomed.2004.01.001
31. Jaganath IB, Crozier A. Overview of health-promoting compounds in fruit and vegetables. In: Tomas-Barberan FA, Gil MI, editors. Improving the Health- promoting Properties of Fruit and Vegetable Products. Cambridge: Woodhead Publishing; 2008. pp. 3-4
32. Seger R, Krebs EG. The MAPK signaling cascade. Federation for American Societies Journal of Experimental Biology. 1995;**9**:726-735. DOI: DOI. org/10.1096/ fasebj.9.9.7601337
33. Jaganath IB, Crozier A. Overview of health-promoting compounds in fruit and vegetables. In: Tomas-Barberan FA, Gil MI, editors. Improving the Health- promoting Properties of Fruit and Vegetable Products. Cambridge: Woodhead Publishing; 2008. pp. 3-4

34. Van Dross RT, Hong X, Essengue S, Fischer SM, Pelling JC. Modulation of UVB-induced and basal cyclooxygenase-2 (COX-2) expression by apigenin in mouse keratinocytes: Role of USF transcription factors. Molecular Carcinogenesis. 2007;**46**(4):303-314. DOI: 10.1002/mc.20281

35. Crespo I, Garcia-Mediavilla MV, Gutierrez B, Sanchez-Campos S, Tunon MJ, Gonzalez-Gallego JA. Comparison of the effects of kaempferol and quercetin on cytokine-induced pro-inflammatory status of cultured human endothelial cells. British Journal of Nutrition. 2008;**100**(5):968-976. DOI: 10.1017/S0007114508966083

36. Park HH, Lee HHS, Son HY, Park SB, Kim MS, Choi EJ, et al. Flavonoids inhibit histamine release and expression of proinflammatory cytokines in mast cells. Archives of Pharmaca Research. 2008;**31**(10):1303-1311. DOI: 10.1007/s12272-001-2110-5

37. Ruiz PA, Braune A, Holzlwimmer G, Quintanilla-Fend L, Haller D. Quercetin inhibits TNF-induced NF- κB transcription factor recruitment to proinflammatory gene promoters in murine intestinal epithelial cells. Journal of Nutrition. 2007;**137**(5):1208-1215

38. Torres-Piedra M, Ortiz-Andrade R, Villalobos-Molina R, Singh N, Medina-Franco JL, Webster SP, et al. A comparative study of flavonoid analogues on streptozotocin- nicotinamide induced diabetic rats: Quercetin as a potential antidiabetic agent acting via 11β-hydroxysteroid dehydrogenase type 1 inhibition. European Journal of Medicinal Chemistry. 2010;**45**(6):2606-2612

39. De SD, Maiuri MC, Simeon V, Grassia G, Soscia A, Cinelli MP, et al. Lycopene, quercetin and tyrosol prevent macrophage activation induced by Gliadin and IFN-γ. European Journal of Pharmacology. 2007;**566**(1-3):192-199. DOI: 10.1016/j.ejphar.2007.03.051

40. Kim GY, Kim JH, Ahn SC, Lee HJ, Moon DO, Lee CM, et al. Lycopene suppresses the lipopolysaccharide- induced phenotypic and functional maturation of murine dendritic cells through inhibition of mitogen-activated protein kinases and nuclear factor-κB. Immunology. 2004;**113**(2):203-211. DOI: 10.1111/j.1365-2567.2004.01945.x

41. Bai SK, Lee SJ, Na HJ, Ha KS, Han JA, Lee H, et al. β-Carotene inhibits inflammatory gene expression in lipopolysaccharide-stimulated macrophages by suppressing redox- based NF-κB activation. Experimental and Molecular Medicine. 2005;**37**(4):323-334. DOI: 10.1038/ emm.2005.42

42. Kim JH, Na HJ, Kim CK, Kim JY, Ha KS, Lee H, et al. The non-provitamin A carotenoid, lutein, inhibits NF-κB- dependent gene expression through redox-based regulation of the phosphatidylinositol 3-kinase/PTEN/ Akt and NF-κB-inducing kinase pathways: Role of H2O2 in NF-κB activation. Free Radical Biology and Medicine. 2008;**45**(6):885-896. DOI: 10.1016/j.freeradbiomed.2008.06.019

43. Rouphael Y, Kyriacou MC, Vitaglione P, Giordano M, Pannico A, Colantuono A, et al. Genotypic variation in nutritional and antioxidant profile among iceberg lettuce cultivars. Acta Scientiarum Polonorum Hortorum Cultus. 2017;**16**:37-45. DOI: 10.24326/ asphc.2017.3.4

44. Mou B. Lettuce. In: Prohens J, Nuez F, editors. Handbook of Plant Breeding-Vegetables, I Asteraceae, Brassicaceae, Chenopodiaceae, and Cucurbitaceae. New York, NY, USA: Springer; 2008. pp. 75-116

45. Wang C, Riedl KM, Schwartz SJ. Fate of folates during vegetable juice processing—Deglutamylation and interconversion. Food Research International. 2013;**53**:440-448. DOI: 10.1016/j.foodres.2013.05.011

46. Gan YZ, Azrina A. Antioxidant properties of selected varieties of lettuce (Lactuca sativa L.) commercially available in Malaysia. International Food Research Journal. 2016;**23**:2357-2362

47. Rouphael Y, Kyriacou MC, Petropoulos SA, De Pascale S, Colla G. Improving vegetable quality in controlled environments. Horticultural Science. 2018;**234**:275-289

48. Kyriacou MC, Rouphael Y. Towards a new definition of quality for fresh fruits and vegetables. Scientia Horticulturae. 2018;**234**:463-469. DOI: 10.1016/j.scienta.2017/09/046

49. Survase SA, Bajaj IB, Singhal RS. Biotechnological production of vitamins. Food Technol. Biotechnol. 2006;**44**(3):381-396. ISSN 1330-9862

50. López A, Javier GA, Fenoll J, Hellín P, Flores P. Chemical composition and antioxidant capacity of lettuce: Comparative study of regular-sized (Romaine) and baby-sized (Little Gem and Mini Romaine) types. Journal of food Composition and Analysis. 2014;**33**:39-48. ISSN: 0889-1575

51. Mou B. Genetic variation of beta- carotene and lutein contents in lettuce. American Socoety for Horticultural Science. 2005;**130**(6):870-876. DOI: 10.21273/JASHS.130.6.870

52. Baslam M, Morales F, Garmendia I, Goicoechea N. Nutritional quality of outer and inner leaves of green and red pigmented lettuces (Lactuca sativa L.) consumed as salads. Scientia Horticulturae. 2013;**151**:103-111. DOI: 10.1016/j.scienta.2012.12.023

53. Kim MJ, Moon Y, Tou JC, Mou B, Waterland NL. Nutritional value, bioactive compounds and health benefits of lettuce (Lactuca sativa L.). Journal of Food Composition and Analysis. **2016;49**:19-34. DOI: 10.1016/j. jfca.2016.03.004

54. Johnson EJ, Hammond BR, Yeum KJ, Qin J, Wang XD, Castaneda C, et al. Relation among serum and tissue concentrations of lutein and zeaxanthin and macular pigment density. American Journal of Clinical Nutrition. 2000;**71**(6):1555-1562. DOI: 10.1093/ ajcn/71.6.1555

55. Kim DE, Shang X, Assefa AD, Keum YS, Saini RK. Metabolite profiling of green, green/red, and red lettuce cultivars: Variation in health benefi-

cial compounds and antioxidant potential. Food Research International. 2018;**105**:361-370. DOI: 10.1016/j.foodres.2017.11.028

56. Mulabagal V, Ngouajio M, Nair A, Zhang Y, Gottumukkala AL, Nair MG. *In vitro* evaluation of red and green lettuce (Lactuca sativa) for functional food properties. Food Chemistry. 2010;**118**:300-306. DOI: 10.1016/j. foodchem.2009.04.119
57. Rouphael Y, Petropoulos SA, El-Nakhel C, Pannico A, Kyriacou MC, Giordano M, et al. Reducing energy requirements in future bioregenerative life support systems (BLSSs): Performance and bioactive composition of diverse lettuce genotypes grown under optimal and suboptimal light conditions. Frontiers in Plant Science. 2019;**10**:1305. DOI: 10.3389/fpls.2019.01305. eCollection 2019
58. Pannico A, El-Nakhel C, Kyriacou MC, Giordano M, Stazi SR, De Pascale S, et al. Combating micronutrient deficiency and enhancing food functional quality through selenium fortification of selectlettuce genotypes grown in a closed soilless system. Frontiers in Plant Science. 2019;**10**:1495
59. Rice-Evans CA, Miller NJ, Paganga G. Antioxidant properties of phenolic compounds. Trends in Plant Science. 1997;**2**(4):152-159. DOI: 10.1016/S1360-1385(97)01018-2
60. Lee JH, Felipe P, Yang YH, Kim MY, Oh YK, Sok DE, et al. Kim MRE_Effects of dietary supplementation with red-pigmented leafy lettuce (Lactuca sativa) on lipid profiles and antioxidant status in C57BL/6J mice fed a high-fat high-cholesterol diet. British Journal of Nutrition. 2009;**101**:1246-1254
61. Nicolle C, Cardinault N, Gueux E, Ja_relo L, Rock E, Mazur A, et al. Health effect of vegetable-based diet: Lettuce consumption improves cholesterol metabolism and antioxidant statusin the rat. Clinical Nutrition. 2004;**23**(4):605-614. DOI: 10.1016/j.clnu.2003.10.009
62. Neugart S, Baldermann S, Hanschen FS, Klopsch R, Wiesner-Reinhold M, Schreiner M. The intrinsic quality of brassicaceous vegetables: How secondary plant metabolites are affected by genetic, environmental, and agronomic factors. Scientia Horticulturae. 2018;**233**:460-478. DOI: org/10.1016/j. scienta.2017.12.038
63. Martínez-Sánchez A, Gil-Izquierdo A, Gil MI, Ferreres FA. comparative study of flavonoid compounds, vitamin C, and antioxidant properties of baby leaf Brassicaceae species. Journal of Agriculture and Food Chemistry. 2008;**56**(7):2330-2340. DOI: 10.1021/jf072975+
64. Mageney V, Baldermann S, Albach DC. Intraspecific variation in carotenoids of brassica oleracea var. sabellica. Journal of Agriculture Food and Chemistry. 2016;**64**(16):3251-3257. DOI: 10.1021/acs.jafc.6b00268
65. Zheng YJ, Zhang YT, Liu HC, Li YM, Liu YL, Hao YW, et al. Supplemental blue

light increases growth and quality of greenhouse pak choi depending on cultivar and supplemental light intensity. Journal of Integrative Agriculture. 2018;**17**:2245-2256. DOI: 10.1016/S2095-3119(18)62064-7

66. Khoo HE, Azlan A, Tang ST, Lim SM. Anthocyanidins and anthocyanins: Colored pigments as food, pharmaceutical ingredients, and the potential health benefits. Food and Nutrition Research. 2017;**61**(1):1361779. DOI: 10.1080/16546628.2017.1361779
67. Akpaso MI, Atangwho IJ, Akpantah A, Fischer1 VA. Igiri AO and Ebong PE. Effect of combined leaf extracts of *Vernonia amygdalina* (Bitter Leaf) and *Gongronema latifolium* (Utazi) on the Pancreatic β-Cells of Streptozotocin-induced diabetic rats. Journal of Advances in Medicine and Medical Research. 2011;**1**(1):24-34. DOI: 10.9734/BJMMR/2011/215
68. Huffman MA, Seifu M. Observation on the illness and consumption of a possibly medicinal plant Vernonia amygdalina (Del.), by a wild chimpanzee in the Mahale Mountains National Park, Tanzania. Primates. 1989;**30**:51-63. DOI: 10.1007/ BF02381210
69. Igile GO, Olezek W, Jurzysata M, Burda S, Fafunso M, Fasanmade AA. Flavonoids from Vernonia amygdalina and their antioxidant activities. Journal of Agricultural and Food Chemistry. 1994;**42**(11):2445-2448. DOI: 10.1021/ jf00047a015
70. Iwu MM. Empirical investigation of dietary plants used in Igbo- Ethnomedicine. In: Iwu MM, editor. Plants in Indigenous Medicine and Diet. Nina Etkined Redgrove. 1st ed. New York: Publishers Co; 1986. pp. 131-150. eBook ISBN: 9781315060385
71. Nwanjo HU. Efficacy of aqueous leaf extract of *Vernoniaamygdalina* on plasma lipoprotein and oxidative status in diabetic rat models. Nigerian Journal of Physiological Sciences. 2005;**20**(1-2):39-42
72. Gutpa S, Shukla R, Prabhu KM, Agarwal S, Rusia U, Murthy PS. Acute and chronic toxicity studies on partially purified hypoglycemic preparation from water extract of bark of Ficus bengalensis. Indian Journal of Clinical Biochemistry. 2002;**17**(**1**):56-63. DOI: 10.1007/BF02867943
73. Atangwho IJ, Ebong PE, Egbung GE, Eteng MU, Eyong EU. Effect of *Vernonia amygdalina* Del. on liver function in alloxaninduced hyperglycaemic rats. Journal of Pharmacy and Bioresources. 2007a;**4**(1):25-31. DOI: 10.4314/jpb.v4i1.32107
74. Jisaka M, Ohigashi H, Takegawa K, Hirota M, Irie R, Huffman MA, Koshmizu K. Steroid gluccosides from *Vernonia amygdalina,* a possible chimpanzee medicinal plant. Phytochemistry. 1993a; 34 (2): 409-413. ISSN : 0031-9422
75. Izevbigie EB. Discovery of water- soluble anticancer agents (edotides) from a vegetable found in Benin City, Nigeria. Experimental Biolology and Medicine. 2003;**228**(3):293-298. DOI: 10.1177/153537020322800308
76. Farombi EO, Owoeye O. Antioxidative and chemopreventive properties of

vernonia amygdalina and garcinia biflavonoid. International Journal of Environmental Research and Public Health. 2011;**8**(6):2533-2555. DOI: 10.3390/ijerph8062533

77. Oyugi DA, Luo X, Lee KS, Hill B, Izevbigie EB. Activity markers of the anti-breast carcinoma cell growth fractions of Vernonia amygdalina extracts. Experimental Biology and Medicine. 2009;**234**(4):410-417. DOI: 10.3181/0811-RM-325
78. Izevbigie EB, Byrant JL, Walker A. A novel natural inhibitor of extracellular signal-regulated kinases and human breast cancer cell growth. Experimental Biology and Medicine. 2004;**229**(2):163-169. DOI: 10.1177/153537020422900205
79. Gbile ZO. Ethnobotany, taxonomy and conservation of medicinal plants. In: Sofowora A, editor. The State of Medicinal Plants Research in Nigeria. Ibadan, Nigeria: University of Ibadan Press; 1986
80. Oboh G. Hepatoprotective property of ethanolic and aqueous extracts of *Telfairia occidentalis* (Fluted Pumpkin) leaves against garlic-induced oxidative stress. Journal of Medicinal Food. 2005;**8**:560-563
81. Kayode AAA, Kayode OT, Odetola AA. *Telfairia occidentalis* ameliorates oxidative brain damage in malnorished rats. International Journal of Biological Chemistry. 2010;**4**(1):10- 18. DOI: 10.3923/ijbc.2010.10.18
82. Tikkiwal M, Ajmera RL, Mathur NK. Effect of zinc administration on seminal zinc and fertility of oligospermic males. Indian. J. Physiol. Pharmacol. 1987;**31**(1):30-34
83. Salman TM, Olayaki LA, Oyeyemi WA. Aqueous extract of Telfairia occidentalis leaves reduces blood sugar and increases haematological and reproductive indices in male rats. African Journal of Biotechnology. 2008;**7**(14):2299-2303. eISSN: 1684-5315
84. Ogbe RJ, Adoga GI, Abu AH. Anti-anaemic potentials of some plant extracts on phenyl hydrazine-induced anaemia in rabbit. Journal of Medicinal Plants Research. 2010;**4**(8):680-684. DOI: 10.5897/JMPR09.487
85. Onajobi FD. Smooth muscle contracting lipid soluble principles in chromatographic fractions of ocimum gratissimum. Journal of Ethnopharmacology. 1986;**18**(1):3-11. DOI: 10.1016/0378-8741(86)90038-3
86. Zeghichi SS, Kallithkara, Simopoulos AP. Nutritional composition of molehiya (*Corchorus olitorius*) and Stamnagathi (*Cichorium spinosum*). In: Simopoulus AP, Gopalan C, editors. Plants in human health and nutrition policy. Basel: Karger; 2011. pp. 1-22
87. Aiyeloja AA, Bello OA. Ethnobotanical potentials of common herbs in Nigeria: A case study of Enugu state. Educational Research and Reviews. 2006;**1**(1):16-22. ISSN-1990-3839

88. Fasinmirin JT, Olufayo AA. Yield and water use efficiency of jute mallow *Corchorus olitorius* under varying soil water management strategies. Journal of Medicinal Plants Research. 2009;**3**(4):186-191. ISSN 1996-0875

89. Salawu SO, Akindahunsi AA. Protective effect of some tropical vegetables against CCl4 -induced hepatic damage. Journal of Medicinal Food. 2007;**10**(2):350-355. DOI: 10.1089/jmf.2006.212

90. Ugochukwu NH, Babady NE. Antihyperglycemic effect of aqueous and ethanolic extracts of *Gongronema latifolium* leaves on glucose and glycogen metabolism in livers of normal and streptozotocin-induced diabetic rats. Life Sciences. 2003;**73**(15):1925-1938. DOI: 10.1016/s0024-3205(03)00543-5

91. Burkill HN. The Useful Plants of West Tropical Africa, Kew, published by Royal Botanic Gardens. 2nd ed. Kew: University Press of Virginia; 1985, 40 (176. pp. 456-596. DOI: 10.1007/ BF02859140

92. Atangwho IJ, Ebong PE, Eyong EU, Williams IO, Eteng MU, Egbung GE. Comparative chemical composition of leaves of some antidiabetic medicinal plants: Azadirachta indica, Vernonia amygdalina and Gongronema latifolium. African Journal of Biotechnology. 2009b;**8**(18):4685-4689

93. Kubec R, Svobodova M, Velisek J. Distribution of *S*-Alk(en)ylcysteine sulfoxides in some *Allium* Spe- cies. Identification of a New flavour precursor: *S*-ethyl- cysteine sulfoxide (ethiin). Journal of Agriculture and Food Chemistry. 2000;**48**(2):428-433. DOI: 10.1021/jf990938f

94. Kubec R, Svobodova M, Velisek J. Gas-chromatographic determination of S-Alk(eny)lylcysteine sulfoxide. Journal of Chromatography. 1999;**862**(1):85-94. DOI: 10.1016/s0021-9673(99)00902-4

95. Ip C, Lisk DJ. Enrichment of selenium in allium vegetables for cancer prevention. Carcinogenesis. 1994;**15**(9):1881-1885. DOI: 10.1093/ carcin/15.9.1881

96. El-Bayoumy K. The role of selenium in cancer prevention. In: DeVita VT, Hellman S, Rosen-berg SA, editors. Cancer Principles and Practices of Oncology. Philadelphia; 1991. pp. 1-15

97. Wang H, Kruszewki A, Brautigan DI. Cellular chromium activation of insulin receptor kinase. Biochemistry. 2005;**44**(22):8167-8175. DOI: org/10.1021/ bi0473152

98. Ritsema T, Smeekens S. Fructans: Beneficial for Plants and Humans. Current Opinion in Plant Biology. 2003;**6**(**3**):223-230. DOI: 10.1016/ S1369-5266(03)00034-7

99. Kruse HP, Kleessen B, Blaut M. Effects of inulin on faecal bifidobacteria in human subjects. The British Journal of Nutrition. 1999;**82**(5):375-382. DOI: 10.1017/s000711459900162

100. Kilian S, Kritzinger S, Rycroft C, Gibson GR, Du Preez J. The effects of the

novel bifidogenic trisac- charide, neokestose, on the human colonic microbiota. World Journal of Microbiology and Biotechnology. 2002;**18**(7):637-644. DOI: 10.1023/A:1016808015630

101. Scholz-Ahrens KE, Schaafsma G, Van Den Heuvel EGHM, Schrezenmeir J. Effects of prebiotics on mineral metabolism. The American Journal of Clinical Nutrition. 2001;**73**(2):459-464
102. Jackson KG, Taylor GR, Clohessy AM, Willieams CM. The effect of the daily intake of inulin on fasting lipid, insulin and glucose concentrations in middle-aged men and women. The British Journal of Nutrition. 1999;**82**(1):23-30. DOI: 10.1017/ s0007114599001087
103. Srinivasan K. Plant foods in the management of diabetes mellitus: Spices as beneficial antidiabetic food adjuncts. International Journal of Food Science Nutrition. 2005;**56**(6):399-414. DOI: 10.1080/09637480500512872
104. Dorant E, Van Den Brandt PA, Goldbohm RA, Sturnmans F. Comsumption of onions and a reduced risk of stomach carcinoma. Gastroenterology. 1996;**110**(1):12-20. DOI: 10.1053/gast.1996.v110.pm8536847
105. Grant WB. A Multicountry ecologic study of risk and risk reduction factors for prostate cancer mortality. European Urology. 2004;**45**(3):271-279. DOI: 10.1016/j.eururo.2003.08.018
106. You WC, Zhang L, Gail MH, Ma JL, Chang YS, Blot WJ, et al. Helicobacter pylori infection, garlic intake and pre- cancerous lesions in a chinese population at low risk of gastric cancer. International Journal of Epidemiology. 1998;**27**(6):941- 944. DOI: 10.1093/ije/27.6.941
107. Melino S, Sabelli R, Paci M. Allyl sulfur com-pounds and cellular detoxification system: Effects and perspectives in cancer therapy. Amino Acids. 2011;**41**(1):103-112. DOI: 10.1007/s00726-010-0522-6
108. Shon MY, Choi SD, Kahng GG, Nam SH, Sung NJ. Antimutagenic, antioxidant and free radical scavenging activity of ethyl acetate extracts from white, yellow and red onions. Food Chemical Toxicology. 2004;**42**(4):659- 666. DOI: 10.1016/j.fct.2003.12.002
109. Yang J, Meyers KJ, Van der Heide J, Liu RH. Varietal differences in phenolic content and antioxi- dant and antiproliferative activities of onions. Journal of Agriculture and Food Chemistry. 2004;**52**(22):6787-6793. DOI: 10.1021/ jf0307144
110. Osmont KS, Arnt CR, Goldman IL. Temporal aspects of onion-induced antiplatelet activity. Plant Foods for Human Nutrition. 2003;**58**(1):27-40. DOI: 10.1023/A:1024062330700
111. Prasoona J, Kumari BA, Sarkar S, Kiran VK, Swamy R. Carrot and beetroot greens: An underutilized green leafy vegetables with health benefits. Just

Agriculture. 2021;**1**(7):2582-2822

112. Clinton S. Lycopene: Chemistry, biology and implication for human health and disease. Nutrition Reviews. 1998;**56**(2):35-51. DOI: 10.1111/j.1753-4887.1998.tb01691.x
113. Scott KJ, Hart DJ. Development and evaluation of an HPLC method for the analysis of carotenoids food and the measurement of the carotenoid content vegetables and fruits commonly consumed in the UK. Food Chemistry. 1995;**54**(1):101-111. DOI: 10.1016/ 0308-8146(95)92669-b
114. Rao AV, Waseem Z, Agarwal S. Lycopene contents of tomatoes and tomato products and their contribution to dietary lycopene. Food Research International. 1998;**31**(10):737-741. DOI: 10.1016/S0963-9969(99)00053-8
115. Albushita AA, Daood HG, Biacs PA. Change in carotenoids and antioxidant vitamins in tomato as a function of varietal and technological factors. Journal of Agriculture and Food Chemistry. 2000;**48**(6):2075-2081. DOI: 10.1021/ jf990715p
116. Albushita AA, Hebshi EA, Daood HG, Biacs PA. Determination of antioxidant vitamins in tomato. Food Chemistry. 1997;**60**(2):207-212
117. Stewart AJ, Bozonnet S, Mullen W, Jenkins GI, Lean ME, Crozier MEA. Occurrence of flavonols in tomatoes and tomato-based products. Journal of Agriculture and Food Chemistry. 2000;**48**(7):2663-2669. DOI: 10.1021/ jf000070p
118. Kim HS, Bowen P, Chen L. Effects of tomato sauce consumption on apoptotic cell death in prostate benign hyperplasia and carcinoma. Nutrition and Cancer. 2003;**47**(1):40-47. DOI: 10.1207/s15327914nc4701_5
119. Giovannucci E. Tomatoes, tomato- based products, lycopene, and cancer: Review of the epidemiological literature. Journal of the National Cancer Institute. 1999;**91**(4):317-331. DOI: 10.1093/jnci/91.4.317
120. Briviba K, Schnabele K, Rechkemmer G, Bub A. Supplementation of a diet low in carotenoids with tomato or carrot juice does not affect lipid peroxidation in plasma and feces of healthy men. The Journal of Nutrition. 2004;**134**(5):1081- 1083. DOI: 10.1093/jn/134.5.1081
121. Bose KS, Agrawal BK. Effect of long-term supplementation of tomatoes (cooked) on levels of an-tioxidant enzymes, lipid peroxidation rate, lipid profile and glycated haemoglobin in type 2 diabetes mellitus. West Indian Medical Journal. 2006;**55**(4):274-278. DOI: 10.1590/ s0043-31442006000400010
122. Upritchard JE, Sutherland WH, Mann JI. Effect of supplementation with tomato juice, Vitamin E and Vitamin C on LDL oxidation and products of inflammatory activity in type 2 diabetes. Diabetes Care. 2000;**23**(6):733-738. DOI: 10.2337/diacare.23.6.733
123. O'Kennedy CNL, Whelan S. Effects of tomato extract on platelet function: A double-blinded crossover study in healthy humans. The American Journal of Clinical Nutrition. 2006;**84**(3):561-569
124. Bosland PW. Capsicums: Innovative uses of an ancient crop. In: Janick J, ed-

itor. Progress in New Crops. Arlington: ASHS Press; 1996. pp. 479-487

125. Suzuki T, Iwai K. Constituents of red pepper species: Chemistry, biochemistry, pharmacology, and food science of the pungent principle of *Capsicum* species. In: Brossi A, editor. The Alkaloids. San Diego: Academic Press; 1984. pp. 227-299. DOI: 10.1016/ S0099-9598(08)60072-3
126. Szallasi A, Blumberg PM. Vanilloid (Capsaicin) receptors and mechanisms. Pharmacological Reviews. 1999;**51**(2):159-211
127. Frei B, Lawson S. Vitamin C and cancer revisited. Proceedings of the National Academy of Sciences (USA). 2008;**105**(32):11037-11038. DOI: org/10.1073/ pnas.0806433105
128. Matsuzoe N, Yamaguchi M, Kawanobu S, Watanabe Y, Higashi H, Sakata Y. Effect of dark treatment of the eggplant on fruit skin color and its anthocyanin components. Journal of the Japanese Society for Horti-cultural Science. 1999;**68**(1):138-145. DOI: 10.2503/JJSHS.68.138
129. Aen-Amos A, Fishler R. Analysis of carotenoids with emphasis on 9-*cis* *β*-carotene in vegetables and fruits commonly consumed in Israel. Food Chemistry. 1998;**62**(4):515-520. DOI: 10.1016/S0308-8146(97)00196-9
130. Wood R. The Whole Foods Encyclopedia. New York: Prentice- Hall Press; 1988
131. Jorge PA, Neyra LC, Osaki RM. Effect of eggplant on plasma lipid levels, lipidic peroxidation and reversion of endothelial dysfunction in experimental hypercholesterolemia. Arquivos. Brasileiros de Cardiologia. 1998;**70**(2):87-91. DOI: 10.1590/s0066-782x1998000200004
132. Guimarães PR, Galvão AM, Batista CM, Azevedo GS, Oliveira RD, Lamounier R, et al. Eggplant (*Solanum melongena*) infusion has a modest and transitory effect on hypercholesterolemic subjects. Brazilian Journal of Medical and Biological Research. 2000;**33**(9):1027-1036
133. Kwon YI, Apostolidis E, Shetty K. *In vitro* studies of eggplant (*Solanum melongena*) phenolics as in-hibitors of key enzymes relevant for type 2 diabetes and hypertension. Bioresource Technology. 2008;**99**(8):2981-2988. DOI: 10.1016/j. biortech.2007.06.035
134. Berrino F, Villarini A. Fruit and vegetables and cancer. In: Tomas-Barberan FA, Gil MI, editors. Improving the Health-Promoting Properties of Fruit and Vegetable Products. Cambridge: Woodhead Publishing; 2008. pp. 76-94
135. Harnly JM, Doherty RF, Beecher GR, Holden JM, Haytowitz DB, Bhagwat S, et al. Flavonoid content of U.S. fruits, vegetables, and nuts. Journal of Agriculture Food and Chem. 2006;**54**(26):9966-9977. DOI: 10.1021/ jf061478a
136. Bohm V, Puspitasari-Nienaber NL, Ferruzi MG, Schwartz SJ. Trolox equivalent antioxidant capacity of different geometrical isomers of Alpha- carotene, Beta-carotene, Lycopene, and Zeaxanthin. Journal of Agriculture Food and Chem. 2002;**50**(1):221-226. DOI: 10.1021/jf010888q

137. Brown JE, Khodr H, Hider RC, Rice-Evans, C. A. Structural dependence of flavonoid interactions with Cu2 + Ions: Implications for their antioxidant properties. Biochemistry Journal 1998; **330** (30): 1173-1178. DOI: 10.1042/bj3301173
138. Damon M, Zhang NZ, Haytowitz DB, Booth SL. Phylloquinone (Vitamin K1) content of vegetables. Journal of Food and Compound Analysis. 2005;**18**(8):751-758. DOI: 10.1016/j.jfca.2004.07.004
139. Chen WJ, Song JR, Guo P, Wen ZY. Butein, a More effective antioxidant than α-tocopherol. Journal of Molecular Structure. 2006;**736**(1-3):161-164. DOI: 10.1016/j.theochem.2005.12.035
140. Normen L, Johnsson M, Anderson H, van Gameren Y, Dutta P. Plant sterol in vegetables and fruits commonly consumed in Sweden. European Journal of Nutrition. 1999;**38**(2):84-89. DOI: 10.1007/ s003940050048
141. Yoshida Y, Niki E. Antioxidant effects of phytosterol and its component. Journal of nutritional Science and Vitaminology. 2003;**49**(4):277-280. DOI: 10.3177/ jnsv.49.277
142. Rhodes CJ, Dintinger TC, Moynihan HA, Reid ID. Radio labelling studies of free radical reactions using muonium (The Second Hydrogen Radioisotope): Evidence of a direct antioxidant role for vitamin K in repair of oxidative damage to lipids. Magnetic Resonance in Chemistry. 2000;**38**(8):646-649. DOI: 10.1002/1097-458X
143. Fahey JW, Zalcmann AT, Talalay P. The chemical diversity and distribution of glucosinolates and isothiocyanates among plants. Phytochemistry. 2001;**56**(*1*):5-51. DOI: 10.1016/s0031-9422(00)00316-2
144. Gürbüz N, Uluişik S, Frary A, Doğanlar S. Health benefits and bioactive compounds of eggplant. Food Chemistry. 2018;**268**:602-610. DOI: 10.1016/j.foodchem.2018.06.093
145. van Breda SGJ, de Kok TMCM. Smart combinations of bioactive compounds in fruits and vegetables may guide new strategies for personalized prevention of chronic diseases. Molecular Nutrition and Food Research. 2018;**62**(1):1700597. DOI: 10.1002/ mnfr.201700597
146. El-Sayed MA, Humayoun A, Khalid Z, Rashida A. Dietary sources of lutein and zeaxanthin carotenoids and their role in eye health. Nutrients. 2013;**5**(4):1169-1185. DOI: 10.3390/ nu504116
147. Matsubara K, Kaneyuki T, Miyake T, Mori M. Antiangiogenic activity of nasunin, an antioxidant anthocyanin, in eggplant peels. Journal of Agriculture, Food and Chemistry. 2005;**53**:6272-6275. DOI: 10.1021/ jf050796r
148. Han SW, Tae J, Kim JA, Kim DK, Seo GS, Yun KJ, et al. The aqueous extract of *Solanum melongena* inhibits PAR2 agonist-induced inflammation. Clinica Chimica Acta. 2003;328:39-44. DOI: 10.1016/S0009-8981(02)00377-7

3 Chapter

VEGETABLE PROTEINS: NUTRITIONAL VALUE, SUSTAINABILITY, AND FUTURE OUTLOOK

1. INTRODUCTION: AN OVERVIEW

Protein is an essential nutrient; however, not all protein-rich foods are equally represented, and your actual requirements may be less than you assume. Protein, together with carbohydrates and fats, constitutes one of the three necessary components. Protein comprises over 20 amino acids that facilitate the creation and repair of muscle fibers, thereby earning the designation of the "building blocks" of the body. Adults should intake a minimum of 0.8 g of protein per kilogram of body weight daily, equating to slightly over 7 g for every 20 lb of body weight. In 2005, it was estimated that 20% of children under five years old in the lowlands were overweight [1]. Although Southern Asia possesses the largest population of children (74 million), East Africa and Southeast Asia exhibit the greatest rates of stunting. When nations with a stunting prevalence of at least 20% were considered, around 36 accounted for 90% of all cognitively impaired young children. [1].

Reports from Germany, the United Nations, and France in the previous century indicate that acute malnutrition was more prevalent among hospitalized pediatric patients in Germany, the United Kingdom, the United States, and France [2–4]. Pawellek and colleagues found that 24% of children in a prominent German government hospital were underweight (below the 90th percentile for weight relative to height), with 17.9% exhibiting mild malnutrition, 4.4% moderate malnutrition, and 1% severe malnutrition, according to Waterlow's criteria. Undernutrition was observed in 40% of patients with chondrogenesis, 34% of patients with infections, 33.3% of patients with cystic fibrosis, 28.6% of patients with heart illness, 27.3% of patients undergoing cancer treatment, and 23.6% of patients with gastrointestinal diseases [3]. Protein is essential for optimal growth and the prevention of malnutrition.

The global population is projected to double by 2063, increasing from approximately 6.5 billion to 13 billion. As proteins are the sole macronutrient among the three in our diet (the others being carbs and fats), it is imperative to provide sufficient protein availability to sustain the global population. Proteins consist of various amino acid residues essential for tissue growth, repair, and regeneration. The cost of protein is escalating due to societal changes, including higher earnings, population growth, and enhanced living standards, alongside the acknowledged need of protein for good aging and the growing awareness of its dietary necessity. Economic urban development is producing notable population dynamics affecting nutritional status, especially in moderate and high-income nations, with developing countries representing the majority of the global surge in demand for animal-based meals. [5].

2. VEGETABLE PROTEIN

Vegetables are not only the most energy effective in the green house, but they are also a major source of energy, nutrients, elements, protein, omega-3 fatty acids, and widely accessible energy for global agricultural production. Over 3000 plant species have been utilized for food by humans over history, with at least 50 species being farmed for business reasons. Nevertheless, approximately 20 different vegetable crops support the majority of the world's population. Vegetables produce two-thirds of all dietary protein on the planet. **Table 1** shows that cereal crops, especially, provide a significant amount of protein to the global protein supply. The data are derived from FAO and Agro state sites [7].

Table 1. Worldwide relative importance of various food groups and per capita intake.

	Protein (g)	Protein (%)	Energy (Kcal)	Energy (%)
Total plant	46.1	65	2277	84
Cereals	33.7	47	1385	51
Pulses, nuts	6.0	8	109	4
Starchy roots	2.0	3	141	5
Other vegetables	2.5	4	46	2
Fruits	0.8	1	65	2
Total animal	25.0	35	433	16
Total	71.1	100	2710	100

Table 2 illustrates the amounts of necessary amino acids in various dietary sources, with the most restricting amino acids in protein sources meals shown in bold. The necessary protein lysine is significantly lower in vegetable-food protein classes than in animal nutrition, as can be shown. Children can thrive as well as recover from acute malnutrition if carefully prepared meals based exclusively on vegetable food sources are supplied. As a consequence, vegetable meals can supply the critical components needed for optimal health and function in the correct ratios and combinations. Protein sources meal combinations have the potential to be nutrient dense. Despite the soybean's reduced sulfur content, soya, peanut

and sesame flour, and cereal grains, for instance, are mostly poor in lysine. This suggests that oil-seed proteins, especially soy proteins, might be used efficiently for most cereal grains to enhance protein properties. Data from the FAO and the US Department of Agriculture are included in **Table 2** [8].

Table 2. The presence of the amino acid content of different food protein source.

Food source	Threonine	Tryptophan	Lysine	Sulfur amino acids
mg/g protein				
Legumes	38 ± 3	12 ± 4	64 ± 10	25 ± 3
Cereals	32 ± 4	12 ± 2	31 ± 10	37 ± 5
Nuts, seeds	36 ± 3	17 ± 3	45 ± 14	46 ± 17
Fruits	29 ± 7	11 ± 2	45 ± 12	27 ± 6
Animal foods	44 ± 6	12 ± 6	85 ± 9	38 ± 7

3. PROTEIN REQUIREMENT

This is entirely dependent on the individual's age, exercise level, weight, medical history, and desired outcomes. People who are sedentary have distinct needs, adults have different requirements, and athletes possess different needs. Assessments of protein requirements, as according to FAO/WHOIUNU [9], refer to metabolic activities that last for a long duration. Though protein and amino acid requirements are usually expressed as daily rates, this is not always the case, and this does not mean that they must be ingested every day. To preserve muscle mass and function, as well as to fight osteoporosis, the aged require a larger protein consumption than younger people [10, 11]. Although current guidelines recommend a protein consumption of 0.8 grams per kilogram body weight per day (g/kg/d) [12], it is been shown that fit and healthy seniors need a protein intake of 1.2 g/kg/d to avoid age-related weight gain and function [13, 14]. There is presently no evidence on the usual protein consumption of healthy and active older adults, including any potential differences between males and females, to our knowledge. **Figure 1** shows how protein requirements differ depending on the individual [15].

Because our bodies are unable to manufacture critical amino acids, we must obtain them from food. Some amino acids found in proteins could not be used by our bodies. **Figure 2** shows many forms of amino acids [15].

When a person consumes vegetable-based protein, their nutritional needs are higher than when they consume animal protein. **Figure 3** depicts the protein requirements for the plant. Vegetable-based proteins that include essential amino acids and branched-chain amino acids include soy, pea, and rice. These are complete proteins that are hypoallergenic and gluten free. Proteins are essential components of human diets because they aid in the maintenance of muscle mass, the regulation of immune responses, the healing of cells, and the enhancement of communication. Proteins are important components in functional meals because of their helpful activities in terms of delivering different proteins in the human diet, such as stiffening and meshing capability, emulsion, fizzing, water retention, and body fat [16, 17].

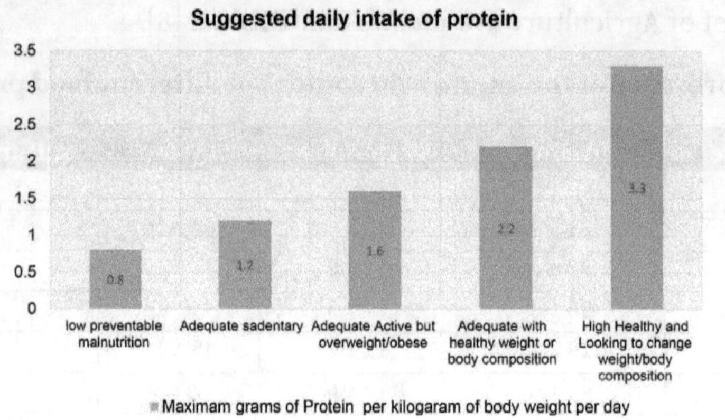

Figure 1. Protein intake requirements on a daily basis.

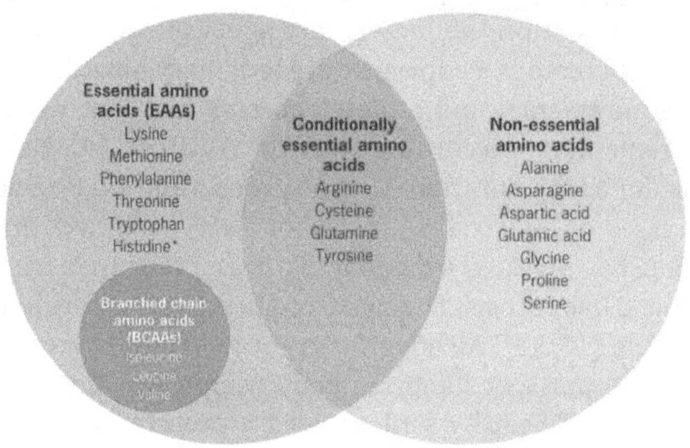

Figure 2. Types of amino acids with their importance.

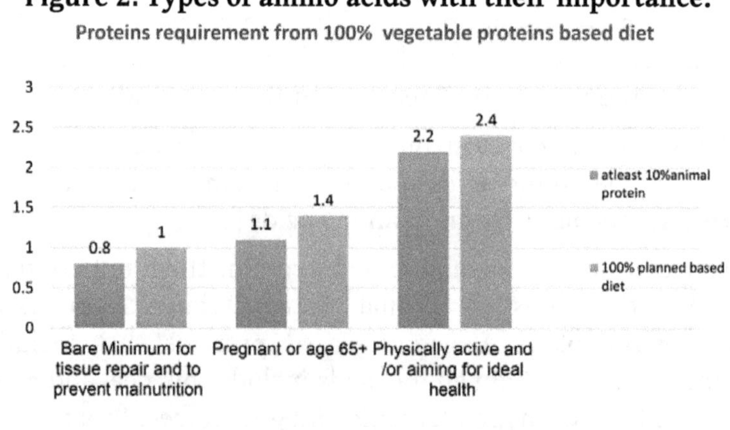

Figure 3. The protein requirements including essential amino acids and branched-chain amino acids.

4. VEGETABLE PROTEINS AND THEIR QUALITY

Figure 4 shows a wide range of vegetable-based proteins derived from several sources. These can be extracted from low-cost and ecological sources such as agricultural wastes and crop and oil industry by-products, which can help to achieve food security [16].

Because of the varied metabolic requirements of certain tissues, rats, and humans have varying amino acid requirements. The fact that the protein-energy ratio of animal proteins (or a combination of proteins) does not reach its maximum value but rises when methionine is added supports the notion that rodents have

a higher sulfur amino acid need than humans. The consumption of a protein in people can be measured by monitoring the fecal matter and urine nitrogen (N) losses; it is predicated on the nitrate adjustment study premise. These studies indicated that some vegetable proteins, particularly beans and wheat, are underappreciated. Wheat's net protein utilization was calculated to be 41% when compared to egg protein. In humans, data show that most vegetable sources of protein have true digestibility in the 80–90% range, with references having lower digestibility (e.g., "rice, cereal": 75% and "rice, polished": 88%) and others having higher digestibility (e.g., "wheat, refined": 96%, "soy protein isolate": 95%). When protein percentage is purer, vegetable protein is much more digestive. Instead of looking at total nitrogen levels, per specific amino acid's digestion should be studied, as this has been suggested for many years. Because some protein acids personal digestion fluctuates or is lower than that of others for several reasons, average digestion is an insufficient proxy with each protein bioavailability.

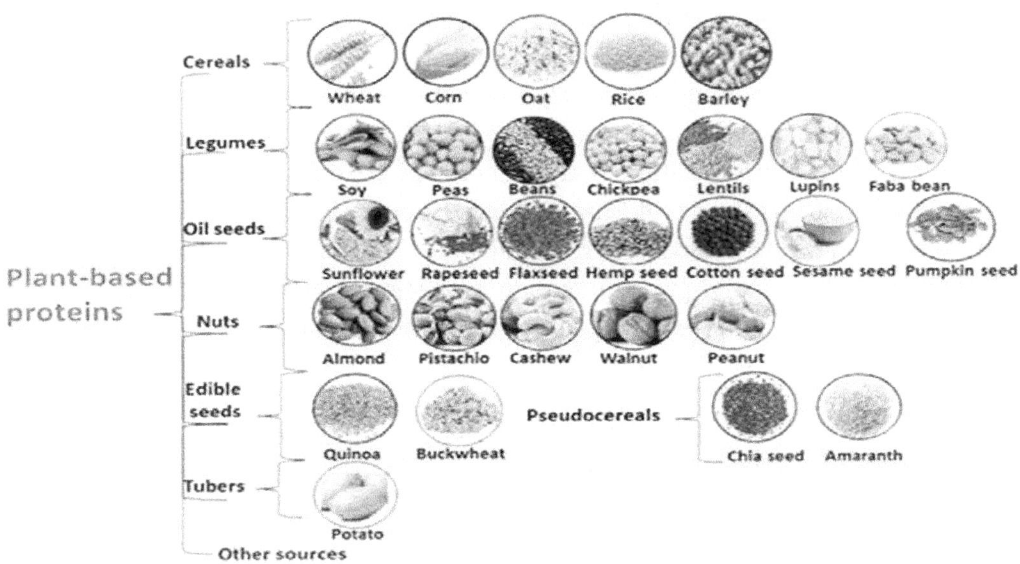

Figure 4. Major plant-based protein sources.

The digestibility of various dietary proteins was found to be between 89 and 95 % in investigations using advanced techniques. The findings were 89 %–92 % for soy protein isolate, pea protein flour or isolate, wheat flour, and lupine flour, which were close to those reported for eggs (91%) and meat (90 –94 %), but somewhat lower than those published for milk protein isolate (95 %). When absorbed at a level that fulfills the entire protein needs (e.g., 0.66 g/kg weight), a dietary protein with an amino acid score offers precisely the exact quantities of proteins that satisfy metabolic requirements and at a ratio in which no amino acid limits utilization of others. Lysine in grains and sulfur amino acids in beans typically emerge as restricted AAS stands for Amino acid score of vegetable protein. A closer analysis of the literature discloses that lysine rates are reduced or near zero in cereals such as sorghum, wheat, triticale (47, 49, and 62%), and rice (80%), and also low in few more both these sources (walnuts and almonds: 60%). In contrast, it is high or extremely high in other sources, notably legumes (e.g., pea: 168 %; fava bean: 152 %; soybean: 134 %) and animal proteins (milk: 168 %; beef: 193 %; egg: 160 %) [18].

5. CHALLENGES FOR VEGETABLE PROTEIN UTILIZATION

Vegetable proteins can also be used to generate bioactive peptides. Most vegetable-based proteins, however, are insoluble in water due to their poor wettability, intricacy, and vulnerability to pH, osmotic pressure, and temperature, limiting their application. Flaxseed, soy, and pea proteins are examples of vegetable proteins with varied percentages. The presence of antinutrients with in form of individual plant residuals is a disadvantage of vegetable proteins. These compounds are produced by vegetables and have biological functions such as protecting vegetables from bugs, infections, fungi, and other organisms. Some of these changes can also assist to mitigate the negative effects of antinutrients [19]. **Figure 5** shows plant protein utilization issues and challenges. To overcome challenges of vegetablebased proteins to date, modifications tactics have been used to circumvent these difficulties. Protein modification is the process of changing a protein's chemical composition or a few chemical units using particular procedures to enhance its techno-functionality and bioactivity. The advantages and disadvantages of each of the physical techniques of protein change used yet for vegetable proteins will be examined separately. Plant-based protein utilization and challenges are shown in **Figure 5.**

6. MODIFICATION OF VEGETABLE PROTEINS TO IMPROVE DIGESTIBILITY AND QUALITY

Heat is among the most common methods for significantly modifying the structure and sensory attributes of vegetable proteins. Protein unwinding is aided by a low-temperature environment, resulting in an intermediary molten globule state with improved functioning. Extreme heat stability, on the other hand, causes permanent changes in protein structures, resulting in hydrolysis and aggregating *via* various bonds such as disulfide, hydrophobic, and electrostatic, resulting in a loss of functional characteristics [18]. Vegetable proteins have been shown to be an effective technique of decreasing or eliminating anti-nutritional compounds using heat [19]. In chickpeas and soybeans, heat treatment has been used to inactivate trypsin inhibitors, and the combination of heating and met bisulfite as a reductant eliminated the trypsin inhibition up to 99.4%. Trypsin inhibitors are among the anti-nutritional factors limiting its use of lentils because they help prevent the action of pancreatic trypsin and chymotrypsin in the gut, causing a range of issues such as complicated digestive processes, poor protein intake, pancreatic expansion, stunted growth, and muscle loss [19]. Heat treatment of vegetable proteins can also enhance their digestion and nutritional properties. The digestibility is increased to 87.55% and the availability of essential amino acids is increased by heating the albumin protein isolate at 100°C for 30 min [20]. Ohmic heating can cause unfolding, denaturation, and the creation of uniform-sized protein aggregates with different techno-functional characteristics by delivering rapid and uniform heating as well as electrical effects. As a consequence, standard heating methods could be replaced with this electro-heating technology that has minimal negative impacts on protein quality and amino acid concentration [21].

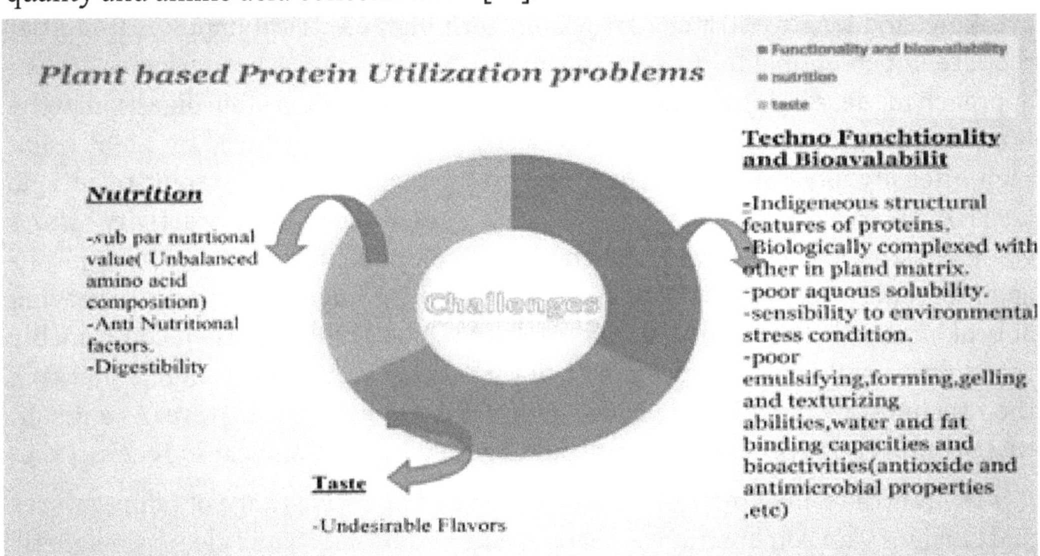

Figure 5. Plant-based protein utilization issue and challenges.

Microwave frequency is less than chemical bond rate [22], and the approach can change the protein without disrupting its core structure, making it attractive as a classification scheme before further physical methods. The microwave approach has also been used to regulate the immune system of vegetable proteins. A significant drop (24.7%) is found in soya mutagenicity by microwave heating at 600 W for 10 min [23]. The activity of chemicals released by dipolar and ionic movement in the existence of an RF field impacts protein function. Radiofrequency (RF) like the microwave is premised on heat production and impacts protein function *via* the action of free radicals generated by dipolar and ionic flexibility in the presence of an RF field. Both RF and microwave heating technologies can solve the issues of low- heating rate impacts that are popular heating techniques due to their heat production [22]. RF heating was discovered to have significant effects on the structure of soy protein by breaking disulfide linkages and increasing surface hydrophobicity [24]. High hydrostatic pressure (HPP) is a non-thermal technique that uses hydrostatic pressure varies from 100 to 800 MPa for a few seconds [25]. HHP treatment influenced the denaturation, aggregation, and interactions of a variety of vegetable proteins. HHP treatment frequently promotes muscle hydrophobicity and lowers solubility, resulting in aggregation, thanks to its capability to expose hidden sulfhydryl groups following unwinding and inhibition [26]. HHP was also investigated as a means of increasing the nutritional value of vegetable proteins. When contrasted to other treatments such as enhanced ultrasonic, microwaving, and elevated homogenization, for example, H. HHP was shown as being the most efficient in lowering the allergen concentration of soy isolate for use in newborn formula [23]. Sonication is a one-of-a-kind, ground-breaking, and lengthy strategy associated with high electrical impulses (>16 kHz) that are imperceptible to the human ear [27]. Sonication has been proposed as an approach for decreasing antinutrients and increasing plant protein digestion in the literature [23]. The inhibitory activity of soy protein was reduced by 18.9% using high-intensity ultrasound [28]. According to the researchers, ultrasonic technique (25 kHz, 400 W, 1–16 min) decreased soymilk protein ace inhibitory activity by 52% while boosting digestion [29]. The combined components contribute to vigorous agitation and severe mechanical forces induced by a huge rotating screw moving at peak pressures (1.5–30.0 MPa) and heats (90–200 C) [30]. Extrusion could enable the molecules of vegetable protein to unfurl, denaturant, and realign, enhancing their techno-functionality while also giving them a meat-like texture. As a result, these contoured vegetable proteins could be used in place of meat in recipes [31].

It is the fourth form of matter could be generated at a wide range of temperatures and tensions *via* a combination of heat, physical, radioactive, and elec- tromagnetic sources of energy. Malnourishment is described as just a structural mismatch in between availability of food and fuel as well as the body's requirement for any of them to ensure growth, repair, and certain activities, as per the World Health Organization [32]. Other dietary deficits often occur with protein-energy malnutrition. Malnourishment has also been linked to a reduction in the number of cells, connections, dendritic arborizations, and myelinations, each of which leads to a shrinking of the mind. Brain growth is retarded and the cerebral cortex gets reduced. Protein-energy malnutrition has been linked to impairments in a global function, brain function, and cognition, with newborns and babies, is now the most vulnerable, considering the newborn brain's adaptability [33]. As hunger advances, development is slowed, resulting in stunted, and some other organs are affected, including hairs, epidermis, fingernails, mucosa, as well as other organs. Because poor nutrition, specifically vitamins and minerals inadequacies are widespread in malnutrition individuals, most of them will show symptoms of them.

Table 3. The utilization of vegetable and animal proteins in disease status.

Vegetable proteins	Animal proteins
Adjustable for CKD patients	Non-adjustable
Lower cholesterol	Higher in cholesterol
Reduced blood pressure	Increase blood pressure
Lower the obesity	Increase risk of obesity

Malnourishment, micronutrient deficiency, kwashiorkor, and miasmic kwashiorkor can all be aided by sufficient protein with correct modification. Decreased thoracic lean muscle, a lower metabolic activity, and electrolyte imbalances can all lead to a reduction in respiratory rate, affecting the ventilator's capacity to act to hypoxia. The utilization of vegetable and animal proteins in disease status is presented in **Table 3.**

7. VEGETABLE PROTEIN SUSTAINABILITY

The expected increase in the world population to 9 billion people (United Nations Population Department, 2000, 2050—medium version) defines the growth plan. As a result, not only does the quantity of food necessary change, and so will the food groups required, as well as their major contributors to nutrition. Predicted protein consumption seems to be of key significance, with forecasts that global demand for animal protein will quadruple by 2050 [34], raising concerns about food management and sustainable development. This is partial because it is widely acknowledged that animal meals emit more greenhouse gasses than vegetables, which are affected by temperature. The fact that rapid growth for animal protein is likely to rise land stress due to the requirement to generate more animal feed adds to the problem. As a result, the clearance of land, lakes, and natural grassland to farming land would rise, negatively impacting greenhouse gasses emissions, diversity, and other essential natural ecosystems [35]. Excess supply of proteins is fueled by social and economic developments including such rising incomes, increase in urbanization, and population growth, in which the importance of proteins to health and longevity is commonly understood [5, 6], as well as awareness of the function of protein in a balanced diet. Economic growth and urbanization are causing large shifts in demographic nutritional status in middle to low nations especially in developing nations accounting for such bulk of worldwide increases in costs for animal-based meals [6]. Protein isolate synthesis, on the other hand, has a smaller impact on the environment but is a long-term alternative since it consumes less fuel, produces less pollution, requires less land, and uses less water. Animals are not efficient kilo converters of the proteins they ingest; thus, manufacturers should provide vegetables to animals to make animal proteins. It is the typical switching frequency of veggie to animal protein is 10 to 1, which indicates that 10 pounds of feed protein is required to create 1 pound of animal protein. Foods high fiber and low in fat are common in vegetable-based diets. A high-fiber, low-fat diet has been demonstrated in multiple trials to reduce the risk of certain malignancies, including colon, breast, and prostate cancers.

Our ever-increasing world population necessitates the availability of low-cost protein. As per world population projections, we will have to deliver protein to nearly nine billion by 2050. Accessibility to low-cost protein sources is critical for sustaining the growing world population while reducing environmental harm.

Vegetable proteins have obvious health advantages. Vegetable proteins have been shown to reduce our risk of cardiovascular disease. Incorporating soy protein into your diet may also aid in the prevention of heart disease. Packaged beef should be designated as a "human carcinogen," according to a research published by the International Agency for Research on Cancer. Meat, on the other side, has been deemed "possibly carcinogenic to humans." While the exact evidence is inconclusive, iron absorption is assumed to play a key role in N-nitroso-compound processing, the formation of lipid oxidation, and a probable cytotoxic impact. The demand for agricultural production would likely be higher to population growth and the moral imperative to provide enough, healthy, and inexpensive food for everyone. By 2050, simply supplying 2900 kcal mean food production with 50% of the proteins coming from animal protein will roughly treble grain consumption. Furthermore, economic expansion, coupled with someone who can afford its desire for an excess of animal goods and poor eat conversion, may drive up prices even more. The purposeful decrease of animal protein intake as well as the replacement of animal proteins with protein sources could be used as a temporary fix to lessen the social change caused by the existing agricultural system. The worldwide market for protein foods is expected to be different combinations of that of the United States. The industry for protein isolate components is growing due to a number of factors. Food producers are reacting to the rising costs for meat-restricted, protein foods among the wellness Boomer Generation and other consumers. Animal food prices are high and variable, placing pressure on global food makers to come up with new ways to reduce expenses.

Our world's expanding population necessitates the availability of low-cost protein. As per population projections, by 2050, we would be supplying protein to roughly 9 billion individuals. The availability of low-cost protein sources is critical for feeding the world's growing population while reducing environmental stress. Vegetables are popular on the market right now. As per market research firm Frost & Sullivan, the U.S. protein-ingredients market alone is expected to produce about $4.5 billion in sales in 2008 (43.3% vegetable proteins and 56.7% animal proteins) and is expected to increase at an annual growth rate of nearly 5%. Barriers are there in shifting from animals to vegetable protein. The transition from animal to vegetable meals is not without its challenges, or at least four have been identified [36]. Change is difficult because social factors oppose it beef has a high social position and that it is the average human's wish to consume high up food given the correct financial circumstances. Lobby groups inside the meat manufacturing chain are significant, and economic concerns against reform are considerable. The advanced technologies required to generate innovative protein sources foods are lacking.

Because the animal protein supply chain has been designed for using all by-products for generations, eliminating meat on a broad scale might just have a significant impact on linked production systems, perhaps canceling out most of the putative environmental benefits.

8. CONCISE SUMMARY

Vegetable proteins are more sustainable and adequate to satisfy the needs of a constantly expanding population. Vegetable proteins have greater adaptability and can be modified in response to health concerns compared to animal proteins. This chapter emphasizes vegetable protein due to the rising protein demand associated with global population growth. The global population is projected to double by 2063, increasing from 6.5 billion to 13 billion individuals. Accelerated urbanization and heightened protein intake have intensified the issue, resulting in a rise in malnutrition among youngsters. Consequently, we must fulfill our protein requirements to promote health and battle malnutrition. We shall emphasize that individual protein requirements, along with the needs for both vegetable and animal proteins, are distinct. Animal protein sources are unsustainable as the global population increases; therefore, we cannot fulfill our protein requirements only from these sources. This chapter examined the stability and adequacy of vegetable protein sources in fulfilling our protein needs. Vegetable diets generate more greenhouse gas emissions than animal-based meals due to their sensitivity to temperature fluctuations. The anticipated rapid expansion of animal protein production is likely to induce land stress due to the necessity of increasing animal feed production, hence exacerbating the issue. In reaction, the conversion of land, lakes, and natural grasslands to agricultural land would increase, endangering greenhouse gas emissions, biodiversity, and other vital natural ecosystems. Vegetable proteins are advocated as more sustainable, adaptable to health conditions, and economically advantageous compared to animal proteins. We transitioned from animal-based to vegetable-based protein. Numerous obstacles must be addressed in the utilization of vegetable proteins. Consequently, we examined various techniques for utilizing vegetable protein and its derivatives to address amino acid shortages, including heating, ohmic heating, microwave, radiofrequency, and extrusion, among others.

REFERENCES

1. Black RE, Allen LH, Bhutta ZA, et al. Maternal and child undernutrition: Global and regional exposures and health consequences. Lancet. 2008;**371**:243-260
2. Joosten KFM, Hulst JM. Prevalence of malnutrition in paediatric hospital patients. Current Opinion in Pediatrics. 2008;**20**:590-596
3. Pawellek I, Dokoupil K, Koletzko B. Prevalence of malnutrition in paediatric hospital patients. Clinical Nutrition. 2008;**27**:72-76
4. Hendricks KM, Duggan C, Gallagher L, et al. Malnutrition in hospitalized pediatric patients. Current prevalence. Archives of Pediatrics & Adolescent Medicine. 1995;**149**:1118-1122
5. Delgado CL. Rising consumption of meat and milk in developing countries has created a new food revolution. The Journal of Nutrition. 2003;**133**:3907S-3910S
6. Popkin BM, Adair LS, Nq SW. Global nutrition and the pandemic of obesity in developing countries. Nutrition Reviews. 2012;**70**:3-21
7. Food and Agriculture Organization/ Agrostat. Computerized Information Series No 1. Food Balance Sheets. Rome: Food and Agriculture Organization; 1991
8. Young VR, Pellett PL. Plant proteins in relation to human protein and amino acid nutrition. The American Journal of Clinical Nutrition. 1994;**59**(5):1203S-1212S. DOI: 10.1093/ ajcn/59.5.1203S
9. Food and Agriculture Organization/ World Health Organization! United Nations University. Energy and protein Requirements. Report of Joint FAO/ WHO/UNU Expert Consultation (WHO Tech rep ser no 724). Geneva: World Health Organization; 1985
10. Houston DK, Nicklas BJ, Ding J, Harris TB, Tylavsky FA, Newman AB, et al.

Dietary protein intake is associated with lean mass change in older, community-dwelling adults: The Health, Aging, and Body Composition (Health ABC) Study. The American Journal of Clinical Nutrition. 2008;**87**(1):150-155

11. Volpi E, Campbell WW, Dwyer JT, Johnson MA, Jensen GL, Morley JE, et al. Is the optimal level of protein intake for older adults greater than the recommended dietary allowance? The Journals of Gerontology Series A. Biological Sciences and Medical Sciences. 2013;**68**(6):677-681. DOI: 10.1093/gerona/gls229
12. World Health Organization. Protein and Amino Acid Requirements in Human Nutrition: Report of a Joint WHO/FAO/UNU Expert Consultation. Report 935. Geneva: WHO Press; 2007
13. Gaffney-Stomberg E, Insogna KL, Rodriguez NR, Kerstetter JE. Increasing dietary protein requirements in elderly people for optimal muscle and bone health. Journal of the American Geriatrics Society. 2009;**57**(6):1073- 1079. DOI: 10.1111/j.1532-5415. 2009.02285.x
14. Deutz NE, Bauer JM, Barazzoni R, Biolo G, Boirie Y, Bosy-Westphal A, et al. Protein intake and exercise for optimal muscle function with aging: Recommendations from the ESPEN Expert Group. Clinical Nutrition. 2014;**33**(6):929-936. DOI: 10.1016/j. clnu.2014.04.007 https://www.precisionnutrition. com/plant-based-protein
15. International Agency for Research on Cancer. Consumption of Red Meat and Processed Meat. Lyon, France: IARC Working Group; 2018
16. Nasrabadi MN, Doost AS, Mezzenga R. Modification approaches of plant-based proteins to improve their techno-functionality and use in food products. Food Hydrocolloids. 2021;**118**:106789. DOI: 10.1016/j. foodhyd.2021.106789. ISSN 0268-005X. Available from: https://www. sciencedirect.com/science/article/pii/ S0268005X21002058
17. Aryee ANA, Agyei D, Udenigwe CC. 2—Impact of processing on the chemistry and functionality of food proteins. In: Yada RY, editor. Woodhead Publishing Series in Food Science, Technology and Nutrition, Proteins in Food Processing. 2nd ed. Woodhead Publishing; 2018. pp. 27-45. DOI: 10.1016/B978-0-08-100722-8.00003-6. ISBN 9780081007228
18. Avilés-Gaxiola S, Chuck-Hernández C, del Refugio Rocha-Pizaña M, García-Lara S, López-Castillo LM, Serna-Saldívar SO. Effect of thermal processing and reducing agents on trypsin inhibitor activity and functional properties of soybean and chickpea protein concentrates. LWT. 2018;**98**:629-634. DOI: 10.1016/j.lwt.2018.09.023. ISSN 0023-6438
19. Mir NA, Riar CS, Singh S. Structural modification in album (*Chenopodium album*) protein isolates due to controlled thermal modification and its relationship with protein digestibility and functionality. Food Hydrocolloids.

2020;**103**:105708. DOI: 10.1016/j. foodhyd.2020.105708. ISSN: 0268-005X

20. Mesías M, Wagner M, George S, Morales FJ. Impact of conventional sterilization and ohmic heating on the amino acid profile in vegetable baby foods. Innovative Food Science & Emerging Technologies. 2016;**34**:24-28. DOI: 10.1016/j.ifset.2015.12.031. ISSN 1466-8564

21. Ji H, Dong S, Han F, et al. Effects of dielectric barrier discharge (DBD) cold plasma treatment on physicochemical and functional properties of peanut protein. Food and Bioprocess Technology. 2018;**11**:344-354. DOI: 10.1007/s11947-017-2015-z

22. Lee H, Yildiz G, dos Santos LC, Jiang S, Andrade JE, Engeseth NJ, et al. Soy protein nano-aggregates with improved functional properties prepared by sequential pH treatment and ultrasonication. Food Hydrocolloids. 2016;**55**:200-209. DOI: 10.1016/j.foodhyd.2015.11.022. ISSN 0268-005X

23. Li R, Cui Q, Wang G, Liu J, Chen S, Wang X, et al. Relationship between surface functional properties and flexibility of soy protein isolate-glucose conjugates. Food Hydrocolloids. 2019;**95**:349-357. DOI: 10.1016/j.foodhyd.2019.04.030. ISSN 0268005X

24. Hite BH. The effect of pressure in the preservation of milk: A preliminary report. West Virginia Agricultural and Forestry Experiment Station Bulletins. 1899;**58**. Available from: https:// researchrepository.wvu.edu/ wv_agricultural_and_forestry_ experiment_station_bulletins/58

25. Queiros RP, Saraiva JA, JAL d S. Tailoring structure and technological properties of plant proteins using high hydrostatic pressure. Critical Reviews in Food Science and Nutrition. 2018;**58**(9):1538-1556

26. Gharibzahedi SMT, Smith B. The functional modification of legume proteins by ultrasonication: A review. Trends in Food Science & Technology. 2020;**98**:107-116. DOI: 10.1016/j. tifs.2020.02.002. ISSN 0924-2244

27. Vanga SK, Wang J, Raghavan V. Effect of ultrasound and microwave processing on the structure, in-vitro digestibility and trypsin inhibitor activity of soymilk proteins. Lebensmittel-Wissenschaft und-Technologie. 2020;**131**:109708

28. Jin J, Okagu OD, Yagoub AEA, Udenigwe CC. Effects of sonication on the in vitro digestibility and structural properties of buckwheat protein isolates. Ultrasonics Sonochemistry. 2021;**70**:105348

29. Sedaghat Doost A, Nikbakht Nasrabadi M, Kassozi V, Dewettinck K, Stevens CV, Van der Meeren P. Pickering stabilization of thymol through green emulsification using soluble fraction of almond gum—Whey protein isolate nanocomplexes. Food Hydrocolloids. 2019;**88**:218-227

30. Zahari I, Ferawati F, Helstad A, Ahlstrom C, Ostbring K, Rayner M, et al. Development of high-moisture meat analogues with hemp and soy protein

using extrusion cooking. Food. 2020;**9**(6):772

31. de Onis M, Monteiro C, Clugston G. The worldwide magnitude of protein energy malnutrition: An overview from the WHO global database on child growth. Bulletin of the World Health Organization. 1993;**71**(6):703-712
32. Georgieff MK. Nutrition and the developing brain: Nutrient priorities and measurement. The American Journal of Clinical Nutrition. 2007;**85**:614S-6120S
33. Westhoek and Colleagues. Available from: http://www.fao.org/fileadmin/user_upload/animalwelfare/Protein_ Puzzle_web_1.pdf [Accessed: July 17, 2017]
34. Van Zanten HHE, Mollenhorst H, Klootwijk CW, van Middelaar CE, de Boer IJM. Global food supply: Land use efficiency of livestock systems. International Journal of Life Cycle Assessment. 2016;**21**:747-758
35. Helms M. Food sustainability, food security and the environment. British Food Journal. 2004;**106**(5):380-387. DOI: 10.1108/00070700410531606

35. [illegible]

4 Chapter

THE HEALTH ADVANTAGES OF VEGETABLES; PREVENTIVE IMPLICATIONS FOR CHRONIC NON-COMMUNICABLE DISEASES

1. INTRODUCTION: AN OVERVIEW

Malnutrition affects 33% of individuals in South Asia and 21% in Sub-Saharan Africa, contributing to 60% of the 10.9 million yearly fatalities among children under five years old[1]. Malnutrition adversely impacts health and wellbeing, leading to physical and psychological harm, which in turn results in suboptimal educational outcomes and detrimental effects on adult life [2]. It compromises the immune system, increases the risk of both non-communicable and communicable diseases, diminishes productivity, and engenders various negative social and economic repercussions for individuals, households, communities, and nations[3]. Approximately 805 million individuals are projected to be chronically undernourished from 2012 to 2014, a reduction of over 100 million in the past decade and 209 million fewer than the figures from 1990 to 1992. [4].

It is imperative for a healthy human diet to have a varied assortment of foods that supply all necessary nutrients for sustaining life. Vegetables, specifically, may aid in the prevention of non-communicable diseases (NCDs) and the management of micronutrient deficiencies in economically disadvantaged populations[7–10]. Vegetables, being abundant in vitamins, minerals, dietary fiber, and low in calories, are essential for the proper functioning of the human body[11]. Adequate consumption of vegetables has been epidemiologically associated with a diminished risk of numerous non-communicable diseases. [12–16].

Micronutrient deficiencies linked with malnutrition are the most common nutritional deficiencies in the world, causing significant developmental difficulties, and they are a key indication of poor nutrition and health [17]. Low birth weight, sex, age, rural residency, infectious illness, malnutrition, poor wealth status, and household food insecurity were all linked to it [18]. Iron, vitamin A, zinc, and iodine are critical micronutrients for children's growth, development, and survival, making them critical in terms of global public health [19]. Both iron deficiency anemia and iodine deficiency diseases have been shown to impair cognitive development in children, resulting in reduced educational attainment and, as a result, a detrimental influence on economic growth [20]. Vitamin A deficiency also affects immune function and leads to an increased risk of morbidity [21]. Zinc deficiency may lead to growth failure and impaired immune function [22]. Both vitamin A and zinc deficiencies increase the risk of mortality in children. Children under five years of age are particularly vulnerable to micronutrient deficiencies due to the relatively high requirements of micronutrients for rapid growth and development [23]. Micronutrient deficiencies are currently being addressed by strengthening nutrition information systems, monitoring, and evaluating mechanisms [24] that conceptualize optimal nutrient intake using a variety of foods across and within food groups [25]. Lately, the governments of developing countries have sought to promote diversification of production and exports away from traditional commodities in order to accelerate economic growth, expand employment opportunities, and reduce rural poverty. Market-oriented production could allow households to increase their income by producing output with higher returns on land and labor and using the income generated from sales to purchase goods for consumption [26].

NCDs account for the majority of the global disease burden, disproportionately affecting low and lower-middle income countries [27]. According to recent estimates, innovative, cost-effective, and feasible interventions could prevent 82% of NCD-related premature deaths. CVDs, type 2 diabetes, cancer, and chronic obstructive pulmonary disease were the most common NCDs [27]. This rising burden adds to the stress on already overburdened health systems, disproportionately hurting the poorest people and impeding countries' social and economic growth. According to the WHO, Ethiopia's annual mortality rate from NCDs was 34% in 2008 [28]. Cardiovascular diseases constituted 15% of total fatalities, with cancers and respiratory illnesses each at 4%, diabetes at 2%, injuries at 9%, and other non-communicable diseases at 9%, as per the study. Misganaw et al. [29] performed a study in Addis Ababa examining the etiology of death (51%) attributable to non-communicable diseases (NCDs). A separate study revealed that under non-communicable diseases (NCDs), cardiovascular disease (CVD) was the predominant cause of mortality at 24%, succeeded by malignant neoplasms at 10%, respiratory tract disorders at 9%, and diabetes at 5%.

The study revealed a significant rise in mortality from non-communicable diseases (NCDs) among those aged 44 to 74 years. A separate population-based survey conducted in Jimma, Ethiopia, revealed an overall prevalence of non-communicable diseases (NCDs) at 8.9%, comprising 3.1% for diabetes, 9.3% for hypertension, 3.0% for cardiovascular diseases (CVDs), 1.5% for asthma, and 2.7% for mental illness. [30].

2. DRIVING BARRIERS TO VEGETABLE CONSUMPTION

Many of the empirical literature reviews revealed that people consumed few vegetables. Accordingly, the vegetable consumption trend was reported in Ethiopia and Uganda studies with a prevalence of 20 and 12.7%, respectively [31]. There is a variation in household consumption patterns, depending on specific geographical and socio-cultural characteristics. A high percentage of caloric consumption comes from cereals, and per capita intake of calories is relatively higher in rural areas than in urban areas [32, 33]. In Ethiopia, cereals, together with Kocho (Ensete), and vegetables, contribute 85% of the total food intake per person per day [33]. A study conducted in the rural Sima area by Kebebu [34] reported that there was negligible consumption of meat, fish, or vitamin A-containing vegetables. Consumer preferences influence whether or not they consume anything. Poor households have no choice but to rely on cheap sources of energy such as grains and starchy staples until they meet physiological demands to satisfy hunger. Households begin to expand their meals by incorporating animal food sources, dairy products, and vegetables once their fundamental energy demands have been met [35].

Due to several factors, nutrient-dense food crops such as pulses and vegetable types are not consumed [34]. Most published research on the drivers of food demand has always concentrated on the impact of income and prices throughout the world [36, 37], while only a few have looked into the demand features (both economic and noneconomic variables) of healthy foods like vegetables [37–41]. The following variables were found in several systematic studies focused specifically on factors affecting vegetable consumption [37, 39, 42–45]. Biological determinants, such as gender, age and food properties; economic determinants, such as income and price; physical determinants, such as time, cooking skills, accessibility, availability (production) and living area; social determinants, such as marital status, having children, education, family, peers, cultural values, habits and meal patterns and also other social factors (preparation, storage and handling of food).

Psychosocial factors at the consumer or individual levels, such as individual taste preferences, consumer awareness, quality consciousness, intra-household decision making power, working status of women, social support, intention, attitudes, beliefs, and traditions, as well as geographical and environmental stage of change, motivation, and knowledge, all interact in a complex manner to shape dietary consumption patterns of an individual and household.

2.1 The physical environment

Lack of time is frequently mentioned as a barrier to vegetable consumption, as well as convenience and knowledge of how to prepare and cook vegetables [46]. According to the FAO, food availability is defined as the availability of sufficient quantities of food of appropriate quality, supplied through domestic production or imports. Accessibility is an important physical factor influencing food choice, reflecting the fact that people's living environments have an impact [39, 47]. The quantity of vegetables consumed by a population can vary depending on where they live. In all 11 nations (Bangladesh, Congo, Ivory Coast, Ecuador, Kenya, Paraguay, Philippines, Tunisia, Ukraine, and Zambia), those residing in urban areas were more likely to have a poor vegetable intake [48]. Overall vegetable intake was greater in urban regions compared to rural ones, according to a meta-analysis of household spending surveys done in 10 Sub-Saharan African nations [49].

Similarly, rural Americans in the United States were less likely than their urban counterparts to consume the required number of daily vegetable servings [50]. A research done in middle-income nations found no variations in vegetable intake between urban and rural residents [51].

Furthermore, 42% of households did not produce any of the common vegetables, respectively. It explained that while their own production of vegetables was significantly better in rural areas, their market availability was significantly better in urban areas [52]. Furthermore, Tadesse [35] conducted an analysis of changes in food consumption patterns in urban Ethiopia, which confirms that urban household consumption has begun to shift away from staple food grains and toward high-value food products. The relative availability of vegetables has decreased in most countries and is still well below the recommended level in both developed and developing countries [53]. The proportion of households who reported having cultivated vegetables in their gardens even once over the year is considered negligible [52]. In a study of Dutch adults, Kamphuis et al. [54] discovered that the availability of a wide range of vegetables throughout the year was positively related to vegetable consumption [39], especially among individuals with higher socioeconomic levels. Previous research in high income nations found that the season had an influence on vegetable intake. Indeed, decreased vegetable consumption was related to the winter season [43]. The seasonal availability of several vegetables limits their use in low-income SubSaharan African nations [49]. According to studies, having a vegetable garden and eating veggies has a beneficial link. As a consequence, people with at least one family member engaged in a community garden program were more likely to consume more vegetables, according to research performed in the United States [55].

The proportion of households that produce adequate vegetables in their gardens is negligible [52, 56]. Home gardening is increasingly being used as a strategy for improving food quality, reducing food insecurity, saving money, and building a more sustainable food system because it provides immediate access to fresh food for households [57, 58]. The study in Ethiopia found that the majority of vegetable producers' households have been found to be more food secure than their nonproducing counterparts [52, 56, 59]. A study also indicates that home gardens benefit households by providing better access to a diversity of plant and animal food items, leading to an overall increase in dietary intake [60] and improving the bioavailability/absorption of essential nutrients [61]. More recently, a study conducted amongst US older adults suggested that compared to non-gardeners, gardeners were more likely to consume vegetables [62]. According to central statistics agency (CSA) [63], the total area under vegetables is about 12,576 hectares, which is insignificant as compared to other food crops; vegetables constitute 2.7% of the total area of all crops. Production decisions are partly determined by consumption needs, and that consumption partly depends on production opportunities [33, 49]. Vegetables are highly perishable, so the cost of getting them to or from the market will be high for households in remote rural areas [49, 64]. Thus, the consumption of some vegetables may be constrained by whether or not they can be grown by the household. Access to water, seeds, and information on horticultural production methods may limit both production and consumption of vegetables [33, 49]. Information from the FAO [24] statistical database suggests that the total supply of fruit and vegetables available is 173 kg per person per year, which divided into 111.6 kg of vegetables.

2.2 Psychosocial aspects

2.2.1 Consciousness, desire, beliefs and attitudes, phases of change, and purpose

A review of studies on the psychosocial predictors of vegetable consumption among adults, primarily conducted in Europe and the United States, found that selfefficacy (also known as perceived behavioral control, which refers to people's perception of their ability to perform a given behavior) was the strongest predictor of vegetable intake [39]. Other characteristics, including obstacles, attitudes and beliefs, stage of change, and intention, according to research, might also influence vegetable intake, but to a lesser extent. Nonetheless, this was less consistent among trials [45]. Recent studies, one on obese Canadian adults [65] and one on US students [66], found that perceived behavioral control was a major predictor of intention to consume vegetables. Furthermore, research of US males and immigrants found that reduced perceived obstacles as well as an advanced stage of transformation were related to increased vegetable intake [67]. The majority of these psychosocial elements are often employed in psychosocial models such as the Social Cognitive Theory, the Health Belief Model, the TransTheoritical Model, and the Theory of Planned Behavior [42]. The goal of such models is to forecast either the intention to consume vegetables or the intake of vegetables. Ajzen's Theory of Planned Behavior is one of the most often utilized models (1991).

2.2.2 Sensory properties and preference

Taste is a key impact on food choice, and individual preferences typically determine consumer decisions about what they eat. In a study of older Irish individuals,

Appleton et al. [68] found that higher vegetable consumption was related to a higher liking for vegetables in a study of older Irish individuals. A similar finding was reached in young adults in the United States [69]. In a systematic review, Pollard et al. [44] published data from two studies focused on the connection between flavor and vegetable intake, one performed in the Netherlands and one in the United States.

2.2.3 Cultural and social support

The diet is based on locally available produce and local agricultural practices; it goes beyond that to the extent that food habits are culturally defined. Food preferences are, in turn, strongly determined by culture and tradition because, generally, what is acceptable as food is dictated to a large extent by cultural norms. A study based on dietary habits learned during childhood seems to be predictive of vegetable intakes in adulthood [54]. Therefore, individuals who ate a lot of vegetables during their childhood usually remain good consumers in adulthood. Social support was significantly associated with vegetable intake [39, 45, 70]. Moreover, there is a misconception among most of the population that the monotonous diet is adequate and superior to vegetables [33, 35, 52]. Eating at fast food restaurants was related to eating significantly less vegetables [71]. Similarly, a study conducted among young Australian adults reported that subjects eating takeaway food at least twice a week were less likely to meet the dietary recommendations for vegetables [72]. Several research studies on children and teenagers found that meal habits, particularly dining together as a family and viewing TV during meals, were associated with vegetable consumption [73, 74]. These studies found that viewing television while eating was linked with lower vegetable intakes in both children and adolescents, but dining together as a family was associated with higher vegetable consumption [75].

The World Health Survey 2002–2003 conducted among adults reported that gender-specific worldwide prevalence of low vegetable consumption, mainly in low and middle-income countries, showed that 78.0% (77.6% of men and 78.4% of women) consumed less than the minimum recommended five daily servings of vegetables.

Indeed, in five countries (Comoros, Dominican Republic, Guatemala, Morocco, and Paraguay), women consume fewer vegetables than men, whereas in the remaining ten countries (Czech Republic, Estonia, Hungary, Slovakia, Slovenia, Spain, Swaziland, Ukraine, Uruguay, and Vietnam), women consume more vegetables than men [48]. Hall et al. [48] found that Men's low vegetable intake ranged from 36.6% (Ghana) to 99.2% (Pakistan), whereas women's consumption ranged from 38.0% (Ghana) to 99.3% (Pakistan) [48].

But many studies specifically indicate that women consume more vegetables than men [44]. Furthermore, most studies investigating the relationship between vegetable consumption and age conclude that the amount of vegetables consumed increases with age [50, 76], and in Canada among adults [65]. On the contrary, in Iran, Esteghamati et al. [51] reported that older adults were more likely to be low consumers. In Canada, a national representative survey reported that middleaged adults consumed fewer vegetables compared to younger and older adults [77].

Low vegetable consumption was associated with age and tended to increase with age and a decrease in prevalence from the 40–49 year age to the 50–59 year age. The oldest age stratum had a higher risk of eating fewer vegetables compared to younger adults in 26 countries, compared to one or more younger age groups in all countries. But in Brazil and Uruguay, the youngest age group has a higher risk of low vegetable consumption [48]. But in Nigeria, demand for vegetables is higher among house- holds with younger members, compared to households with older members [78].

The World Health Organization Survey reported that Ethiopian women (64.1%) had a higher risk of low vegetables than men (57.7%) [48]. Women are more concerned about a healthy diet and more often classify foods according to the assumed nutrient content than men, i.e. the share of products of vegetable origin is higher in women's diets [79]. The European Journal of Public Health published a survey report on gender differences in the consumption of meat and vegetables from Western European countries such as Finland and the Baltic countries, showing that men ate meat more often while women ate vegetables [79]. Other studies done in diverse circumstances found that women eat more vegetables than males in Europe [76, 80–82], Iran [51], Canada [77], and the United States [50].

2.3 Economic access and the household's income

Households' ability to purchase food depends on the households' income and the price of food. Regardless of the source of food, households must have the means to acquire appropriate foods, and economic access therefore refers to the affordability of food for the consumer. In some of the lowest income countries, households may spend as much as 60% of their income on food, versus 15% or less for households in high-income countries [49, 64]. A meta analysis of 52 mainly low-and middle- income countries showed that the prevalence of low vegetable consumption decreased with increasing income. The poorest community had the highest prevalence of low vegetable consumption, whereas the richest community had the lowest prevalence [48] (Burundi, Ethiopia, Ghana, Guinea, Kenya, Malawi, Mozambique, Rwanda, Tanzania, and Uganda). From the study in sub-Saharan Africa, researchers investigated a positive trend of increasing consumption of vegetables as gross domestic product per capita goes up, even if the trend is not fully linear [49].

Vegetables account for a small proportion of the food share budget, ranging from less than 4.5% in Ethiopia to 15.6% in Rwanda [49]. In studies conducted in several countries, such as Australia [83], Canada [77], Finland [84, 85], France [76], and the United States, higher socioeconomic status was associated with higher vegetable consumption. Several studies have found that those living in higher-income neighborhoods tend to consume more vegetables, while people living in the most disadvantaged regions tend to consume considerably fewer vegetables than people living in the most advantaged areas [86]. Research from the U.S. indicates that those in low-income households are less likely to meet the U.S. Department of Agriculture guidelines for vegetable consumption [87]. Only Kocho and vegetables cover 55% of the daily intake in the low income group, whereas low (15%) in the high income group. The demand for high-value food items (e.g. meat, milk, vegetables, and fruits) increases with income, which implies that poor households are unlikely to access them [63]. Households with income near to subsistence level consume large quantities of grains and starchy staples and few fruits, vegetables, meat, milk, and milk products.

The most often mentioned impediment to vegetable intake is cost [44, 46]. As a result, the exorbitant cost is mostly related to a person's income or socioeconomic position, which is often based on income, education, and work; hence, income and cost are inextricably intertwined and will not be handled independently. The 2002–2003 World Health Survey discovered a substantial link between vegetable intake and income in 33 of 52 nations. With the exception of Ghana, the number of poor vegetable eaters fell as wealth increased in all of these nations [48]. However, a study conducted on the effect of distance and cost on vegetables Consumption in Rural Texas revealed that neither access nor cost was related to vegetable consumption among white respondents. Among blacks, the cost was also not associated with rural consumption [88]. Only 15% of breastfeeding children aged 6–23 months consumed vitamin A-rich vegetables or foods made from roots and tubers [63]. The consumption patterns are determined by the combination of four main factors: the production level in the community garden, income level, preferences of the house- hold, and market prices. Accessibility and availability are core [39, 49, 89].

2.4 Social

2.4.1 Children and marital status

There is a compromise when it comes to marital status and vegetable consumption, with married people being more likely to consume more vegetables. However, this link is less obvious when it comes to having children. Indeed, numerous studies [39, 43, 44] found that being married was linked with higher vegetable intakes than being single in systematic reviews focused on a variety of factors of vegetable consumption. Two European research studies found that married status was a greater predictor of vegetable intake in males than in women [54, 81]. Having children has mixed correlations, according to Kamphuis et al. [43]. Indeed, studies have found a negative relationship between having children and vegetable consumption, i.e. parents consume less vegetables [90]; others have found that this relationship varies depending on ethnicity in the US population; and others conducted in the UK concluded that median vegetable intakes were not significantly different between women who had or did not have children under the age of 16 years [91].

2.4.2 Food knowledge and skills

People with a greater level of education consume substantially more veggies. This connection is frequently influenced by income, as more education is generally associated with a larger salary. There is research that suggests a link between education and vegetable intake. Positive associations between vegetable consumption and education have also been reported for Irish adults [81], French adults [76], US adults [50], Finnish adults [92], and Swedish adults [93]. Studies conducted in Canada reported that higher education was associated with purchasing greater amounts of vegetables [94] and with higher intakes of vegetables [77]. A study that looked at the relationship between income and vegetable intake at different levels of education found that Finnish people with low education reported greater vegetable intakes if their income was higher than individuals with intermediate or high education [85]. A comprehensive analysis of the connection between food consumption and dietary knowledge discovered that consumption of vegetables was positively associated with knowledge among people living in highincome nations [45]. Some studies, the majority of which were conducted in highincome countries, found that a high level of nutrition knowledge, specifically knowledge about the health benefits of high vegetable consumption and knowledge about the associations between diet and diseases, was associated with higher vegetable intakes [45, 67, 95–97]. The only study conducted in lowand middle-income countries discovered the opposite, namely that there was no relationship between diet knowledge and vegetable intake in South African Black adults [98].

Baker and Wardle [80] found that older individuals with higher awareness of the link between vegetables and illnesses ate considerably more vegetables, in both men and women, in a study of older adults in England. Other authors who focused on procedural nutrition knowledge, which is defined as knowledge of how to eat a balanced diet, found that Swiss men who answered correctly consumed more vegetables [99]. Other studies conducted in Africa with low income also showed that the level of education attained and knowledge gained were positively correlated with more consumption of vegetables [15, 49, 100, 101]. In this literature review, it is also recognized that a level of education attained positive associated nutritional value knowledge of vegetables. Knowledge alone will not ensure adequate health intake practice. The appropriate practice of vegetable consumption was, however, not as high as knowledge. The major reasons for this lower proportion of practice as compared to knowledge are the cost of food items and seasonal availability [101].

Knowledge of the nutritional value of vegetables was found to be statistically associated with sex, with females having higher knowledge than their male peers [49, 101]. The only study conducted in low and middle-income countries found the opposite, namely that there was no association between diet knowledge and vegetable intakes in South African Black adults [98].

Furthermore, climate and weather conditions, harvesting and handling techniques, packaging, storage and transportation facilities, market situation, diseases and pests were identified as major causes of post-harvest [111] loss in Dirre Dewa, Ethiopia, with post-harvest losses ranging from 20 to 50% between marketing and consumption [102]. Seid et al. [103] found that the severe postharvest loss and quality deterioration of vegetables mainly occurred during harvesting followed by marketing, transporting and storage [103]. Another study done in Akaki, Ethiopia showed that cabbage post-harvest loss in the supply chain was 58.9% [104]. This might be due to storage/transportation temperature, mechanical damage during transportation, disease infection, poor quality of irrigation water, poor storage mechanisms. In 2015, postharvest loss of vegetables occurred at the retail level (56% for bananas) due to rotting before reaching consumers in Ethiopia [105]. Besides, postharvest loss assessment was conducted in Jimma town, Ethiopia and the result indicated that the postharvest losses of mango and banana were 35.5% and 40.0%, respectively [106].

3. VEGETABLE NUTRITIONAL VALUE AND HEALTH MECHANISM

Based on the evidence of the role of vegetables in the prevention of many health problems, the FAO and WHO have recommended that people eat at least 240 g of vegetables per person per day, or alternatively, three servings a day. The standard portion size is assumed to be 80 g and the studies reviewed showed that, on average, this is appropriate. Actual portions, on the other hand, tend to be lower for vegetables and higher for fruits: at least two servings (160 g) of fruits and three servings (240 g) of vegetables, with at least one serving of vegetables involving dark green and leafy or orange vegetables (nutrient-rich vegetables), or approximately 146 kg per capita (per person) per year of fruit and vegetables combined (excluding potatoes and other starchy tubers, cassava, and sugar). This recommendation was based on a dose-response effect, which indicates an increased risk of disease at <200 g/day; yet little benefit >240 g/day. This quantity is believed to provide l2sufficient micronutrients and to reduce the risk of chronic diseases by ensuring adequate supply, availability and consumption of vegetables around the world [107–109].

Also, they create opportunities for smallholder farmers, a source of a better life for the economy as well as significant for improving feeding behavior for people [24]. Low vegetable intake is the main contributor to micronutrient deficiencies in the developing world, especially in populations with low intakes of nutrient-dense animal source foods such as meat and dairy products, which are expensive for low-income individuals [15, 40].

Accordingly, sufficient intake of vegetables has been related epidemiologically to a reduced risk of many non-communicable diseases [10, 28]. There are mainly three arguments that explain the health benefits of vegetables. Firstly, the large contribution of vegetables to micronutrients (especially provitamin A carotenoids, vitamin C and E, folate, and minerals, such as potassium or magnesium, which are probably involved in beneficial health effects, i.e., a decreased risk of NCDs. Secondly, the protective effect, due to certain antioxidants, such as vitamin C, carotenoids and polyphenols, against NCDs such as cardiovascular diseases, neurodegenerative and metabolic diseases and certain cancers. Five of the ten leading global disease burden risk factors identified by the World Health Report [19] are among the major causes of these diseases: high blood pressure, high cholesterol, obesity, physical inactivity, and insufficient consumption of vegetables. Four factors in the epidemiology of these diseases, such as poor diet and nutrition, physical inactivity, and tobacco and alcohol use, are of devastating importance to public health [10]. Diet and nutrition are recognized as key factors in the promotion and maintenance of good health throughout the entire life course [53]. Dietary behaviors like intakes of diets low in vegetables and high in sodium contribute significantly to the non-communicable disease burden in Ethiopia [110]. Vegetables also contain minerals such as potassium, magnesium, and, less importantly, iron and calcium, which are hard for the body to absorb and use [109]. A cross-sectional study in Gilgel Gibe found that 9% of the population is affected by chronic diseases [111]. The Ethiopian nationwide study reported a strong association between nutritional impairment and the development of chronic diseases such as cardiovascular diseases, cancer, and diabetes [112]. The survey reported the prevalence of chronic disease and its risk factors in the rural Jimma area of Ethiopia and found that the overall prevalence of chronic disease was 8.9%, and 80% of the rural subjects studied had at least one risk factor for chronic disease [111].

Recent research has shown that dietary calcium intake from vegetable consumption aids in bone mass formation in children aged 8–20 years [113–115]. Many studies have found that eating fewer vegetables than recommended is a risk factor for edentulism [114, 115], which is the condition of having lost all of one's natural teeth [116]. Indeed, numerous studies have shown evidence that nonstarchy veggies may lower the incidence of cancers of the mouth, larynx, pharynx, esophagus, and stomach. Furthermore, some foods, such as allium vegetables, may help prevent stomach cancer, and garlic may help prevent colon cancer [25]. A beneficial effect of consumption of vegetable-origin food is based on the presence of nutrients and phytochemicals including energy, fiber, B vitamins, vitamin C, antioxidants, potassium, flavonoids, and other identified and unidentified food compounds that provide good health for both children and adults. Health benefits include antihistamine, anti-inflammatory, diuretic, and especially antioxidant properties, and they have a proven preventive role in cancer processes [117]. Much research has focused on the possible mechanisms of action of phytochemicals in preventing or treating NCDs, cancer, and heart disease [118, 119]. Phytochemicals treat cancer by neutralizing free radicals, inhibiting enzymes that activate carcinogens, and activating enzymes that detoxify carcinogens. They protect against cancer by blocking or suppressing active carcinogens or tumor promoters from reaching target tissue [118–120]. The blocking actions could include activating enzymes that detoxify carcinogens, trapping carcinogens, or inhibiting cellular events linked to tumor formation. Study findings suggest that phytochemicals have also been promoted for the prevention and treatment of heart disease and may reduce the risk of coronary heart disease by preventing the oxidation of low-density lipoprotein (LDL) cholesterol, reducing the synthesis or absorption of cholesterol, normalizing blood pressure and clotting, and improving arterial elasticity [119, 120]. Some of these phytochemicals, such as phenolic compounds and sulfides, are marked by a broad spectrum of health-promoting functions [121] as cited by Charles et al. [118].

Dietary fiber (from fruits, vegetables, cereals, and pulses) has been shown to delay the absorption of carbohydrates after a meal and thereby decrease the insulinemic response to dietary carbohydrates [118, 119, 122, 123]. Since the water retention capacity of fiber adds bulk to fecal bolus, solidifies feces, aids bowel movements and brings beneficial physiological effects such as improving glucose intolerance, gastrointestinal functioning through keeping water, lowering total and LDL in the blood cholesterol level and prevention of some forms of cancer, especially colon cancer [12, 14–16, 122, 123]. Recent studies have concentrated on a specific kind of oligosaccharide: fructo- oligosaccharides (FOS), which have appealing physiological features such as low caloric values (approximately 2 kcal/gram), sweetness, limited potential for producing caries, and effects similar to those of dietary fiber. It is regarded as a prebiotic due to its ability to modify colon flora by promoting the selective growth of groups or individual species of bacteria that discourage the settlement of pathogenic bacteria.

Additionally, FOS fermentation acidifies the environment, thereby reducing carcinogen production [117, 122]. Fiber increases satiety, reduces hunger, and decreases energy intake by reducing the absorption of carbohydrates, thus reducing possible hyperglycemia and serum insulin levels from carbohydrate-rich foods and increasing insulin sensitivity through which they might have a positive effect on preventing one of the risks for type II diabetes, hypertension, and obesity. Because most vegetables are low in energy density, due to high water and low fat content and usually being fiber rich. Given that obesity is perhaps the most potent risk factor in the development of type 2 diabetes, the evidence indicated that the effects of dietary fiber due to its role in weight management were prevented by weight gain [118, 119]. According to the WHO, raising individual vegetable consumption up to the theoretical minimum-risk distribution may lower the global burden of disease for ischaemic heart disease by 30% for men and 31% for women and for ischemic strokes by 18% for men and 19% for women. The possible reduction in illness attributed to an increase in vegetables was 19% for stomach and esophageal cancers, and 19% for gastric cancer. Attributable risk fractions were lower for lung and colorectal cancers, at 12% and 2%, respectively. Overall, it is estimated that up to 2.7 million lives could potentially be saved each year if vegetable consumption were sufficiently increased [107].

Vegetables are the most important sources of vitamin A, a nutrient important for several metabolic activities in the body, in addition to its role as antioxidant; vegetables provide folate and potassium that are known to prevent birth defects, cancer, heart disease, hypertension and stroke; vegetables are good sources of minerals such as iron, zinc, calcium, potassium, and phosphorus [124–128]. Although concerns with respect to the bioavailability of vitamin A from green leafy vegetables have been raised [129, 130], consumption of cooked and pureed green leafy vegetables was shown to have a beneficial effect on improving vitamin A status [131–135]. Based on World Health Organization [27] recommendations suggesting unavailability of dark green leafy vegetables for more than six months in an area as indicative of increased risk to Vitamin A deficient, Inadequate intake of foods containing vitamin A is basically due to lack of access to vitamin A rich foods, which is a function of inadequate production, inadequate availability in the market or inability to purchase the foods.

4. CONCISE SUMMARY

Vegetables are incorporated into dietary requirements due to their substantial content of dietary fiber, vitamins, minerals, and phytochemicals. Consumption of vegetables helps safeguard against certain severe and costly non-communicable diseases, including cardiovascular disease, type II diabetes, some malignancies, and obesity. Notwithstanding the numerous health benefits of vegetables, underprivileged countries often do not meet the daily need of 240 grams for vegetable intake. Efforts must be intensified to enhance vegetable consumption and combat cost-effective, preventable noncommunicable illnesses. Moreover, the promotion of vegetables is more likely to be economically viable for governments and the food sector than measures that limit the use of unhealthy foods. Although the prevalence of noncommunicable diseases is rising, health systems and public health policies have predominantly concentrated on managing infections and addressing dietary inadequacies. This review indicates that low vegetable consumption is associated with economic factors, age, gender, psychological issues, family dynamics, educational level, inadequate nutritional knowledge, cultural influences, environmental conditions, social factors, and vegetable-related practices. The presence of nutrients and phytochemicals, including energy, fiber, B vitamins, vitamin C, vitamin A, antioxidants, potassium, flavonoids, and other known and unknown dietary components, enhances the beneficial effects of vegetable consumption in both children and adults. Phytochemicals combat cancer by neutralizing free radicals, blocking enzymes that activate carcinogens, activating enzymes that detoxify carcinogens, and obstructing active carcinogens or tumor promoters from accessing target tissue. Certain phytochemicals, including phenolic compounds and sulfides, possess a variety of health-enhancing attributes. Vegetables are crucial for postponing carbohydrate absorption post-meal, thereby reducing the insulinemic response to dietary carbohydrates. The water retention capacity of fiber increases fecal bulk, solidifies stool, facilitates bowel movements, and provides advantageous physiological effects, including the enhancement of glucose intolerance, gastrointestinal function through water retention, reduction of total and LDL cholesterol levels in the blood, and prevention of certain cancers, notably colon cancer. Fiber enhances feelings of fullness. Fiber enhances satiety, diminishes hunger, and reduces energy intake by inhibiting carbohydrate absorption, consequently decreasing the likelihood of hyperglycemia and serum insulin levels from carbohydrate-dense foods while improving insulin sensitivity, potentially mitigating risks associated with type II diabetes, hypertension, and obesity. Consequently, nutrition education via multisectoral strategies should be augmented to increase vegetable intake and mitigate the prevalence of non-communicable diseases (NCDs).

REFERENCES

1. Mekides W. Determinants of underweight, stunting and wasting among schoolchildren. BMC Public Health. 2015;**15**(8):1-9
2. Disha A, Rawat R, Subandoro A, Menon P. Infant and young child feeding (IYCF) practices in Ethiopia and Zambia and their association with child nutrition: Analysis of demographic and health survey data. African Journal of Food, Agriculture, Nutrition and Development. 2012;**12**(2)
3. Masresha T. Feeding patterns and stunting during early L8childhood in rural communities of Sidama South Ethiopia. Pan African Medical Journal. 2013;**14**:75

4. UNICEF. Nutrition, Improving Child, the Achievable Imperative for Global Progress. New York, NY; 2013
5. Ekesa BN, Walingo MK, Onyango MO. Accessibility to and consumption of indigenous vegetables and fruits by rural households in Matungu division, western Kenya. African Journal of Food, Agriculture, Nutrition and Development. 2009;**9**(8):1726-1728
6. Slavin JL, Lloyd B. Health benefits of fruits and vegetables. Advances in Nutrition. 2012;**3**(4):506-516
7. Bazzano LA. The high cost of not consuming fruits and vegetables. Journal of the American Dietetic Association. 2006;**106**:1364-1368
8. Lora I, Kenda C, Marie R. Improving Diet Quality and Micronutrient Nutrition. Homestead Food Production in Bangladesh. International Food Policy Research Institute; 2009
9. Tohill BC, Seymour J, Serdula M, Kettel KL, Rolls BJ. What epidemiologic studies tell us about the relationship between fruit and vegetable consumption and body weight. Nutrition Review. 2004;**62**(10):365-374
10. WHO. Diet, nutrition and the prevention of chronic diseases. World Health Organ Tech Rep Series. 2004
11. Uusiku NP, Oelofse A, Duodu KG, Bester MJ, Faber M. Nutritional value of leafy vegetables of sub-Saharan Africa and their potential contribution to human health: A review. Journal of Food Composition and Analysis. 2010;**23**(6):499-509
12. Adebawo O, Salau B, Ezima E, Oyefuga O, Ajani E, Idowu G, et al. Fruits and vegetables moderate lipid cardiovascular risk factor in hypertensive patients. Lipids in Health and Disease. 2006;**5**(5):14
13. WHO. Indicators for Assessing Infant and Young Child Feeding Practices; Part 1 Definitions; World Health Organization Report. Geneva, Switzerland; 2008
14. Abajobir AA, Jake MN, Gail W, Lane S, Alexandra C, Steve K. Substantiated childhood maltreatment and young adulthood cannabis use disorders: A pre-birth cohort study. Psychiatry Research. 2017;**256**:21-31
15. Banwat ME, Lar LA, Daber J, Audo J, Lassa S. Knowledge and intake of fruit and vegetable consumption among adult in urban community in North Central Nigeria. The Nigeria Health Journal. 2012;**12**(1):12-15
16. Darmon N, Drewnowski A. Contribution of food prices and diet cost to socioeconomic disparities in diet quality and health: A systematic review and analysis. Nutrition Reviews. 2015;**73**(10):643-660
17. WHO. The World Health Report for Reducing Risks, Promoting Healthy Life. World Health Organization Report. Geneva, Switzerland; 2008
18. Pasricha SR, Black J, Muthayya S. Determinants of anemia among young children in rural India. Pediatrics. 2010;**126**(1):140-149

19. WHO. The World Health Report for Reducing Risks, Promoting Healthy Life. World Health Report. Geneva, Switzerland; 2002
20. Stoltzfus RJ, Chway HM, Montresor A, Tielsch JM, Jape JK, Albonico M, et al. Low dose daily iron supplementation improves iron status and appetite but not anemia, whereas quarterly anthelminthic treatment improves growth, appetite and anemia in Zanzibari preschool children. The Journal of Nutrition. 2004;**134**(2):348-356
21. Rice AL, West KP, Black RE. Vitamin A deficiency. In: Ezzati MLA, Rodgers A, Murray CLJ, 422 Food and Nutrition Bulletin 38(3), editors. Comparative Quantification of Health Risks; 2004
22. Ploysangam T, Heubi JE, Eisen D, Balistreri WF, Lucky AW. Cutaneous Crohn's disease in children. Journal of the American Academy of Dermatology. 1997;**36**:697-704
23. Black RE, Victora CG, Walker SP. Maternal and child undernutrition and overweight in low income and middleincome countries. Lancet. 2013;**382**(9890):427-451
24. FAO. Food and Agricultural Organization of the United Nations. Children's Work in the Livestock Sector: Herding and beyond Rome. Rome, Italy: FAO; 2013
25. WHO. Millennium Development Goals: Progress towards the Health- Related Millennium Development Goals. Fact Sheet Levels and Trends in Child. Geneva, Switzerland; 2012
26. Solomon A, Bekele S, Franklin S. Does technology adoption promote commercialization? Evidence from Chickpea Technologies in Ethiopia. Food Policy. 2010;**37**(3):283-295
27. WHO. Global Action Plan for the Prevention and Control of Noncommunicable Diseases 2013-2020. World Health Organization; 2014. p. 102
28. WHO. Nutrition Landscape Information System Country Profile Indicators Interpretation Guide. Geneva, Switzerland: WHO; 2010
29. Misganaw A, Haregu TN, Deribe K, Tessema GA, Deribew A, Melaku YA, et al. National mortality burden due to communicable, non-communicable, and other diseases in Ethiopia, 1990- 2015: Findings from the Global Burden of Disease Study 2015. In: Popul Health Metr. Vol. 15(29). London: BioMed Central; 2017
30. Ayalew MT, Haileamlak A, Tessema F, Fessahaye A, Kifle W, Makonnen A, et al. Population based survey of chronic non-communicable diseases at Gilgel Gibe Field research center, Southwest Ethiopia. Ethiopian Journal of Health Sciences. 2012;**22**:7-18
31. Murphy GAV, Asiki G, Ekoru K, Nsubuga RN, Nakiyingi-Miiro J, Young EH, et al. Sociodemographic distribution of noncommunicable disease risk factors

in rural Uganda: A cross-sectional study. International Journal of Epidemiology. 2013;**42**(6):1740-1753

32. Degye G, Belay K, Mengistu K. Measuring diet quantity and quality dimensions of food security in rural Ethiopia. Journal of Development and Agricultural Economics. 2013;**5**(5):174-185
33. Guush B, Zelekawork P, Kibrom T, Seneshaw T. Foodgrain consumption and calorie intake patterns in Ethiopia. Ethiopia Strategy Support Program II. 2011;**23**
34. Kebebu A, Whiting SJ, Dahl WJ, Henry CJ, Kebede A. Formulation of a complementary food fortified with broad beans (*viciafaba)* in southern Ethiopia. African scholar science communication trust. African Journal of Food, Agriculture, Nutrition and Development. 2013;**13**(3):1-15
35. Tadesse K. Analysis of changes in food consumption patterns in urban Ethiopia. In: Ethiopian Development Research Institute, Working Paper 002. Addis Ababa, Ethiopia; 2010
36. Gallet CA. The income elasticity of meat: A meta-analysis. Australian Journal of Agricultural and Resource Economics. 2010;**54**:477-490
37. Yazew T, Daba A. Heath benefits of fruit and vegetables consumption: Preventive implications for non- communicable diseases in Ethiopia. Advanced Techniques in Biology and Medicine. 2020;**8**(3):275
38. Dunn RA, Dean WR, Johnson CM, Leidner A, Sharkey JR. The effect of distance and cost on fruit and vegetable consumption in rural Texas. Journal of Agricultural and Applied Economics. 2012;**44**(4):491-500
39. Gesessew K. Food choice behaviors of lactating women: Association with body mass index and fruits and vegetables intake in Central Amhara Region, Ethiopia—An observational study. Journal of Nutrition and Metabolism. 2021;**2021**:13
40. Meng T, Florkowski WJ, Sarpong D, Resurreccion AVA, Chinnan M. Determinants of Fresh Vegetables, Fresh Fruits, and Fresh Peanut Products Expenditure in Urban Household in Ghana. Washington DC: AAEA Joint meeting CAES; 2013
41. Niu L, Wohlgenant M. Subsidizing for fruits and vegetables by income group: A two-stage budgeting approach. In: Selected Paper Prepared for Presentation at the Agricultural and Applied Economic Association Joint Meeting CAES; Washington, DC, USA. 2013
42. Guillaumie L, Godin G, Vezina-Im LA. Psychosocial determinants of fruit and vegetable intake in adult population: A systematic review. International Journal of Behavioral Nutrition and Physical Activity. 2010;**7**(1):12

43. Kamphuis CBM, Giskes K, De Bruijn GJ, Wendel-Vos W, Brug J, Van Lenthe FJ. Environmental determinants of fruit and vegetable consumption among adults: A systematic review. The British Journal of Nutrition. 2006;**96**:620-635

44. Pollard J, Kirk SFL, Cade JE. Factors affecting food choice in relation to fruit and vegetable intake: A review. Nutrition Research Reviews. 2002;**15**:373-387

45. Shaikh AR, Yaroch AL, Nebeling L, Yeh MC, Resnicow K. Psychosocial predictors of fruit and vegetable consumption in adults a review of the literature. American Journal of Preventive Medicine. 2008;**34**(6):535-543

46. Yeh MC, Ickes SB, Lowenstein LM, Shuval K, Ammerman AS, Farris R, et al. Understanding barriers and facilitators of fruit and vegetable consumption among a diverse multi- ethnic population in the USA. Health Promotion International. 2008;**23**(1):42-51

47. FAO. Food and Agricultural Organization of the United Nations. Rome: The State of Food and Agriculture: Innovation in Family Farming; 2014

48. Hall JN, Spencer M, Sam BH, John WL. Global variability in fruit and vegetable consumption. American Journal of Preventive Medicine. 2009;**36**(5):402-409

49. Marie T, Nicholas M, Lisa S. Patterns and Determinants of Fruits and Vegetables Consumption in Sub Saharan African. A Multicounty Comparison. Washington DC, USA: International Food Policy Research Institute; 2005

50. Centers for Disease Control and Prevention. Sates-specific trends in fruit and vegetable consumption among adults-United States 2000-2009. Morbidity and Mortality Weekly Report. 2010;**59**(35):1125-1130

51. Esteghamati A, Noshad S, Nazeri A, Khalilzadeh O, Khalili M, Nakhjavani M. Patterns of fruit and vegetable consumption among Iranian adults: A SuRF-NCD-2007 study. British Journal of Nutrition. 2011;**108**(1):177-181

52. Tsegaye D, Ahmed A, Dilnesaw Z. Availability and consumption of fruits and vegetables in nine regions of Ethiopia with special emphasis to vitamin A deficiency. Ethiopian Journal of Health Development. 2009;**23**(3):216-222

53. FAO/WHO. Workshop on Fruit and Vegetables for Health, Held at the WHO Centre for Health Development. Kobe, Japan; 2004

54. Kamphuis CB, van Lenthe FJ, Giskes K, Brug J, Mackenbach JP. Perceived environmental determinants of physical activity and fruit and vegetable consumption among high and low socioeconomic groups in the Netherlands. Health and Place. 2007;**13**(2):493-503

55. Alaimo K, Packnett E, Miles RA, Kruger DJ. Fruit and vegetable intake among Urban Community gardeners. Journal of Nutrition Education and Behavior. 2008;**40**(20):94-101

56. Rolien W, André de J. Business Opportunities in the Ethiopian Fruit and Vegetable Sector. Wageningen University and Research Centre LEI; 2009
57. Ariel M, Jeff SS, Jill KB, Clark, Kristi SL. Barriers to home fruit and vegetable cultivation in Ohio [MSc thesis]. Ohio: School of Environment and Natural Resources at the Ohio State University; 2013
58. John RB, Mohammed A, Fernandez AB, Megan EC, Guillou M, et al. Achieving food security in the face of climate change: Summary for policy makers from the Commission on Sustainable Agriculture and Climate Change. 2011
59. Getu GE. Contributions of Vegetable Production to the Household Food Security: The Case of Kebeles Surrounding Holeta Town. Ethiopia: School of Graduate Studies Addis Ababa University; 2011
60. Galhena DH, Freed R, Maredia KM. Home gardens: A promising approach to enhance household food security and wellbeing. Agriculture & Food Security. 2013;**2**(8)
61. Akrofi S, Struik PC, Price LL. HIV and orientation of subsistence and commercial home gardens in rural Ghana: Crop composition, crop diversity and food security. African Journal of Agricultural Research. 2010;**5**(18):2593-2607
62. Sommerfeld AJ, McFarland AL, Waliczek TM, Zajicek JM. Growing minds: Evaluating the relationship between gardening and fruit and vegetable consumption in older adults. Horticulture Technology. 2010;**20**(4):711-717
63. Central Statistical Agency and ORC Macro. Ethiopia Demographic and Health Survey, Addis Ababa, Ethiopia and Calverton, Maryland. USA: CSA and ORC Macro; 2011
64. FAO. Food and Agricultural Organization of the United Nations. Rome: The State of Food Insecurity in the World High Food Prices and Food Security Threats and Opportunities; 2008
65. Godin G, Amireault S, Bélanger-Gravel A, Vohl MC, Pérusse L, Guillaumie L. Prediction of daily fruit and vegetable consumption among overweight and obese individuals. Appetite. 2010;**54**(3):480-484
66. Blanchard CM, Fisher J, Sparling PB, Hunt Shanks T, Nehl E, Rhodes RE, et al. Understanding adherence to 5 servings of fruits and vegetables per day: A theory of planned behavior perspective. Journal of Nutrition Education and Behavior. 2009;**41**(1):3-10
67. Wolf RL, Lepore SJ, Vandergrift JL, Wetmore-Arkader L, McGinty E, Pietrzak G, et al. Knowledge, barriers and stage of change as correlates of fruit and vegetable consumption among urban and mostly immigrant Black men. Journal of the American Dietetic Association. 2008;**108**(8):1315-1322
68. Appleton KM, McGill R, Neville C, Woodside JV. Barriers to increasing fruit and vegetable intakes in the older population of Northern Ireland: Low lev-

els of liking and low awareness of current recommendations. Public Health Nutrition. 2010;**13**(4):514-521

69. Larson N, Laska MN, Story M, Neumark-Sztainer D. Predictors of fruit and vegetable intake in young adulthood. Journal of the Academy of Nutrition and Dietetics. 2012;**112**(8):1216-1222
70. Watters JL, Satia JA, Galanko JA. Associations of psychosocial factors with fruit and vegetable intake among African-Americans. Public Health Nutrition. 2007;**10**(7):701-711
71. Keihner A, Adkins S, Scruggs V. Out-of-Home Eating Relates to Fruit and Vegetable Consumption among African Americans. Sacramento, CA: California Department of Health Services; 2012
72. Smith KJ, McNaughton SA, Gall SL, Blizzard L, Dwyer T, Venn AJ. Takeaway food consumption and its associations with diet quality and abdominal obesity: A cross-sectional study of young adults. International Journal of Behavioral Nutrition and Physical Activity. 2009;**6**(29)
73. Feldman S, Eisenberg ME, Neumark SD, Story M. Associations between watching TV during family meals and dietary intake among adolescents. Journal of Nutrition Education and Behavior. 2007;**39**(5):257-263
74. Fitzpatrick E, Edmunds LS, Dennison BA. Positive effects of family dinner are undone by television viewing. Journal of the American Dietetic Association. 2007;**107**(4):666-671
75. Boutelle KN, Birnbaum AS, Lytle LA. Associations between perceived family meal environment and parent intake of fruit, vegetables and fat. Journal of Nutrition Education and Behavior. 2003;**35**:24-29
76. Estaquio C, Druesne-Pecollo N, Latino-Martel P, Dauchet L, Hercberg S, Bertrais S. Socioeconomic differences in fruit and vegetable consumption among middle-aged French adults: Adherence to the 5 a day recommendation. Journal ofAmerican Dietetic Association. 2008;**108**(12):2021-2030
77. Azagba S, Sharaf M. Disparities in the frequency of fruit and vegetable consumption by socio-demographic and lifestyle characteristics in Canada. Nutrition Journal. 2011;**10**(1):118
78. Kolawole O, Sadiat FA. Determinants of household demand for fresh fruit and vegetable in Nigeria: A double hurdle approach. Quarterly Journal of International Agriculture. 2016;**52**(3):199-216
79. Prättälä R, Paalanen L, Grinberga D, Helasoja V, Kasmel A, Petkeviciene J. Gender differences in the consumption of meat, fruit and vegetables are similar in Finland and the Baltic countries. European Journal of Public Health. 2006;**17**(5):520-525
80. Baker AH, Wardle J. Sex differences in fruit and vegetable intake in older adults. Appetite. 2003;**40**(3):269-275

81. Friel S, Newell J, Kelleher C. Who eats four or more servings of fruit and vegetables per day? Multivariate classification tree analysis of data from the 1998 Survey of Lifestyle, Attitudes and Nutrition in the Republic of Ireland. Public Health Nutrition. 2005;**8**(2):159-169
82. Boffetta P, Couto E, Wichmann J, Ferrari P, Trichopoulos D, Bueno-de-Mesquita HB, et al. Fruit and vegetable intake and overall cancer risk in the European prospective investigation into cancer and nutrition. Journal of the National Cancer Institute. 2010;**102**(8):529-537
83. Ball K, Crawford D, Mishra G. Socio-economic inequalities in women's fruit and vegetable intakes: A multilevel study of individual, social and environmental mediators. Public Health Nutrition. 2006;**9**(5):623-630
84. Lallukka T, Laaksonen M, Rahkonen O, Roos E, Lahelma E. Multiple socioeconomic circumstances and healthy food habits. European Journal of Clinical Nutrition. 2007;**61**(6):701-710
85. Lallukka T, Pitkaniemi J, Rahkonen O, Roos E, Laaksonen M, Lahelma E. The association of income with fresh fruit and vegetable consumption at different levels of education. European Journal of Clinical Nutrition. 2010;**64**(3):324-327
86. Shohaimi S, Welch A, Bingham S, Luben R, Day N, Wareham N, et al. Residential area deprivation predicts fruit and vegetable consumption independently of individual educational level and occupational social class: A cross sectional population study in the Norfolk cohort of the European Prospective Investigation into Cancer. Journal of Epidemiology and Community Health. 2004;**58**(8):686-691
87. Guenther P, Dodd K, Reedy J, Krebs-Smith S. Most Americans eat much less than recommended amounts of fruits and vegetables. Journal of the American Dietetic Association. 2006;**106**:1371-1379
88. Richard AD, Joseph RS, Scott H. The effect of fast-food availability on fast-food consumption and obesity among rural residents: An analysis by race/ethnicity. Economics and Human Biology. 2012;**10**(1):1-13
89. Ethiopia Nutrition Profile. Somali Region, Ethiopia; 2018
90. Wandel M. Dietary intake of fruits and vegetables in Norway: Influence of life phase and socio-economic factors. International Journal of Food Sciences and Nutrition. 1995;**46**(3):291-301
91. Pollard T. Changes in mental well-being, blood pressure and total cholesterol levels during workplace reorganization: The impact of uncertainty. Work and Stress. 2001;**15**(1):14-28
92. Paalanen L, Prattala R, Palosuo H, Laatikainen T. Socio-economic differences in the consumption of vegetables, fruit and berries in Russian and Finnish Karelia: 1992-2007. The European Journal of Public Health. 2007;**21**(1):35-42

93. Elfhag K, Tholin S, Rasmussen F. Consumption of fruit, vegetables, sweets and soft drinks are associated with psychological dimensions of eating behaviour in parents and their 12-year- old children. Public Health Nutrition. 2008;**11**(9):914-923

94. Ricciuto L, Tarasuk V, Yatchew A. Socio-demographic influences on food purchasing among Canadian households. European Journal of Clinical Nutrition. 2006;**60**:778-790

95. Moynihan PJ, Mulvaney CE, Adamson AJ, Seal C, Steen N, Mathers JC, et al. The nutrition knowledge of older adults living in sheltered housing accommodation. Journal of Human Nutrition and Dietetics. 2007;**20**(5):446-458

96. Wardle J, Steptoe A, Oliver G, Lipsey Z. Stress, dietary restraint and food intake. Journal of Psychosomatic Research. 2000;**48**(2):195-202

97. Beydoun MA, Powell LM, Wang Y. Reduced away-from-home food expenditure and better nutrition knowledge and belief can improve quality of dietary intake among US adults. Public Health Nutrition. 2009;**12**(3):369-381

98. Karl P, Supa P. Knowledge, barriers, and benefits of fruit and vegetable consumption and lay conceptions of nutrition among rural and semi-urban Black south Africans. Psychological Reports. 2004;**94**(3)

99. Dickson-Spillmann M, Siegrist M. Consumers' knowledge of healthy diets and its correlation with dietary behavior. Journal of Human Nutrition and Dietetics. 2011;**24**(1):54-60

100. Bettina MB, Rochelle R, Leam M. Knowledge, Attitude and Practices Related to Fruit and Vegetable Consumption of High School Students; 2011. Available from: www.linkighub. elsevier.com

101. Mathilda EB, Luretalbert L, Jonathan D, Sunday A, Samuel L. Knowledge and intake of fruit and vegetables consumption among adults in an Urban Community in north Central Nigeria; Department of Community Medicine, Jos University Teaching Hospital, Jos, Nigeria. The Nigerian Health Journal. 2012;**2**(1):1-4

102. Mohammed K, Afework B. Post- harvest loss and quality deterioration of horticultural crops in Dire Dawa Region, Ethiopia. Journal of Saudi Society of Agricultural Sciences. 2013;**17**(1):1-9

103. Seid H, Hassen B, Yitbarek WH. Postharvest loss assessment of commercial horticultural crops in South Wollo, Ethiopia challenges and opportunities. Food Science and Quality Management. 2013;**17**:34-39

104. Hagos F, Alemseged F, Balcha F, Berhe S, Aregay A. Application of nursing process and its affecting factors among nurses working in Mekelle zone hospitals, northern Ethiopia. Nursing Research & Practice. 2014;**675212**

105. Mulualem AM, Jema H, Kebede W, Amare A. Determinants of postharvest Banana loss in the marketing; chain of Central Ethiopia. Food Science and Quality Management. 2015;**37**:52-63

106. Adugna D, Gerba D, Diriba B, Kassaye T. Identification of major causes of postharvest losses among selected fruits in Jimma zone for proffering veritable solutions. International Journal of Current Research. 2011;**3**(11):34-40
107. Lock K, Pomerlau J, Causer L, Altmann DR, McKee M. The global burden of disease attributable to low consumption of fruit and vegetables: Implications for the global strategy on diet. Bulletin of the World Health Organization. 2005;**83**:100-108
108. Park HM, Heo J, Park Y. Calcium from plant sources is beneficial to lowering the risk of osteoporosis in postmenopausal Korean women. Nutrition Research. 2011;**31**:27-32
109. WHO/FAO. Diet, Nutrition and the Prevention of Chronic Diseases: Report of a Joint WHO/FAO Expert Consultation, Geneva: WHO Technical Report Series. 2003; 916.
110. Yohannes A, Awoke M, Amare D, Gizachew A, Kebede D, Berhe WS, et al. The impact of dietary risk factors on the burden of non-communicable diseases in Ethiopia: Findings from the Global Burden of Disease study. International Journal of Behavioral Nutrition and Physical Activity. 2016;**13**(122):1-13
111. Prevett M. Chronic non- communicable diseases in Ethiopia—A hidden burden. Ethiopia Journal of Health Science. 2012;**22**(S):1-2
112. Central Statistical Agency. Federal Democratic Republic of Ethiopia: Household Income, Consumption, and Expenditure Survey of 2004/5. Vol. I. Addis: Statistical Bulletin; 2007. p. 394
113. Baxter-Jones A, Vatanparast H, Robert F, Donald B, Susan W. Positive effects of vegetable and fruit consumption and calcium intake on bone mineral accrual in boys during growth from childhood to adolescence: The University of Saskatchewan Pediatric Bone Mineral Accrual Study. The American Journal of Clinical Nutrition. 2005;**82**(3):700-706
114. De Marchi RJ, Hugo FN, Padilha DM, Hilgert JB, Machado DB, Durgante PC, et al. Edentulism, use of dentures and consumption of fruits and vegetables in south Brazilian community dwelling elderly. Journal of Oral Rehabilitation. 2011;**38**:533-540
115. Tsakos G, Herrick K, Sheiham A, Watt RG. Edentulism and fruit and vegetable intake in low-income adults. University of Wisconsin-Stout Menomonie, WI 54751 University, 301 HS, NSCI, and Stillwater, OK 74078, USA. Public Health Nutrition. 2010;**89**:462-467
116. Sussex PV. Edentulism from a New Zealand perspective a review of the literature. New Zealand Dental Journal. 2008;**104**:84-96
117. FAO. World Agriculture: Towards 2030/2050. Prospects for Food, Nutrition, Agriculture and Major Commodity Groups. Interim Report. Rome: Global Perspective Studies Unit, Food and Agriculture Organization of the United Nations; 2005

118. Charles L, Wendy L, Guire M. What do we know about health care team effectiveness? A review of the literature. Medical Care Research and Review. 2006;**63**(3):263-300
119. Miller LB. Phytochemicals as a motivational tool to change fruit, vegetable and whole grain consumption [MSc thesis]. Food Science and Nutrition; 2002
120. American Cancer Association. Special Section: Cancer-Related Pain. Cancer Facts and; Figures, Atlanta, Georgia; 2007.
121. Schreiner M. Vegetable crop management strategies to increase the quantity of phytochemicals. European Journal of Nutrition. 2005;**44**(2):85-94
122. David A, Virginie B, Françoise B, Veronique B, Stéphanie H, Emmanuelle BS, et al. High cumulative risks of cancer in patients with *PTEN* hamartoma tumor syndrome. Journal of Medical Genetics. 2013;**50**:255-263
123. Tosh SM, Yada S. Dietary fiber in pulse seeds and fractions: Characterization, functional attributes, and applications. Food Research International. 2010;**43**:450-460
124. Thomson HC, Kelly WC. Vegetable Crops. 5th ed. New Delhi: McGraw Hill Publishing Co. Ltd.; 1990
125. Hirvonen K, Hoddinott J. Agricultural Production and children's Diets: Evidence from Rural Ethiopia. Washington (DC): International Food Policy Research Institute; 2014
126. Prior R, Cao G. Antioxidant phytochemicals in fruits and vegetables; diet and health implications. HortScience. 2000;**35**:588-592
127. Quebedeaux B, Bliss F. Horticulture and human health: Contributions of fruits and vegetables. Proc.1st Intl Symp, Hort and Human Health Prentice. 1998
128. Wargovich M. Anticancer properties of fruits and vegetables. Hort Science. 2000;**35**:573-575
129. De Pee S, West CE, Muhilal KD, Hautvast JGA. Lack of improvement in vitamin a status with increased consumption of dark-green leafy vegetables. Lancet. 1995;**46**(8967):75-81
130. Khan NC, West CE, De Pee S. The contribution of plant foods to the vitamin A supply of lactating women in Vietnam: A randomized controlled trial. American Journal of Clinical Nutrition. 2007;**85**:1112-1120
131. Flyman MV, Afolayan AJ. The suitability of wild vegetables for alleviating human dietary deficiencies. South African Journal of Botany. 2006;**72**:492-497
132. Haskell MJ, Jamil KM, Hassan F. Daily consumption of Indian spinach (*Basella alba*) or sweet potatoes has a positive effect on total-body vitamin A stores in Bangladeshi men. The American Journal of Clinical Nutrition. 2004;**80**:705-714

133. Haskell MJ, Pandey P, Graham JM, Peerson JM, Shrestha RK, Brown KH. Recovery from impaired dark adaptation in night-blind pregnant Nepali women who receive small daily doses of vitamin A as amaranth leaves, carrots, goat liver, vitamin A-fortified rice, or retinyl palmitate. The American Journal of Clinical Nutrition. 2005;**81**:461-471
134. Odhav B, Beekrum S, Akula U, Baijnath H. Preliminary assessment of nutritional value of traditional leafy vegetables in KwaZulu-Natal, South Africa. Journal of Food Compositio Analls. 2007;**20**:430-435
135. Takyi EEK. Children's consumption of dark green, leafy vegetables with added fat enhances serum retinol. The Journal of Nutrition. 1999;**129**:1549-1554

5 Chapter

VEGETABLE AND HERBAL EXTRACTS: A PATHWAY TO PREVENTIVE AND THERAPEUTIC REGIMENS

1. INTRODUCTION: AN OVERVIEW

Ayurveda, the traditional healthcare system of India, is regarded as the oldest medical system. Historically, plants (vegetables and fruits) have served as medicinal agents for the treatment and prevention of diseases in both traditional and contemporary contexts. In India, Ayurveda has been utilized for decades as a natural approach to disease prevention and treatment, with plants playing a crucial role in this process [1]. Ayurveda, the science of longevity, encompasses a range of therapeutic approaches and delineates many ailments along with natural remedies to address them. Traditional herbal treatments are considered to have existed for a minimum of 3000 years. Medicines or extracts derived from whole plants or specific plant components are commonly referred to as phytomedicines or phytoextracts.

These chemicals exhibit significant antioxidant effects. Antioxidants are molecules or chemicals that inhibit the process of oxidation. Key elements in plant products include flavonoids, alkaloids, terpenes, bisbenzyls, and coumarins, which provide potential therapeutic usefulness [2]. Currently, much emphasis is directed on natural antioxidants derived from plant products due to their accessibility, cost-effectiveness, and non-toxic characteristics in contrast to manufactured medicines [3]. Diverse plant compounds are endorsed for their antioxidative, anticarcinogenic, anti-tumorigenic, healing, and chelating properties [4]. Research indicates that cultures consuming diets abundant in natural antioxidants exhibit a reduced prevalence of disorders associated with oxidative damage and cancer [5].

Diverse plant species have been examined with encouraging outcomes. A plethora of vitamins and minerals is present in botanical products. They are inherently abundant in antioxidants, which safeguard the body against aging and diseases such as cancer and cardiovascular conditions. Medicinal plants naturally manufacture secondary metabolites such as alkaloids, flavonoids, glycosides, lactones, resins, sterols, and sterol derivatives. A diverse array of bioactive chemicals has been uncovered in contemporary science, facilitating the development of novel pharmaceuticals derived from plants. Various phytochemicals have demonstrated potential effectiveness against life-threatening disorders, necessitating additional investigation into their mechanisms of action, preclinical studies, and clinical trials. Approximately 80% of Africa's population relies on traditional medicines for their healthcare needs. Studies have demonstrated the use of medicinal plants in the treatment and prevention of chronic disorders [6].

This chapter examines the advantageous properties of several medicinal plants and their specific components. Medicinal plants and Ayurvedic remedies are gaining popularity due to the soaring costs of allopathic medications[7]. The most useful components of aloe vera are the gel and latex derived from its leaves, while in turmeric, the rhizome exhibits the majority of its therapeutic characteristics [8, 9]. The leaves of the tea tree are utilized, while the leaves, blossoms, and seeds of the Ashwagandha plant serve medicinal reasons; the entire plant of Centella asiatica is beneficial, and the leaves of Holy Basil are recognized for their therapeutic properties in treating many ailments [10–13]. Moringa oleifera contains several phytoconstituents in its leaves, flowers, seeds, roots, and stem bark [14]. The berries and seeds of Sea Buckthorn are highly beneficial; conversely, the fruit of Amla is the most advantageous[15, 16]. Garlic is a bulbous plant whose bulb possesses significant utility. These plant components can be utilized directly as medications or for the production of therapeutic agents. Consequently, the current chapter concentrates on the application of herbal items to mitigate the detrimental effects of various xenobiotics. It is crucial to investigate and ascertain the significance of natural medicinal systems and adapt them to contemporary requirements.

2. ALOE VERA (ALOE BARBADENSIS MILLER)

Aloe vera is a perennial succulent plant native to the plant family Asphodelaceae (**Figure 1**). Ayurveda, Siddha, Unani, and homeopathy are just a few indigenous systems of medicine that use homeopathic remedies. Among the 250 species of Aloe, Aloe vera is one of the 250 species of Aloe barbadensis miller [17]. Aloes grow in various climates, including deserts, grassland, coastal, and even alpine areas.

They are often thought to only increase in hot, dry climates.

Figure 1. Aloe Vera.

2.1 Phytochemistry

There are 75 active compounds in aloe vera which falls under vitamins, enzymes, minerals, sugars, lignin, saponins, salicylic acids, and amino acids. It contains antioxidant vitamins A, E and C. Furthermore, it contains folic acid, vitamin B_{12}, and choline (**Figure 2a** and **b**).

The system includes eight enzymes: alkaline phosphatase, brady kinase, catalase, carboxypeptidase, cellulase, amylase, lipase, and peroxidase [18]. Excessive inflammation is reduced by topically applied brady kinase, whereas the others aid in breakdown of sugars and fats.

It has high concentrations of various minerals as calcium, chromium, copper, selenium, magnesium, manganese, potassium, sodium, and zinc [19, 20]. The plant provides polysaccharides glucomannan and polymannose as well as monosaccharides glucose and fructose. Mucopolysaccharides originate in the plant's mucilage layer. Monosaccharides that dominate are mannose-6-phosphate and glucomannans [beta (1,4)-acetylated mannan] [20]. Aloe vera gel has been shown to contain anti-inflammatory glycoprotein and a novel anti-allergic substance, C-glucosyl chromone [19]. It comprises 12 anthraquinones, which are phenolic compounds traditionally used as laxatives. Aloin and other hydroxy anthracene derivatives such as emodin are responsible for the laxative effect of Aloe vera. It also contains four fatty acids, cholesterol, campesterol, sitosterol and lupeol. Lupeol is also antiseptic, and analgesic and all of these compounds have anti-inflammatory properties [20].

2.2 Medicinal and nutritional value

Aloe vera has long been used to treat burns, commonly known as the burn tree and first aid plant. Aloe vera is anti-inflammatory, antibacterial, antiviral, increases immune activity, and has a lower histamine activity than other organic compounds, which accelerates burn wounds' healing. The medicinal properties of Aloe vera

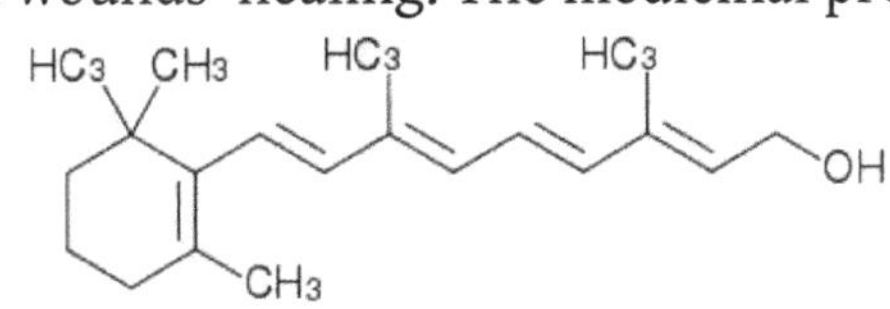

Vitamin A (a)

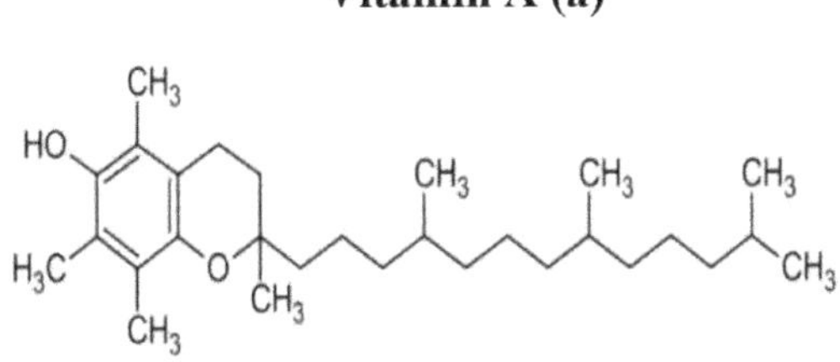

Vitamin E (b)

Figure 2. Structure of potent vitamins. Vitamin A (a). Vitamin E (b).

have led to its increased popularity. It contains amino acids essential to wound healing [21]. Additionally, it contains several inorganic electrolytes, such as iron, potassium, magnesium, chromium, copper, sodium, calcium, and zinc, essential for wound healing [22]. By releasing growth factors, it produces antibodies and promotes wound healing. Aloe vera gel, has anti-inflammatory properties, healing effects, stimulation of mucus, and regulates gastric secretions, which can help prevent and treat gastric ulcers [22]. As a result of the presence of anthraquinones and chromone, aloe vera gel has strong anti-inflammatory properties [23]. Aloe vera also relieves joint pain because it has an anti-inflammatory effect. Inflammation is caused by the body's response to various injuries, which produces bradykinin (a mediator of inflammation). Aloe vera extracts stored for three to 10 days in the dark produce active compounds prostanoids from the glycoprotein and polysaccharide fractions. Chronic bronchial asthma can be effectively treated with these compounds. Several types of cancer are non-resistant to the chemo-preventive power of aloe vera due to its glycoproteins and polysaccharides [24]. These agents stimulate the immune system to fight cancer [22]. Researchers examined the cytotoxicity of barbarol, aloe-emodein, and aloesin extracted from Aloe vera against acute myeloid leukemia (AML) and acute lymphocytic leukemia (ALL) [21].

3. TURMERIC (*CURCUMA LONGA*)

It is one of the most useful herbal medicinal plants which belongs to Zingiberaceae family (**Figure 3**). It is known for its medicinal properties for 4000 years [25].

Figure 3. Turmeric.

It is used widely as a spice all over the world but mostly in eastern region. It is broadly cultivated in Asian countries like India and China. In ancient times, it was used in many ways including its presence in ointments due to its great healing properties, it is used to improve digestion, intestinal problems, to strengthen organs such as gall bladder and liver [26]. In 1870s, chemists found that the yellow-orange color of turmeric changes to reddish brown in alkaline medium. This led to the formation of turmeric paper which is used to test alkalinity [9].

3.1 Phytochemistry

Turmeric powder contains 69.4% carbohydrates, 6.3% protein, 5.1% fat and 3.5% minerals. The yellow color is due to the presence of curcumin (3–4%), which consists of curcumin I (94%), curcumin II (6%) and curcumin III (0.3%) **(Figure 4)** [27].

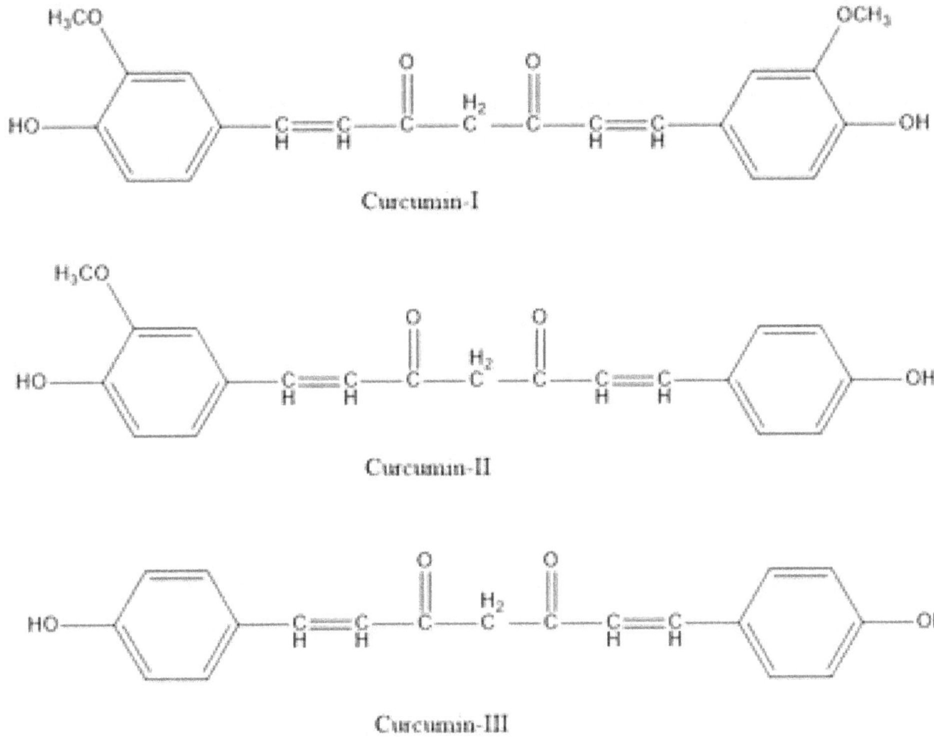

Figure 4. Curcumin derivatives.

Curcumin derivatives such as dimethoxy and bis-dimethoxy have also been isolated from turmeric. Several substances occur in the rhizomes, including tumerone a, tumerone b, curzerenone, curdione, monoethoxycurcumin, and diethoxycurcumin. Its essential oils are derived from its leaves [28].

Turmeric has been reported to contain multiple chemical compounds such as linalool, caryophyllene, geraniol, sabinene, myrcene, and α-phellandrene.

3.2 Medicinal and nutritional value

Although turmeric has been around for thousands of years, it has recently become popular due to its medicinal properties. Traditionally, turmeric was used for its digestive and carminative properties [29]. Free radicals generated by oxygen can be scavenged by curcumin. In terms of antioxidant activity, it is comparable to vitamins C and E [30]. It prevents oxidative damage to lipids and hemoglobin.

Macrophages that are activated by it are significantly less likely to produce reactive oxygen species such as H_2O_2, superoxide ions, and nitrite radicals [31]. Antioxidant properties are also present in its derivatives, bisdemethoxycurcumin and demethoxycurcumin. Researchers have shown that curcumin pre-treatment decreases the oxidative stress and improves ischemic conditions [32]. In an in vitro study, curcumin was found to enhance cellular resistance to oxidative stress in response to an inducible stress protein [33]. Researchers have found that turmeric can treat a number of diseases, and it can also prevent certain forms of cancer [34]. Skin cancers and precancerous conditions can also be treated with turmeric [34]. Turmeric is antispasmodic and can reduce digestive cramps, menstrual cramps and pains caused by osteoarthritis due to its ability to reduce pain and disability [35].

Curcumin shows strong potential for scavenging free radicals such as superoxide radicals, hydrogen peroxide and nitric oxide [36]. It reduces iron complex and inhibit lipid peroxidation [37].

Turmeric contains curcumin, a powerful anti-inflammatory ingredient that works by inhibiting inflammatory molecules in the body [35]. Researchers have shown that curcumin is beneficial to people suffering from conditions, including rheumatoid arthritis, inflammatory bowel disease. Several studies have suggested applicability of curcumin against cancer of many organs [38]. It induces apoptosis, inhibit progression of cell cycle and prevent growth of cancer cells [39]. curcumin induces apoptosis in different cell types such as human bladder cancer cells and arrests cancer cells [40]. It takes part in various signaling pathways and inhibit signaling through NF-κB which regulates expression of many genes responsible for malignancy [41]. Owing to its anti-inflammatory and anti-oxidant property it was also used against Alzheimer's disease [42]. It is also known to have metal chelating properties thus may reduce amyloid aggregation and reduce oxidative neurotoxicity [43]. Curcumin can be a future drug for the therapy of various neurological disorders especially depression and diabetic neuropathy [44].

4. GARLIC (*ALLIUM SATIVUM*)

Garlic (Fam. Liliaceae) is widely distributed and grown around the world (**Figure 5**). It is considered to be a valuable preventive remedy, folk food and spice. As the primary antidote to an epidemic, garlic has been used as a remedy throughout history for typhus, dysentery, cholera and influenza [45].

Figure 5. Garlic.

Garlic has long been appreciated by physicians from different nations as a remedy in the ancient and Middle Ages as well as during the modern period [46]. Garlic has been used as a remedy in ancient China since 2700 BC. The ancient Indians used garlic as a tonic, to treat cases of lack of appetite, skin problems, rheumatism, hemorrhoids, and more. In addition to being known as Russian penicillin, garlic was also used for years as a remedy for children with respiratory tract diseases via inhalation [47].

4.1 Phytochemistry

Garlic contains various sulfur compounds, vitamins (A, B_1), enzymes and minerals (germanium, calcium, copper, iron, potassium, magnesium, and selenium). Total 17 amino acids are found in garlic, and major are glutamine, glycine, lysine, cysteine, valine, methionine, isoleucine, leucine, tryptophan, and phenylalanine [48].

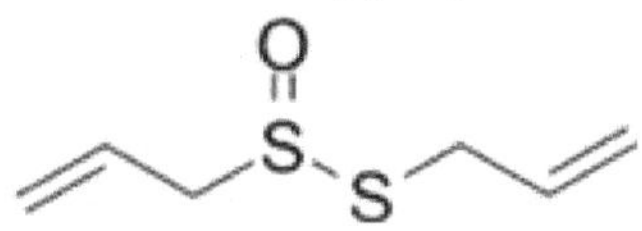

Figure 6. Allicin.

The pungent odor of garlic and many of its therapeutic effects are caused by the high levels of sulfur compounds present in this species of Allium. Allicin (diallyl thiosulfinate or diallyldisulfide) is a highly biologically active compound in garlic (**Figure 6**) [49]. In fresh and dry garlic, alliin (S-allyl cysteine sulfoxide) is the compound producing the highest levels of sulfur. Garlic is commonly prepared by cutting, mincing, crushing, which disturbs S-allyl cysteine sulfoxide and exposes it to the allinase enzymes, which then convert it into diallyl thiosulfinate, the aromatic constituent [48]. Below a pH of 3.5 or upon heating, the allinase enzyme responsible for diallyl thiosulfate conversion becomes inactive. Recent studies have shown that other compounds may play a greater role in anti-oxidant function than allicin [50].

4.2 Medicinal and nutritional value

There have been several studies showing possible cancer-preventive effects of garlic preparations and their constituents in vitro and in vivo. In garlic, a number of potent anticancer compounds have been found, including allylsulfide derivatives. Different garlic derivatives are shown to modulate various cellular mechanisms for promoting cancer, such as DNA adduct formation, mutagenesis, and free radical scavenging. They may also contribute to angiogenesis [46]. A number of studies have concluded that garlic protects liver cells from some toxic compounds [49].

Organosulfur compounds derived from garlic are known to inhibit cancer in various animal models by modifying cytochrome P_{450} which is responsible for the activation of nitrosamines, hydrazine and benzene [51]. The chemical present in garlic extract diallyl sulfide and its oxidation products diallyl sulfoxide and diallyl sulfone inhibit chemical carcinogenesis and mutagenesis [52]. These compounds restrict development of chemically induced cancers by blocking the phase-1 enzymes. These organo sulfur compounds also induce phase-2 enzymes such as glutathione S-transferase, quinone reductase and glucuronate transferase which eventually regulate removal of toxic compounds. Garlic extract is also known to modify -SH containing enzymes and increases their activity via formation of disulfide bond between protein thiol and thiol group of cysteine [53]. This type pf bond formation with proteins is an effective redox method to regulate protein activity.

5. SEA BUCKTHORN (*HIPPOPHAE RHAMNOIDES L.*)

A native plant of Asia and Europe, sea buckthorn is a hardy winter shrub that flourishes throughout the year (**Figure 7**). This species is divided into eight subspecies, which are economically valuable [54]. These subspecies are abundant and commercially cultivated mainly in Asia, where sinensis and Mongolia are found.

Figure 7. Sea Buckthorn.

In addition to drought, salinity and alkalinity, sea buckthorn can also withstand extreme temperatures and salinity.

5.1 Phytochemistry

The fatty acid content of sea-buckthorn fruit oil makes it unique compared to other vegetable oils. This oil contains rare palmitoleic acid (omega-7), a component of skin lipids that helps stimulate regenerative processes in skin cells and promotes wound healing. Sea-buckthorn oil also contains numerous active substances, resulting in multiple effects. There are exceptionally high vitamins A, C, E, F, P, and B complex in the oil [55]. Its regenerative and anti-wrinkle properties are attributed to vitamin A in carotenoids (200 mg/100 g). The vitamin C content of orange fruit (approximately 695 mg/100 g) is 15 times higher than in oranges. Vitamin C has antioxidative and protective properties against ultraviolet A and UVB rays. The walls of capillary blood vessels are strengthened by vitamin E, which is present in the form of tocopherols (approximately 200–600 mg/100 g) and minerals [56]. It also contains sterols, fruit acids (malic acid, citric acid), phenolic compounds, tannins, sugars, synthecanthins, phospholipids, ascorbic acid, selenium, copper, and zinc mineral salts.

5.2 Medicinal and nutritional value

Sea Buckthorn contains important phytochemicals like flavonoids, carotenoids, fatty acids etc. Studies have shown that sea buckthorn has high antioxidant activity. This plant can be used as a natural remedy for cardiovascular diseases and in diseases of skin, liver and stomach as well [56]. There are a lot of phytopharmaceuticals present, mainly phenolic compounds such as curcumin, resveratrol and proanthocyanidins which are known for cancer chemoprevention [57]. It is found that the seed oil of this plant enhances non-specific immunity and has anti-tumor effects. However, clinical studies with Sea Buckthorn need to validate the effects and mechanism on human cancer patients [15]. Sea Buckthorn helps in reducing cholesterol and improve cardiovascular health. It increases the blood circulation and relaxes cardiac muscle tissues which helps in restoring cardiac function [58]. To protect the cells of immune system and balance the inflammation, antioxidants are very important. There are a lot of antioxidants, isorhamnetin, quercetin, kaempferol in Sea Buckthorn. It has carotenoids, vitamin C and E as well which makes this fruit rich in anti-inflammatory properties as well [59]. Furthermore, it prevents gastrointestinal ulcers as it reduces inflammation. Researches have also shown that Sea Buckthorn helps in balancing liver enzymes and protects the liver from harmful effects of toxic chemicals [60]. The oil of this fruit contains palmitoleic acid which is a component of skin. It helps in the treatment of burns and healing of wounds. Sea Buckthorn oil has UV- blocking and emollient properties which help in regeneration of tissues [60].

6. TEA TREE (*MELALEUCA ALTERNIFOLIA*)

Tea tree oil (TTO) is also known as Melaleuca alternifolia (**Figure 8**). The volatile essential oil derived from the Australian native plant offers a variety of medicinal properties. Many cutaneous infections are treated using TTO as its active ingredient due to its antimicrobial properties. There are over-the-counter versions of Melaleuca alternifolia available in Australia, Europe, and North America.

Figure 8. Tea tree.

Herbal products and medicinal uses of this plant are very important [61]. To treat sore throats and skin ailments, tea tree leaves infusions were made from the leaves of the tea tree, which were also used for cough treatment. Traditional medicine of the Aboriginal Bundjalung people of eastern Australia relies on Melaleuca alternifolia leaves crushed and its oil to cure coughs, colds, and to treat wounds [62]. After its anti-microbial properties were reported in a series of papers in the 1920s and 1930s, Melaleuca alternifolia oil became widely known for its benefits. It was Melaleuca alternifolia which was reported to be 11 times more potent than phenol in its antimicrobial properties that led to tea tree oil being used in industry for the first time. In the 1950s, the introduction of antibiotics prompted a decline in the use of natural products in medicine, which had a negative effect on the production of tea tree oil [63]. During the 1970s, there was a general renaissance of interests in natural products such as tea tree oil. A large-scale production of consistent essential oils was achieved through commercial plantations established in the 1970s and 1980s. Melaleuca alternifolia is typically used to extract tea tree oil commercially, but Melaleuca *dissitiflora* and Melaleuca *linariifolia* can also be used to extract tea tree oil [10]. Tea tree oil has been extensively studied for its antibacterial, antifungal and antiviral properties.

6.1 Phytochemistry

The active ingredients in Melaleuca alternifolia are terpene hydrocarbons, which comprise monoterpenes, sesquiterpenes, and their alcohols. Tea Tree Oil has a relative density of 0.885 to 0.906, and is soluble in water and insoluble in non-polar solvents [64]. There is an international standard for tea tree oil that sets maximum and minimum specifications for 14 components of the oil, one of which is terpinen-4-ol (**Figure 9**).

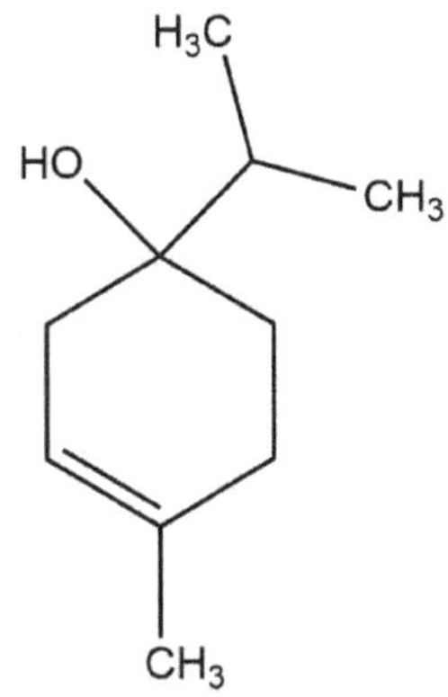

Terpin-1-ol

Figure 9. Terpinen-4-ol.

There are six varieties of M. *alternifolia*, each producing oil with a different chemical composition [62]. There are four 1,8-cineole chemotypes as well as a terpenen-4-ol chemotype and a terpinolene chemotype. Tea tree oil produced commercially contains about 30 to 40% of the terpinen4-ol chemotype, which has antimicrobial properties [63].

6.2 Medicinal and nutritional value

Tea Tree oil is added to many products including moisturizers, body lotions, foot sprays, face cleansers, and shaving powders. It helps in treating dermal infections by inhibiting the growth of a fungi named Malassezia [62]. Tea tree oil shows antiseptic and anti-inflammatory properties as well which helps in the treatment of gingival infections. It kills the bacteria which causes dental problems. Hence, it is used in mouthwashes as well. Tea tree oil is also used for the treatment of acne as it reduces inflammation [64]. In cases of vaginal candidiasis, tea tree oil is found to be a good alternative to be active against both gram positive and negative bacteria. Similarly, it is also effective for the treatment of athlete's foot which is again a fungal infection [62]. Tea tree oil is an essential oil which has a lot of medicinal properties. However, it can be toxic if ingested. It can cause drowsiness, confusion, coma, blood cell abnormalities, diarrhea, severe rashes etc. High doses of tea tree oil can also cause side effects. Hence, it should be used carefully [61].

7. AMLA (*EMBLICA OFFICINALIS*)

The Amla fruit is a natural gift to humanity (**Figure 10**). Ayurveda and Unani consider it an essential part of their systems with amazing medicinal properties. It is known as Amalaki or Dhatriphala in Sanskrit. In the Ayurvedic literature "Charak Samhita" (500 BC), amla is perhaps the most frequently mentioned herb

[65]. Amlaki means sustainer of fruit, or the place where the Goddess of Prosperity resides in Sanskrit.

Figure 10. Amla.

As it provides so much nourishment to mankind that this tree is worshipped as the Earth Mother in Hindu mythology. In addition to cultivated (gramya), there are wild (vanya) varieties of Amla. It is smaller and rougher in wild varieties, whereas cultivated amla is bigger, smoother, and juicier. Its trunk is crooked and its branches tend to spread out in a crooked pattern. It reaches heights of 8 to 18 m. Usually deciduous, small branchlets are glabrous or finely pubescent, 10–20 cm long. It has simple, tightly set leaves that resemble pinnate leaves, which are subsessile and subsedile along branches [65]. Berry harvesting is done in autumn, when the berries reach their mature stage. Despite its fibrous texture, Indian gooseberries have a sharp and bitter flavor and are bitter and astringent [66].

7.1 Phytochemistry

Approximately 80% of the amla fruit's chemical composition is water. Besides protein, carbohydrate, fiber, and mineral contents, it also contains gallic acid, which is an important polyphenol. Amla contains tannins, alkaloids, and phenols. In the whole plant, 28% of the tannins are present in the fruits. Two of the tannins in the fruit, Emblicanin A and B, when hydrolysed yield gallic acid, ellagic acid and glucose. Phyllaemblic compounds, tannins, flavonoids, pectin, and vitamin C, along with polyphenolic compounds, are abundant in this fruit [67]. Vitamin C is reported to be 20 times as abundant in amla fruit as in orange juice.

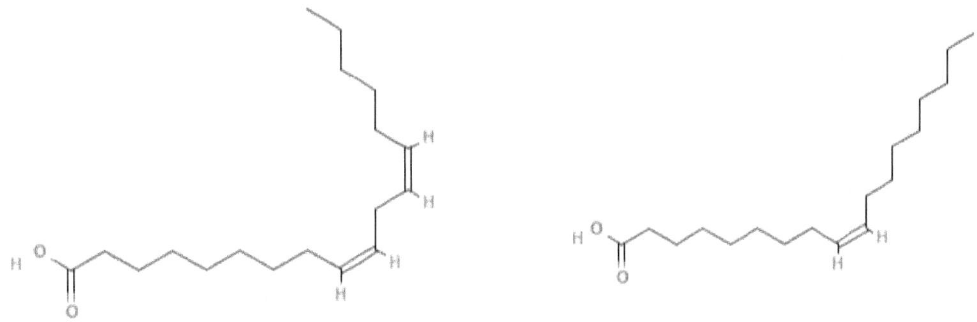

Figure 11. Linoleic acid. Oleic acid.

Ascorbic acid concentration is 160 times higher in the amla fruit tissue than in an apple. Minerals and amino acids are also found in higher concentrations in the fruit than in apples. The ash from amla fruit contains chromium, zinc, and copper. It has been demonstrated that a wide range of phytochemical components, including terpenoids, alkaloids, flavonoids, and tannins, possess useful biological activity. Tannins are abundant in the fruits, leaves, and bark. Ellagic acid and lupeol are found in the root, and leucodelphinidin is found in the bark. Oil is extracted from the seeds (16%) and is brownish-yellow in color. It contains Linolenic Acid (8.8%), Linoleic Acid (44.0%), Oleic Acid (28.4%), Stearic Acid (2.15%), Palmitic Acid (3.0%) and Myristic Acid (1.0%) (**Figure 11**) [68]. It contains a small amount of essential oil, fixed oil, and phosphides in its seeds. It also contains gallic acid, ellagic acid, chebulagic acid and chebulinic acid in their leaves. P. emblica roots provide Phyllaemblic acid, a novel, highly oxygenated norbisabolane.

7.2 Medicinal and nutritional value

Every part of this plant is useful due to its medicinal properties. E. officinalis is known to have antioxidant, anticancerous, anti-inflammatory, anti-diabetic and antimicrobial properties [69]. The extract of its fruit prevents hyperlipidemia, osteoporosis and several other disorders [70]. It shows strong antioxidant properties similar to ascorbic acid due to the presence of various tannins such as emblicanin A, emblicanin B, punigluconin, gallic acid [71]. It quenches various free radicals such as hydroxyl, superoxide and reduces DPPH free radical by reducing Fe^{+3} ions [72]. The antioxidative effect of fruit extracts was investigated against alcohol induced toxicity and it was suggested that the efficacy is due to the presence of polyphenols, flavonoids, and other ascorbic acid related compounds [73]. It strengthens antioxidant defense system by increasing level of GSH and various antioxidant enzymes like superoxide dismutase, catalase, GSH peroxidase, GSH reductase and GSHS-transferase [74]. The anti-diabetic or hypoglycemic potential involves inhibition of α-amylase and glucosidase owing to the presence of high concentration of ellagic acid, ascorbic acid and flavonoids in fruit extract [75].

Significant reduction in serum glucose and triglyceride levels have been reported in STZ-induced diabetic male Long-Evans rats [76]. These fruit extracts are also known to reduce significantly total cholesterol, LDL and VLDL-cholesterol level [77]. It is also reported to improve ischemia–reperfusion induced oxidative stress conditions by reducing myocardial lipid peroxidation and augmenting antioxidant enzymes activity in rat heart [78]. Stress related oxidative damage is a major factor in various physiological disorders. Administration of E. officinalis extracts have been reported to lower down oxidative stress conditions in brain frontal cortex and striatum due to certain external shocks [79]. Its antidepressant activity is also reported due to inhibition of brain biogenic amines, and affinity with serotonin receptors, dopaminergic D2-receptor and GABA-B receptors [80]. Fruits extracts has antiviral properties and also functions as an antibacterial and antifungal agent. *Emblica* significantly reduced UV induced erythema, had excellent free-radical quenching capabilities and chelating properties for iron and copper [68].

8. CONCISE SUMMARY

Numerous studies substantiate the therapeutic benefits of plant-derived phytochemicals. These plant extracts are often employed to avert severe diseases and as a component of healthcare routines. Herbal products have numerous benefits; they are non-toxic and also cost-effective. It is clear that phytochemicals originating from plants exhibit diverse capabilities; therefore, study is essential to identify combinations of desired phytochemicals to tackle intricate healthcare challenges. The inclusion of herbal components in the diet will promote longevity and provide protection against harmful xenobiotics.

REFERENCES

1. Jima TT, Megersa M. Ethnobotanical study of medicinal plants used to treat human diseases in berbere district, bale zone of oromia regional state, South East Ethiopia. Evidence-based Complementary and Alternative Medicine. 2018;**2018**:1-16
2. Way EL. Review of reviews. Annual Review of Pharmacology and Toxicology. 1986;**26**(1):567-576
3. Duh PD, Yen GC. Antioxidative activity of three herbal water extracts. Food Chemistry. 1997;**60**(4):639-645
4. Flora SJS, Mittal M, Mehta A. Heavy metal induced oxidative stress & its possible reversal by chelation therapy. Indian Journal of Medical Research. 2008;**128**(4):501
5. Reddy LALINI, Odhav B, Bhoola KD. Natural products for cancer prevention: a global perspective. Pharmacology & Therapeutics. 2003;**99**(1):1-13
6. Okoye TC, Uzor PF, Onyeto CA, Okereke EK. Safe African medicinal plants

for clinical studies. In: Toxicological Survey of African Medicinal Plants. Amsterdam: Elsevier; 2014. pp. 535-555

7. Manvitha K, Bidya B. Aloe vera: a wonder plant its history, cultivation and medicinal uses. Journal of Pharmacognosy and Phytochemistry. 2014;**2**(5):85-88
8. Christaki EV, Florou-Paneri PC. Aloe vera: a plant for many uses. Journal of Food, Agriculture and Environment. 2010;**8**(2):245-249
9. Bhowmik DC, Kumar KS, Chandira M, Jayakar B. Turmeric: A herbal and traditional medicine. Archives of Applied Science Research. 2009;**1**(2):86-108
10. Carson CF, Hammer KA, Riley TV. Melaleuca alternifolia (tea tree) oil: A review of antimicrobial and other medicinal properties. Clinical Microbiology Reviews. 2006;**19**(1):50-62
11. Singh S, Singh DR, Banu VS, Avinash N. Functional constituents (micronutrients and phytochemicals) and antioxidant activity of Centella asiatica (L.) Urban leaves. Industrial Crops and Products. 2014;**61**:115-119
12. Gohil KJ, Patel JA, Gajjar AK. Pharmacological review on Centella asiatica: A potential herbal cure-all. Indian Journal of Pharmaceutical Sciences. 2010;**72**(5):546
13. Cohen PA. Hazards of hindsight— monitoring the safety of nutritional supplements. New England Journal of Medicine. 2014;**370**(14):1277-1280
14. Brilhante RSN, Sales JA, Pereira VS, Castelo DDSCM, de Aguiar Cordeiro R, de Souza Sampaio CM, et al. Research advances on the multiple uses of Moringa oleifera: A sustainable alternative for socially neglected population. Asian Pacific Journal of Tropical Medicine. 2017;**10**(7):621-630
15. Zeb A. Chemical and nutritional constituents of sea buckthorn juice. Pakistan Journal of Nutrition. 2004;**3**(2):99-106
16. Dasaroju S, Gottumukkala KM. Current trends in the research of Emblica officinalis (Amla): A pharmacological perspective. International Journal of Pharmaceutical Sciences. 2014;**24**(2):150-159
17. Manvitha K, Bidya B. Aloe vera: a wonder plant its history, cultivation and medicinal uses. Journal of Pharmacognosy and Phytochemistry. 2014;**2**(5):85-88
18. Surjushe A, Vasani R, Saple DG. Aloe vera: A short review. Indian Journal of Dermatology. 2008;**53**(4):163
19. Foster M, Hunter D, Samman S. Evaluation of the nutritional and metabolic effects of Aloe vera. In: Herbal Medicine: Biomolecular and Clinical Aspects. 2nd ed. Boca Raton, FL: CRC Press, Taylor & Francis; 2011
20. Raphael E. Phytochemical constituents of some leaves extract of Aloe vera and Azadirachta indica plant species. Global Advanced Research Journal of Environmental Science and Toxicology. 2012;**1**(2):014-017

21. Maan AA, Nazir A, Khan MKI, Ahmad T, Zia R, Murid M, et al. The therapeutic properties and applications of Aloe vera: A review. Journal of Herbal Medicine. 2018;**12**:1-10
22. Hashemi SA, Madani SA, Abediankenari S. The review on properties of Aloe vera in healing of cutaneous wounds. BioMed Research International. 2015;**2015**:1-6
23. Mukherjee PK, Nema NK, Maity N, Mukherjee K, Harwansh RK. Phytochemical and therapeutic profile of Aloe vera. Journal of Natural Remedies. 2013;**14**(1):1-26
24. Joseph B, Raj SJ. Pharmacognostic and phytochemical properties of Aloe vera linn an overview. International Journal of Pharmaceutical Sciences. 2010;**4**(2):106-110
25. Ali RE, Rattan SI. Curcumin's biphasic hormetic response on proteasome activity and heat-shock protein synthesis in human keratinocytes. Annals of the New York Academy of Sciences. 2006;**1067**(1):394-399
26. Nasri H, Sahinfard N, Rafieian M, Rafieian S, Shirzad M, Rafieian- Kopaei M. Turmeric: A spice with multifunctional medicinal properties. Journal of HerbMed Pharmacology. 2014;**3**(1):5-8
27. Heath DD, Khwaja F, Rock CL. Curcumin content of turmeric and curry powders. The FASEB Journal. 2004;**18**:A125
28. Wang YJ, Pan MH, Cheng AL, Lin LI, Ho YS, Hsieh CY, et al. Stability of curcumin in buffer solutions and characterization of its degradation products. Journal of Pharmaceutical and Biomedical Analysis. 1997;**15**:1867-1876
29. Kuwatada JS, Raja M, Sood P. Turmeric: A boon to oral health. International Journal of Oral Care and Research. 2017;**5**(3):338-341
30. Prasad S, Aggarwal BB. Turmeric, the golden spice. In: Herbal Medicine: Biomolecular and Clinical Aspects. 2nd ed. Boca Raton, FL: CRC Press, Taylor & Francis; 2011
31. Mortellini R, Foresti R, Bassi R, Green CJ. Curcumin, an antioxidant and anti-inflammatory agent, induces heme oxygenase-1 and protects endothelial cells against oxidative stress. Free Radical Biology & Medicine. 2000;**28**: 1303-1312
32. Arbiser JL, Klauber N, Rohan R, van Leeuwen R, Huang MT, Fisher C. Curcumin is an in vivo inhibitor of angiogenesis. Molecular Medicine. 1998;**4**(6):376-383
33. Aggarwal BB, Kumar A, Bharti AC. Anticancer potential of curcumin: Preclinical and clinical studies. Anticancer Research. 2003;**23**:363-398
34. Mahady GB, Pendland SL, Yun G, Lu ZZ. Turmeric (Curcuma longa) and curcumin inhibit the growth of Helicobacter pylori, a group 1 carcinogen. Anticancer Research. 2002;**22**:4179-4181

35. Balasubramanyam M, Koteswari AA, Kumar RS, Monickaraj SF, Maheswari JU, Mohan V. Curcumin-induced inhibition of cellular reactive oxygen species generation: novel therapeutic implications. Journal of Biosciences. 2003;**28**:715-721
36. Das KC, Das CK. Curcumin (diferuloylmethane), a singlet oxygen quencher. Biochemical and Biophysics Research Communications. 2002;**295**:62-66
37. Thaloor D, Singh AK, Sidhu GS, Prasad PV, Kleinman HK, Maheshwari RK. Inhibition of angiogenic differentiation of human umbilical vein endothelial cells by curcumin. Cell Growth & Differentiation. 1998;**9**(4):305-312
38. Wright LE, Frye JB, Gorti B, Timmermann BN, Funk JL. Bioactivity of turmeric-derived curcuminoids and related metabolites in breast cancer. Current Pharmaceutical Design. 2013;**19**:6218-6225
39. Gautam SC, Gao X, Dulchavsky S. Immunomodulation by Curcumin. In: Aggarwal BB, Surh YJ, Shishodia S, (eds). The Molecular Targets and Therapeutic Uses of Curcumin in Health and Disease. Advances in Experimental Medicine And Biology. Vol 595. Boston, MA: Springer; 2007. https://doi.org/10.1007/978-0-387-46401-5_14
40. Tong QS, Zheng LD, Lu P, Jiang FC, Chen FM, Zeng FQ, et al. Apoptosis- inducing effects of curcumin derivatives in human bladder cancer cells. Anti-Cancer Drugs. 2006;**17**:279-287
41. Perkins S, Verschoyle RD, Hill K, Parveen I, Threadgill MD, Sharma RA, et al. Chemopreventive efficacy and pharmacokinetics of curcumin in the min/+ mouse, a model of familial adenomatous polyposis. Cancer Epidemiology, Biomarkers & Prevention. 2002;**11**:535-540
42. Baum L, Ng A. Curcumin interaction with copper and iron suggests one possible mechanism of action in Alzheimer's disease animal models. Journal of Alzheimer's Disease. 2004;**6**:367-377
43. Calabrese V, Butterfield DA, Stella AM. Nutritional antioxidants and the heme oxygenase pathway of stress tolerance: Novel targets for neuropr tection in Alzheimer's disease. Italian Journal of Biochemistry. 2003;**52**:17-181
44. Khafif A, Hurst R, Kyker K, Fliss DM, Gil Z, Medina JE. Curcumin: A new radio-sensitizer of squamous cell carcinoma cells. Otolaryngology and Head and Neck Surgery. 2005;**132**:317-321
45. Petrovska BB, Cekovska S. Extracts from the history and medical properties of garlic. Pharmacognosy Reviews. 2010;**4**(7):106
46. Bayan L, Koulivand PH, Gorji A. Garlic: a review of potential therapeutic effects. Avicenna Journal of Phytomedicine. 2014;**4**(1):1
47. Amagase H. Clarifying the real bioactive constituents of garlic. The Journal of Nutrition. 2006;**136**:716S-725S
48. Gebreyohannes G, Gebreyohannes M. Medicinal values of garlic: A review. International Journal of Medicine and Medical Sciences. 2013;**5**(9):401-408

49. Mikaili P, Maadirad S, Moloudizargari M, Aghajanshakeri S, Sarahroodi S. Therapeutic uses and pharmacological properties of garlic, shallot, and their biologically active compounds. Iranian Journal of Basic Medical Sciences. 2013;**16**(10):1031

50. Hahn G. Garlic: The Science and Therapeutic Application of Allium Sativum L and Related Species. 2nd ed. USA: Baltimore Williams and Wilkins; 1996. pp. 1-24

51. Wargovich MJ. Diallylsulfide and allylmethysulfide are uniquely effective among organosulfur compounds in inhibiting CYP2E1 protein in animal models. The Journal of Nutrition. 2006;**136**:832S-834S

52. Yang CS, Chhabra SK, Hong JY, Smith TJ. Mechanisms of inhibition of chemical toxicity and carcinogenesis by diallyl sulfide (DAS) and related compounds from garlic. The Journal of Nutrition. 2001;**131**:1041S-1045S

53. Pinto JT, Krasnikov BF, Cooper AJL. Redox-sensitive proteins are potential target of garlic-derived mercaptocysteine derivatives. The Journal of Nutrition. 2006;**136**:835S-841S

54. Li H, Ruan C, Ding J, Li J, Wang L, Tian X. Diversity in sea buckthorn (Hippophae rhamnoides L.) accessions with different origins based on morphological characteristics, oil traits, and microsatellite markers. PLoS One. 2020;**15**(3):e0230356

55. Kulkarni KV, Ghurghure SM. Indian gooseberry (Emblica officinalis): Complete pharmacognosy review. International Journal of Chemistry Studies. 2018;**2**(2):5-11

56. Olas B, Skalski B, Ulanowska K. The anticancer activity of sea buckthorn [Elaeagnus rhamnoides (L.) A. Nelson]. Frontiers in Pharmacology. 2018;**9**:232

57. Chauhan AS, Negi PS, Ramteke RS. Antioxidant and antibacterial activities of aqueous extract of Sea buckthorn (Hippophae rhamnoides) seeds. Fitoterapia. 2007;**78**(7-8):590-592

58. Cheng J, Kondoa K, Suzuki Y, Ikeda Y, Meng X, Umemura K. Inhibitory effects of total flavones of Hippophae rhamnoides on thrombosis in mouse femoral artery and in vitro platelet aggregation. Life Sciences. 2003;**72**:2263-2271

59. Sayegh M, Miglio C, Ray S. Potential cardiovascular implications of Sea Buckthorn berry consumption in humans. International Journal of Food Sciences and Nutrition. 2014;**65**(5): 521-528

60. Gao Z, Li., X. Gu, F. Cheng and F. Jiang. Effect of sea buckthorn on liver fibrosis: A clinical study. W. Journal of Gastroenterology. 2003;**9**:1615-1617

61. Cox SD, Mann CM, Markham JL, Gustafson JE, Warmington JR, Wyllie SG. Determining the antimicrobial actions of tea tree oil. Molecules. 2001;**6**(2):87-91

62. Yadav E, Kumar S, Mahant S, Khatkar S, Rao R. Tea tree oil: A promising essential oil. Journal of Essential Oil Research. 2017;**29**(3):201-213
63. Yang WJ, Chen HF, Zhu FY, Hu MQ, Jiang DA. Low concentration of bisulfite enhances photosynthesis in tea tree by promoting carboxylation efficiency in leaves. Photosynthetica. 2008;**46**(4):615-617
64. Shah G, Baghel US. Pharmacognostic standardization of the leaf of Melaleuca alternifolia (Maiden & Betche) Cheel. African Journal of Traditional, Complementary, and Alternative Medicines. 2017;**14**(3):1-11
65. Kulkarni KV, Ghurghure SM. Indian gooseberry (Emblica officinalis): Complete pharmacognosy review. International Journal of Chemistry Studies. 2018;**2**(2):5-11
66. Kumar KS, Bhowmik D, Dutta A, Yadav AP, Paswan S, Srivastava S, et al. Recent trends in potential traditional Indian herbs Emblica officinalis and its medicinal importance. Journal of Pharmacognosy and Phytochemistry. 2012;**1**(1):18-28
67. Baliga MS, Dsouza JJ. Amla (Emblica officinalis Gaertn), a wonder berry in the treatment and prevention of cancer. European Journal of Cancer Prevention. 2011;**20**(3):225-239
68. Grover HS, Deswal H, Singh Y, Bhardwaj A. Therapeutic effects of amla in medicine and dentistry: A review. Journal of Oral Research and Review. 2015;**7**(2):65
69. Mirunalini S, Krishnaveni M. Therapeutic potential of Phyllanthus emblica (amla): The ayurvedic wonder. Journal of Basic and Clinical Physiology and Pharmacology. 2010;**21**(1):93-105
70. Patel S, Goyal R. Emblica officinalis Geart: a comprehensive review on phytochemistry, pharmacology and ethnomedicinal uses. Research Journal of Medicinal Plant. 2012;**6**:6-16
71. Bhattacharya A, Ghosal S, Bhattacharya SK. Antioxidant activity of tannoid principles of Emblica officinalis (amla) in chronic stress induced changes in rat brain. Indian Journal of Experimental Biology. 2000;**38**(9):877-880
72. Poltanov EA, Shikov AN, Dorman HJ, Pozharitskaya ON, Makarov VG, Tikhonov VP, et al. Chemical and antioxidant evaluation of Indian gooseberry (Emblica officinalis Gaertn., syn. Phyllanthus emblica L.) supplements. Phytotherapy Research. 2009;**23**(9):1309-1315
73. Scartezzini P, Antognoni F, Raggi MA, Poli F, Sabbioni C. Vitamin C content and antioxidant activity of the fruit and of the ayurvedic preparation of emblica officinalis Gaertn. Journal of Ethnopharmacology. 2006;**104**(1-2):113-118
74. Shivananjappa MM, Joshi MK. Influence of Emblica officinalis aqueous extract on growth and antioxidant defense system of human hepatoma cell line (HepG2). Pharmaceutical Biology. 2012;**50**(4):497-505

75. Nampoothiri SV, Prathapan A, Cherian OL, Raghu KG, Venugopalan VV, Sundaresan A. In vitro antioxidant and inhibitory potential of Terminalia bellerica and Emblica officinalis fruits against LDL oxidation and key enzymes linked to type 2 diabetes. Food and Chemical Toxicology. 2011;**49**(1):125-131
76. Ansari A, Shahriar MSZ, Hassan MM, Das SR, Rokeya B, Haque MA, et al. Emblica officinalis improves glycemic status and oxidative stress in STZ induced type 2 diabetic model rats. Asian Pacific Journal of Tropical Medicine. 2014;**7**(1):21-25
77. Gopa B, Bhatt J, Hemavathi KG. A comparative clinical study of hypolipidemic efficacy of Amla (Emblica officinalis) with 3-hydroxy-3- methylglutaryl-coenzyme-A reductase inhibitor simvastatin. Indian Journal of Pharmacology. 2012;**44**(2):238-242
78. Rajak S, Banerjee SK, Sood S, Dinda AK, Gupta YK, Gupta SK. Maulik, Emblica officinalis causes myocardial adaptation and protects against oxidative stress in ischemic- reperfusion injury in rats. Phytotherapy Research. 2004;**18**(1):54-60
79. Bhattacharya A, Chatterjee A, Ghosal S, Bhattacharya SK. Antioxidant activity of active tannoid principles of Emblica officinalis (amla). Indian Journal of Experimental Biology. 1999;**37**(7):676-680
80. Dhingra D, Joshi P, Gupta A, Chhillar R. Possible involvement of monoaminergic neurotransmission in antidepressant-like activity of Emblica officinalis fruits in mice. CNS Neuroscience & Therapeutics. 2012;**18**(5):419-425

6 Chapter

UTILIZATION OF FLAXSEED AS A FUNCTIONAL FOOD

1. INTRODUCTION: AN OVERVIEW

Flaxseed, or linseed, is the seed of the flax plant, an annual herb that reaches approximately 2 feet in height and is part of the Linaceae family. Linumusitatissimum is the Latin designation for flaxseed, signifying "most useful," and it exists in two varieties: brown and yellow or golden. It is often consumed in one of three forms: whole seeds, powdered powder, or oil. It was likely first cultivated in Egypt, but it is now farmed globally, especially in India, China, the United States, Ethiopia, and Canada. Commercially, it was employed in the production of paper and textiles, including linen, until the 1990s, while flaxseed oil and its derivatives were utilized in animal feed[1]. Flaxseed has been ingested by humans since ancient times. Nonetheless, it has generated fresh interest in dietary consumption and illness research over the past 30 years, due to the substantial health advantages linked to certain bioactive constituents. It comprises around 40–50% oil, predominantly consisting of advantageous polyunsaturated fatty acids. It is rich in omega-3 fatty acids, both soluble and insoluble fibers, phytoestrogenic lignans, proteins, and antioxidants, among other nutrients. In vegetarian diets, it serves as a potential supply of alpha-linolenic acid [3]. Flaxseeds comprise fats (41%), dietary fibers (28%), proteins (20%), water (8%), and ash (3%). The lipid content of flaxseed ranges from 37 to 45 g per 100 g, whereas the carbohydrate amount varies from 25 to 30 g per 100 g. Glutamine, Arginine, Valine, Leucine, Tyrosine, and Phenylalanine. Flaxseed, its meals, and separated protein are rich in insoluble fiber components, including cellulose, hemicellulose, and lignin, whereas soluble fiber consists of mucilage gums. Flaxseed mucilage, mostly composed of water-soluble polysaccharides, has superior water retention capacity and physicochemical properties akin to those of guar gum [5]. Flaxseeds comprise three categories of phenolic compounds: phenolic acids, flavonoids, and lignans. The predominant phenolic acids present in defatted flaxseed are Ferulic acid, Chlorogenic acid, and Gallic acid. Flavone C- and Flavone O-glycosides are the predominant flavonoids present in flaxseeds. Secoisolariciresinol diglycoside is the predominant lignan present in flaxseed. Alpha-linolenic acid constitutes 53% of total lipids, while linoleic acid comprises

17%, oleic acid 19%, stearic acid 3%, and palmitic acid 5%. The ratio of n-6 to n-3 fatty acids in flaxseeds is typically 0.3:1 [6]. Furthermore, it comprises minerals and vitamins, notably Phosphorus (K), Magnesium (Mg), Calcium (Ca), Iron (Fe), Zinc (Zn), and a low quantity of Sodium (Na), along with Vitamins A, C, Thiamine, Riboflavin, Niacin, B6, and E [7]. Figure 1 illustrates the comprehensive nutritional profile of flaxseeds. It comprises anti-nutrient substances such as phytic acid, linatine, and cyanogenic glycosides, with advantageous phytochemicals. Moreover, this plant's capacity to absorb and sequester cadmium from the soil, which forms chelates with thiol-containing plant proteins, presents a significant concern. Fortunately, the literature reports no adverse effects, including food poisoning, associated with Flaxseed ingestion [2, 3].

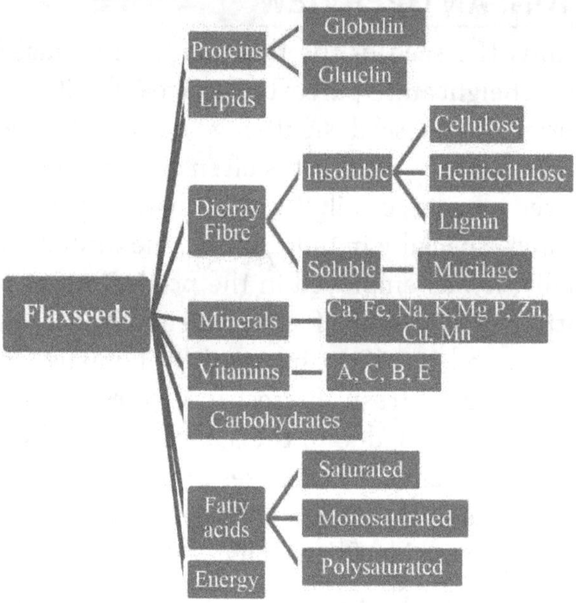

Figure 1. Nutrient profile of flaxseeds.

Flax seeds contain numerous biologically active elements that help to prevent and treat a variety of physiological conditions and non-communicable diseases, including dyslipidemia, obesity, diabetes mellitus, a variety of cancers, kidney and renal failure, irritable bowels, immune function, and more. Their omega-3 fats, Lignans, and Fiber content are primarily responsible for their health benefits. Plant-based ALA fatty acids have been linked to a lower risk of stroke and have been shown to improve heart health. Lignans, which are rich in antioxidants and estrogen, may aid in the prevention of diabetes, breast and prostate cancer, as well as other cancers [8]. **Figure 2** demonstrates major health benefits provided by phytochemicals present in flaxseeds. Seed mucilage consumption via the oro-gastro-intestinal route has been linked to a variety of health benefits, including postprandial glycemic and insulinemic response modulation, hyperlipidemia prevention, satiety enhancement, and gut microbiota function regulation [9].

Diabetes Mellitus, also known as diabetes, is a metabolic condition characterized by raised blood sugar levels. A Hormone called insulin that transports sugar from the bloodstream into cells to be stored or used for energy, is responsible for diabetes. In diabetes, the body either does not produce enough insulin or does not use it effectively. Over the last 50 years, the number of diabetics and obese people has risen considerably in both developed and developing countries [10]. Both the International Diabetes Federation (IDF) and the World Health Organization (WHO) have made significant efforts to keep figures or global predictions of a sharp rise in the number of people with diabetes up to date. According to the IDF, 382 million people had diabetes in 2013, and by 2035, that number will have risen to nearly 600 million. **Figure 3** shows the International Diabetes Federation's global distribution of people with types 1 and 2 diabetes (ranked by country) [11]. Diabetes mellitus is linked to abnormalities in carbohydrate, protein, and lipid metabolism, which can lead to secondary complications. It can harm the heart, blood vessels, eyes, kidneys, and nerves over time. It raises the chances of having a heart attack or a stroke in adults by two to three times. Neuropathy (nerve damage) in the feet, when combined with reduced blood flow, increases the risk of foot ulcers, infection, and eventual limb amputation. Diabetic retinopathy is a common cause of blindness that develops as a result of long-term damage to the retina's small blood vessels. Diabetic nephropathy is one of the most common causes of kidney failure [12]. Basic lifestyle modifications have been shown to restrict the risk of type II diabetes or delay its onset. To aid in the prevention of type II diabetes and its complications, individuals should: attain and maintain healthy body weight; participate in physical activity; consume a nutritious sugar and saturated fat-free diet, and refrain from smoking. Because of the importance of lifestyle prevention, it's critical to look into the protective effects of healthy nutrients and foods. Numerous studies have shown that nutritional therapy plays an important role in preventing or delaying the advancement of these secondary complexities [13].

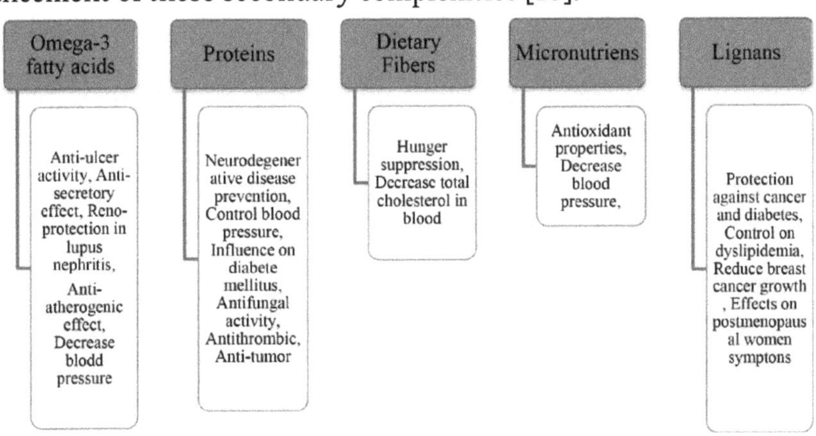

Figure 2. Health benefits of flaxseeds.

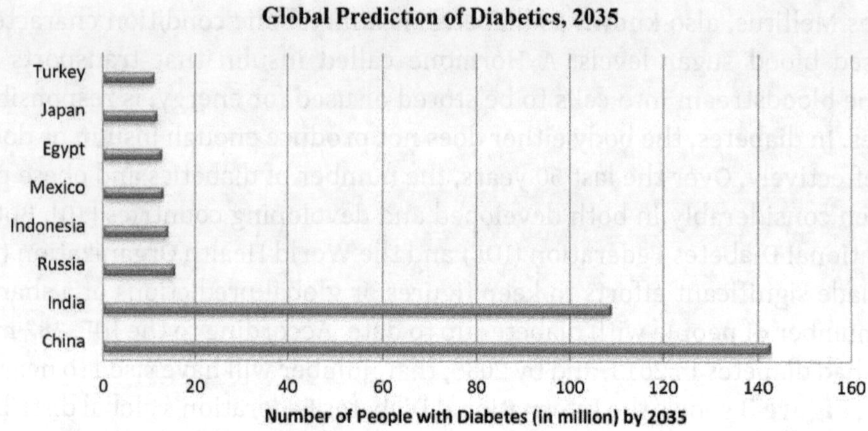

Figure 3. Prediction of growth of diabetic patients, 2035.

A diabetic diet focuses on a general dietary pattern including functional foods such as whole grains, nuts, fruits, leafy green vegetables, herbs, spices, legumes, and seeds that can help in managing blood glucose levels [14, 15]. Functional foods, which can boost health and lower the risk of developing many chronic diseases, have received a lot of attention over the last two decades. The concept "functional foods" was coined to describe foods that have been scientifically proven to have medical benefits [14]. Researchers have been focusing on the characteristics of bioactive substances found in functional foods in the management of multiple facets of diabetes mellitus in recent years; some of the therapeutic potentials of these bioactive substances and food sources have been studied in vitro and in vivo, and numerous analytical studies have even verified the benefits in people with diabetes. Hypoglycemic and Hypolipidemic properties have been discovered in a variety of functional foods [15]. Flaxseeds one of the functional food have been shown in studies to have hypoglycemic and hypolipemic properties, causing foods to have a lower glycemic index. It is high in dietary fibers, omega-3 fatty acids, and antioxidants, as well as being low in carbohydrates, making it an anti-diabetic food. Consequently, the inclusion of flaxseed in several dietary products induces beneficial alterations in the biochemical profiles of diabetics, thus enhancing their metabolic profile [16]. This study consolidates findings on flaxseed-infused meals and their significance in diabetes, as documented in the literature during the past five years.

2. VALUE-ADDED FOOD PRODUCTS PREPARED BY INCORPORATING FLAXSEEDS

After centuries of use as natural medicine, Flaxseed has resurfaced in the functional foods arena as a potential future functional ingredient with a wide variety of clinical benefits. Flaxseed-supplemented food products are becoming increasingly trendy due to their high levels of unsaturated fatty acids, protein, soluble fiber, and phytonutrients. It can be used as roasted and ground seeds, and its oil can be used in a variety of food formulations as vegetable oils, stable emulsions, and micro-encapsulated powder. **Tables 1** and **2** highlights the specifics of flaxseed-supplemented foods. Their impact on the glycemic index is also briefly discussed. The entire information on functional foods is divided into two categories: paper-based information and patent-based information.

Table 1. Information of Flaxseed incorporated functional foods, published since 2017.

S. No.	Author and Year of Publication	Functional Food	Flaxseed Form	Amount of Supplementation	Reference
1.	Seczyk et al., 2017	Wheat Bread	Flaxseed Hulls	1–5%	[17]
2.	Kay et al., 2017	Pudding	Flaxseed Mucilage	2.28 wt%	[18]
3.	Kurek et al., 2018	Wheat Bread	Flax dietary Fiber	10.8%	[19]
4.	Zhu & Li, 2018	Steamed Bread	Flaxseed Flour	2.5–20%	[20]
5.	Zhu & Li, 2018	Noodles	Flaxseed Flour	25gm- 200gm	[21]
6.	Wandersleben et al., 2018	Bread	Flaxseed Flour	10.6%	[22]
7.	Soltanian & Janghorbani, 2018	Cookies	Flaxseed Flour	10g	[23]
8.	Soltanian & Janghorbani, 2018	Cookies	Flaxseed Flour	10g	[24]
9.	Yuksel, 2019	Noodles	Flaxseed Flour	10- 20g	[25]
10.	Hasaniani et al., 2019	Yogurt	Ground Flaxseed	30 g	[26]
11.	de Oliveira Giarola et al., 2019	Sherbet	Golden Flaxseed Flour	1–3%	[27]
12.	Tobias-Espinoza et al., 2019	Extruded products	Flaxseed Flour	6.6–9.3%	[28]
13.	Wirkijowska et al., 2019	Bread	Flaxseed flour & Marc	5–15%	[29]
14.	Geetha et al., 2020	Food Mix	Flaxseed Flour	–	[30]
15.	Sevostyanova et al., 2020	Bread	Flaxseed Flour	–	[31]
16.	Almehmadi et al., 2020	Muffins	Flaxseed Flour	30 g	[32]
17.	Campidelli et al., 2020	Ice-cream	Flaxseed meal Flour	1%	[33]
18.	Vieira et al., 2021	Liquid Thickener	Flaxseed Gum	0.6–1.5%	[34]

2.1 Paper-based information

Published researches that include the food formulation containing flaxseeds have been included in this category.

Table 2. Information about Flaxseed supplemented Functional Foods, patented since 2017.

S. No.	Inventor/s	Title of Patent	Amount of incorporation of Flaxseeds	Patent Application Number	Reference
1.	Vadakkemuri Jolly Mathew	Optimized Nutrient Fatty Acid Composition	Flaxseed oil, 4.2–5.8 Liter	US2017/03600-73A1, 2017	[35]
2.	Haijia Chen, Xiaohu Ge, Fei Wang, Yifei Wang, Jianqiang Cheng, Xiaoyan Wang	One kind is for diabetes B patient's diet and foodstuffs and preparation method thereof	Flaxseed oil, 4 parts by weight	CN107373642A, 2017	[36]
3.	Xiping Li, Xiao Han, YiminGao	A kind of compositions being applicable to diabetes patient and diet therapy soup thereof and preparation method	Flaxseeds powder, 5- 30 g	CN106266429A, 2017	[37]
4.	Jyoti Shirodkar	Food Products and Processes for Preparation Thereof	—	AU2017/1005-71A4, 2017	[38]
5.	Muhua Wang, Peiping Pan, Wei Song, Zhan Deng, Yuhong King, Binnan Su, YinghulCai, Xuxia Li, Xiaoxia Lu, Junfeng Gong	A kind of red yeast rice Flaxseed cake fermentate and its application	—	CN107136183A, 2017	[39]
6.	Chun xiao Li, Hongyu Luo	Diabetes full nutrition formula food and preparation method thereof made of a kind of seaweeds raw material	Flaxseed meal, 1% by weight	CN108522970A, 2018	[40]
7.	Tae Wan Kim	Method of producing a food for preventing and improving diabetes	Flaxseeds, 54.5–71.5% by weight	KR101850-955B1, 2018	[41]

8.	Weiyi Wang	A kind of 3 sour milk powder of omega and its preparation process	Flaxseed oil powder, 15–25% by weight	CN104542974B, 2018	[42]
9.	Yingyan Chen	A kind of 3 solid beverage of omega	Flaxseed oil powder, 60–90% by weight	CN108685006A, 2018	[43]
10.	Vadakkemuri Jolly Mathew, Paul Thomson Kochery	Optimized Nutrient Food	Flaxseeds, 40-50 g	US2019/033-5796A1, 2019	[44]
11.	Kewei Fang, Haijie Zheng, Yanbing Wu	Low-sugar mooncake high in fat and preparation method thereof	–	CN110384123A, 2019	[45]
12.	J. Murray Hockings	Method for treating and reversing Type 2 diabetes	Flaxseed oil, 500 mg	US10765318B2, 2020	[46]
13.	Jianzhang Lu, Shuang Li, Xuli Wu	Low- carbohydrate flaxseed nut wafer and preparation method thereof	Flaxseeds, 40–80 parts by weight	CN111466425A, 2020	[47]
14.	Scott Anderson, John Hall, Mark Yoho	Human dietary supplement and method for treating digestive system and immune-related disorders	–	WO2020081417A1, 2020	[48]
15.	Shiddalingesh Salimath	Enteral or oral nutritional composition	–	WO2020039277A1, 2020	[49]
16.	Jinho Pyo	Preparing method of noodle for improving diabetes, and the noodle obtained thereby	Flaxseeds, 5–30 parts by weight	KR20210012287A, 2021	[50]

Lifestyle choices, such as energy-dense foods and low levels of physical activity, significantly raise obesity rates and adipose tissue stores, leading to an increased risk of Type II Diabetes. Lifestyle changes, such as dietary changes, can help to reduce the risk of T2D. Flaxseed and its byproducts are becoming more popular as functional ingredients in low-glycemic-index foods as a result of their numerous health benefits. Until now, many efforts have been made to ensure the functionality of a wide range of cereal-based products, including bread, pasta, muffins, biscuits, and cakes. In this context, flaxseed hulls were investigated as a functional supplement for wheat bread enrichment and their relationship with the Glycemic Index. Furthermore, the antioxidant capacity, nutritional value, and consumer satisfaction all were investigated in 1–5% flaxseed hull (FH) supplemented wheat bread. Supplementation with 5% FH increased phenolic content by 93 percent, radical scavenging ability by 176 percent, and reducing power by 220 percent when compared to control. It was seen that fortification reduced in vitro starch digestibility slightly but had no effect on the expected glycemic index (eGI); however, it resulted in a reduction in relative protein digestibility of up to 8%. The addition of fortification reduced the volume of the loaf while enhancing the firmness of the crumb. The sensory evaluation results showed that the addition of FH particularly at higher doses had a slightly adverse impact on bread sensory properties but assure satisfactory customer acceptability [17]. According to these findings, flaxseed hulls might be a useful functional food in terms of nutrition, but they were not beneficial to diabetics.

Improving the content of soluble dietary fiber (SDF) in carbohydrate-rich foods is one strategy for lowering the risk of T2D. However, more research is needed into the glycemic response to specific soluble DFs, particularly the role of soluble DF-induced viscosity. Using a randomized, double-blind, crossover design, the acute postprandial glycemic response of puddings containing soluble DF derived from yellow mustard mucilage (YMM), fenugreek gum (FG), and flaxseed mucilage (FM), in 15 people (10 male and 5 female; age 55.1 ± 12.0 years; BMI 29.5 ± 3.4 kg/ m2) who were at extremely high risk of developing T2D was investigated. The nutritional composition of the puddings confirmed that the prepared puddings contained more energy and less fat. YMM incorporated pudding had the highest total dietary fiber content, while FG incorporated pudding had the lowest. Results revealed that all soluble DF puddings significantly reduced blood glucose and plasma insulin at maximum concentration and definite time points when compared to a control pudding, but not differed from one another [18]. This study showed that using viscosity-matched single DF in a pudding matrix can lower acute postprandial peak glucose and insulin levels, as well as the risk of T2D.

Another study looked into the use of oat (OB), flax (FB), and apple (AB) as dietary fiber (DF) fortifiers in wheat bread (WB). In this, wheat bread was compared to flax, oat, and apple bread. The fatty acid profiles were significantly altered when oat and flax fibers were added to bread. Oleic acid (33.83 percent of lipids) and linoleic acid were the most abundant in OB (24.31 percent of lipids). Only FB had a significant amount of Linolenic fatty acid with 18.32 percent. The bioaccessibility trials disclosed that the DF reduced saturated fatty acid intake. In the range of fatty acids, polyunsaturated fatty acids (PUFA) were the least bioaccessible (72 percent in OB to 87 percent in FB). In terms of glycemic load, the commercial bread had the highest value (80.5), which was markedly greater than the values for OB, FB, and AB. The addition of OB, FB, and AB resulted in low glycemic value. In addition, AB had the highest overall phenolic value (897.2 mg/kg), while FB had the least (541.2 mg/kg). In this study, AB had the only significant reduction in caloric values [19].

Chinese steamed bread (CSB) is a fermented basic food popular in China and Asia. Because of the widespread usage and low cost of CSB, there are many chances to create 'nutritious' CSB with added natural products to combat the rising prevalence of chronic illnesses such as diabetes and cancer among the population. Development of CSB fortified with ground flaxseeds at a fortification level of up to 20% was investigated [20]. The resulting CSB was studied for its physicochemical, nutritional, and sensory properties. According to findings, Flaxseed addition reduced the specific volume and spread ratio of CSB while boosting unsaturated lipids, dietary fiber, phenolic content, and hardness in a dose-dependent manner. Flaxseed additions of up to 15% had a slight impact on the total acceptability of CSB. In terms of nutrition, adding flaxseed to CSB increased in vitro antioxidant activities while lowering in vitro starch digestibility and glycemic index from 117 to 97, and the glycemic load from 50 to 32.9 g.

In another study, fresh salted noodles with ground flaxseed flour that replaced wheat flour up to 200 g/kg were formulated. It was noticed that Flaxseed supplementation increased in vitro antioxidant activities while reducing the expected glycemic index of cooked noodles in different dosages. The inclusion of flaxseed increased cooking loss whereas lowered the hardness and thickness of cooked noodles. According to kinetic analysis, adding 200 g/kg flax seed to noodles reduced the eGI from 60 to 52. Flaxseed replacement reduced the glycemic load of the noodles significantly, from 13.4 g in control noodles to 8.5 g in noodles with 200 g/kg flax seed addition in a dose-dependent manner. The noodles developed in this study were found to be beneficial to people suffering from chronic diseases such as diabetes and stroke [21].

Soluble fiber, protein, and resistant starch-rich foods aid digestion and may improve blood sugar response after meals. In this instance, three ingredients in bread production strike equilibrium between nutrition and palatability: lupine grit flour which consists of 60 percent protein, lupine hulls flour, and flaxseed expeller flour which is an excellent source of dietary fiber all were tested for their medical benefits. A customer acceptability test was also conducted on 259 volunteers using the bread deemed best by its balance of sensory and nutritional value. Compared to the control bread the final bread had 125 percent more fiber and 55 percent more protein. In all aspects surveyed, the bread was rated as acceptable by more than 90% of those polled. It was discovered that lupine can replace animal protein very well, and in addition, flaxseed supplemented bread had hypoglycemic properties [22].

In 53 constipated T2D patients, the effects of baked flaxseed cookies versus placebo cookies on constipation symptom scores, body mass, blood glucose, and lipid control were compared. Constipated T2D patients with BMI 20.5–48.9 kg/ m2 were randomly assigned to either 10 g of flaxseed cookies two times a day or placebo cookies, for twelve weeks in a single-blind, randomized controlled trial. Participants in the control group were given sugarless orange-flavored maltodextrin cookies as a placebo. Constipation symptom rating, BMI, fasting blood sugar, glycosylated hemoglobin (HbA1c), and lipid level were measured at the start and end of four, eight, and twelve weeks. A stool diary was used to assess constipation. It was observed that the flaxseed group's constipation symptom scores (2.46), weight (3.8 kg), BMI (1.5 kg/m2), FPG (26.7 mg/dl), cholesterol (37.3 mg/dl), triglycerides (10.4 mg/dl), LDLC (21.0 mg/dl), HDLC (4.7 mg/dl), and cholesterol/ HDLC ratio (1.4 mg/ dl) all decreased significantly from reference point after the 12-week intervention. The differences in absolute change between the flaxseed and placebo groups in all of the studied profiles were statistically significant. This study found that eating 10 g of flaxseed baked into cookies every day for 12 weeks improved constipation symptoms, glycemic and lipid control, BMI, and body weight [23]. Following this study, the effects of baked flaxseed or psyllium versus placebo on constipation symptoms, weight, glycemia, and lipid levels in T2D patients were also compared. Psyllium is a soluble, viscous, gel-forming non-fermented fiber supplement that contains pentoses, hexoses, anduronic acids. It has previously been shown to reduce constipation signs, body fat, glycemic, and lipid profile. In constipated T2D patients, no medical study had done to compare the effects of flaxseed and psyllium on lowering constipation symptoms, mass, glycemic, and lipid status. As a result, 77 patients participated in this clinical trial. 77 constipated T2D patients were randomly assigned to three groups in a singleblinded, randomized controlled trial. For a total of 12 weeks, patients were given either 10 g flaxseed or psyllium supplemented cookies or placebo cookies twice a day. Constipation symptoms (P 1/4 0.002), stool consistency (P 0.001), weight (P 0.001), BMI (P 0.001), FPG (P 1/4 0.004), cholesterol (P 1/4 0.010), LDLC (P 1/4 0.031), and cholesterol/HDLC ratio (P 1/4 0.019) all improved significantly in the flaxseed and psyllium groups compared to the placebo group. It was observed that although both flaxseed and psyllium could help with constipation, weight loss, glycemic control, and lipid levels, flaxseed appears to be better than psyllium [24].

Raw flaxseed-enriched noodles were prepared and their physical properties such as Resistant Starch (RS), Total Starch, Total Dietary Fiber (TDF) content, estimated glycemic index (eGI) value, and nutrient composition of the resultant product was investigated. Flaxseed was added to the noodle blend in 3 different amounts (10, 15, and 20 g/100 g). According to findings, the RS content of the samples was found to be in the range of 0.61–2.14 g/100 g, with the sample containing 10% flaxseed having the highest RS. The noodle sample enriched with 20% flaxseed had a maximum TDF value of 8.28 percent, while the control had a value of 1.68 percent. The flaxseed addition did not affect the estimated eGI of the samples, but it showed very little impact on the mineral content of the noodle samples. Sodium, potassium, magnesium, and phosphorus were the most common mineral compounds found in the samples [25]. It could be concluded that a 10% flaxseed supplemented noodle would be a healthy and nutritious effective alternative to consume, but it would be less effective to diabetics.

An investigation was done to see how flaxseed enriched yogurt affected the elevated glucose and lipid levels, along with hypertension in T2D patients. On 57 patients with T2D, a randomized, open-label, controlled clinical trial was performed. For 2 months, participants were given either 200 g 2.5 percent fat yogurt with 30 g flaxseed or plain yogurt. At the start and end of the study, anthropometric and biochemical parameters were assessed. The findings revealed that Hemoglobin A1c levels in the treatment group were significantly lower than in the control group after 2 months of supplementation. At the end of the trial, there were substantial variations in triglyceride and total cholesterol ($p = 0.04$ and $p = 0.01$), systolic blood pressure, and diastolic blood pressure ($p = 0.02$ and $p = 0.002$, respectively) seen between flaxseed enriched yogurt and control groups. Fasting blood sugar levels were also reduced. Moreover, no differences in low- as well as high-density lipoprotein, body mass, or waist size were found between the two groups [26].

Sherbet is a frozen dessert just like ice cream, but it is made with more fruit pulp and has a more acidic flavor. In addition to glycemic load, two major parameters that should be considered for good quality ice cream, sherbet, or sorbet are sugar crystal- lization and ice recrystallization. The unwanted mechanism of ice recrystallization occurs during the processing and storage of ice creams and sherbets. In this context, analysis of the effects of different concentrations of golden flaxseed flour (GFF) on the glycemic index and ice recrystallization in uvaia (Eugenia pyriformis Cambess.) diet sherbets fortified with iron was studied. In the study, samples were moved from a freezer designed to perform 14 temperature fluctuation phases ranging from 20 °C/12 h to 10 °C/12 h from storage (25 °C) (each cycle, 48 h). Before the tempera- ture cycles, the chemical structure, acidity, and physical parameters such as overrun, rheology of static and transient shear, firmness, thermal properties, and crystallite sizes were evaluated. An optimized descriptive profile was used to evaluate the sensory attributes. The findings revealed that the glycemic load decreased as GFF concentration increased, while the ash content and pH values increased significantly. GFF addition was found to be a viable option for the preparation of an iron-fortified uvaia diet sherbet with a reduced glycemic load and smaller ice crystal sizes. As a result, to obtain a high-quality diet product with improved rheological, thermal, and microstructural properties, GFF should be added at a rate of 1 to 2% [27].

Regardless of the potential benefits of amaranth and flaxseed, there is little information on their combined effect on the physicochemical and functional properties of extruded products. Extrusion of six mixtures in a twin extruder was done with different proportions of amaranth (18.7–33.1%), flaxseed (6.6–9.3%), maize grits (55.6–67.3%), and minor ingredients (4.7%) for the production of extrudates. According to the study, the amount of amaranth and flaxseed used in the development of extruded products had major effects on their functional and physicochemical properties. Extrudates with higher proportions of amaranth and flaxseed had higher insoluble and soluble fiber content. The highest flaxseed proportion had the highest soluble fiber content (1.9 percent). The protein content of the extruded products was high (>12%), which was significantly high than that of commercial breakfast cereals. Aside from these attributes, the extruded products obtained had a healthy caloric density (<5%). Extruded products containing low amounts of starch and high levels of fiber produced extrudates with a low expansion index and more hardness. As a result, extruded products containing amaranth (18.7–22.9 percent), flaxseed (8.6–9.3 percent), and maize grits (63.8–67.3 percent) had acceptable physicochemical and functional properties [28].

Flaxseed is a vital oil crop. By-products of flaxseed oil extraction have a high nutritional value. Therefore, the effectiveness of flaxseed industry by-products (flour (FF) and marc (FM)) as functional raw materials in bread production was highlighted in one study. Flaxseed by-products were added to wheat bread in amounts of 5, 10, and 15%. The effect of adding FF and FM to bread on its quality, chemical composition, and consumer acceptance was studied. According to the outcome of the research, adding FF and FM increased bread yield by 146.6 and 148.4 percent, respectively, for samples with 15% FF and FM addition, compared to the control bread (13.5%). According to texture analysis, Both by-products changed the average values of hardness, chewiness, and firmness, as well as the color of the crumb. Flaxseed by-products improved crumb water content but showed a nega- tive impact on specific volume and pore size distribution, whereas FM has a lesser impact. The introduction of 5% flax components increased the bread's nutritional value while lowering its calorie content thus being beneficial for diabetics. According to the findings, adding 10% flaxseed marc to bread results in bread with high nutritional quality and satisfactory sensory attributes [29].

Diabetes management relies heavily on nutrition, and foods with a low glycemic index are becoming increasingly popular because they slow the release of glucose into the bloodstream. It's critical to create a low-glycemic food mix using locally available ingredients for daily diets. Therefore, an investigation of the glycemic index of traditional recipes (dosa, mudde, and roti) made from developed millet-based food mix, as well as their impact on pre-diabetic subjects was conducted in research. Finger millet, little millet, defatted soy flour, whole green gram, fenugreek seeds, flax seeds, curry leaves, bitter gourd, and skimmed milk powder were used to make a millet-based food mix. Protein (19.41 g/100 g) and dietary fiber (21.11 g/100 g) both were present in the developed millet-based food mix.

With a glycemic load of 11.05, 18.43, and 18.09, the glycemic index of dosa, mudde, and roti was found to be 37, 48, and 53, respectively. All three developed products, however, had a lower glycemic index (<55), as well as a reasonable glycemic load of (<20). Moreover, dietary supplementation on pre-diabetic subjects revealed a significant decrease in FBS (120.50 ± 18.73 to 97.81 ± 20.00) and HbA1c (6.14 ± 0.30 to 5.67 ± 0.40), indicating that they were a wiser choice for diabetes management [30].

The previous studies discovered the impact of diet and various nutrients on metabolism, including the use of low-carb food in the treatment of cardiovascular disease, diabetes, and metabolic syndrome. In this frame of reference, an optimal combination of ingredients for the development of bread, which could be used to improve the course of these diseases, as well as to be consumed by groups of individuals with gluten, milk, and sugar intolerances was studied. Two bread formulations containing flaxseed flour, eggs, vegetable oil, baking soda, salt, and natural apple cider vinegar were formulated and evaluated during the project. The liquid component was represented by oat milk in the first sample and hazelnut drink in the second. The resulting bread was compared to commercially available analogs. The quality analysis revealed that the resultant bread met the quality standards. Both samples had lower calorie (18%), carbohydrate (37.5%), and higher protein content (13.5%). It was discovered that adjusting the nutritional value allowed the product to be integrated into a low-carb diet, which was particularly essential for people with diabetes and metabolic syndrome [31].

Flaxseed has been found to help lower and stabilize blood glucose levels in various investigations. While one study was conducted to investigate whether flaxseed could reduce blood glucose response more successfully when consumed as a single 30 g portion or as a split portion thrice a day. It was a cross-over study with a randomized, repeated measures design. 15 healthy members were provided one of three options: (1) 3 flaxseed muffins containing a total of 30 g flaxseed consumed once in the early hours, (2) 3 flaxseed muffins consumed at 3 distinct time intervals throughout the day (10 g flaxseed per muffin), or (3) 3 control muffins consumed at 3 distinct time intervals throughout the day (0 g flaxseed per muffin). A continuous glucose monitor was used to assess the 24-hour blood glucose reaction. Muffins made were high in crude protein, fatty acids, energy, and fiber profile while low in carbohydrates. The findings showed that flaxseed muffins given thrice a day were more effective than control muffins at reducing and sustaining blood glucose levels over 24 hours. It was observed that Flaxseeds should be consumed in small amounts throughout the day, not all at once [32].

Analysis of the impact of varying sugar, fat, and prebiotic fiber levels in powder preparations used to make mangaba ice cream was studied in one research. P1, P2, and P3 were the first three simple powder mixes for ice creams, and they were also known as powder preparation. Cream milk powder, light milk powder, flaxseed meal powder, corn syrup powder, inulin, FOS, and stabilizers (at different concentrations) were among the ingredients that were required for the production of ice cream. The powdered preparations were tested for solubility, hygroscopicity, water content, and microscopic characteristics. The physicochemical proper-ties, quality attributes, overrun, nutrient profile, mineral absorption simulation in vitro, rheological characteristics, fructooligosaccharide quantitative analysis, and sensory acceptability of the ice cream produced by these mixtures were all examined. As it was discovered that the ice cream mangaba is a source of vitamin C (17–25 mg.100 g-1), dietary fiber (4.9–8.6 g.100 g-1), and prebiotic (FOS content: 3.06–5.68 g.100 g-1), calcium (3.38–3.64 g/kg-1), and contains low sodium (0.48–0.60 g/kg-1) the study concluded that the incorporation of prebiotic fibers and the reduction of sugar and fat levels can provide nutritional benefits. Inulin and FOS, on the other hand, were likely accountable for the increase in solubility (64.22 percent, 71.67 percent, and 79.73 percent) and reduction in hygroscopicity (6.06, 4.23, and 3.14 g/100 g) of powdered ice cream preparations. It was concluded based on findings that incorporating ingredients such as mangaba pulp, flaxseed, prebiotic fibers (inulin and FOS), and lowering sugar and fat levels can strengthen the technological and nutritional characteristics of ice cream [33].

Oropharyngeal dysphagia is a serious type of dysphagia, which refers to difficulty swallowing - moving food from the mouth to the stomach, requires more effort than usual. This mechanical disorder has been linked to aging, neurological issues, brain and neck cancer, cerebral palsy, Parkinson's disease, and benign esophageal stricture, among other things. Oropharyngeal dysphagia patients are given a viscosity-modified diet with thickeners to avoid swallowing problems. Although most commercial products contain xanthan and starch, flaxseed gum (FG) is a potential thickener for liquid foods that also has health benefits. Concerning the use of flaxseed gum (FG) in thickening methodologies for dysphagia patients, one investigation was done to check the relationship between the rheological properties of thickeners and their advantages in terms of nutrition such as enhancing fibers, and phenolic content while reducing glucose absorption. FG was combined with modified starch (MS) and Xanthan gum (XG), with the concentrations of the bio-polymers varied according to a central composite rotational design, with the evaluated responses being rheological and color properties in water. Furthermore, the amount of glucose released following in vitro digestion was assessed and compared to a commercial MS-based thickener. It was observed that the shear time-independent and shear-thinning behavior of all formulations was primarily influenced by the concentrations of XG and MS. The formulations exhibited a predominant elastic character in oscillatory measurements, which was attributed to MS and primarily to XG, which, despite its lower concentration, imposed a similar influence on this rheological property. The most significant factor influencing viscosity, however, was the increase in FG concentration. Surprisingly, the commercial formulation had a significantly higher glucose release than the suggested FG/MS/XG-based formulations. Hence, FG-based thickened beverages could have an advantage over those based on MS in this due consideration, enhancing the quality of life of dysphagia patients and lowering the amount of glucose released in vitro after digestion [34].

2.2 Patent based information

Patented inventions that include the food formulation containing flaxseeds have been included in this category, and it was also investigated whether these foods can be included in the diabetic patient's diet in the information.

Over the last century, human dietary habits have changed dramatically due to the advancement of civilization & industrialization. The dietary intake of linoleic acid, an essential fatty acid of the omega 6 fatty acid group, increased as the production and consumption of oil seeds and oil rose. This resulted in a shift in the omega 6:3 fatty acid ratio in human diets from around 1:1 to 10–40:1. This shift plays a critical role in the treatment of so-called "lifestyle diseases," such as cardiovascular disease, hypertension, type 2 diabetes, and viral infections. In this context, one research was conducted and patented under application number, US0360073A1. The primary goal of this invention was to develop an optimized nutrient fatty acid composition that can be used to resolve a dietary imbalance in essential fatty acid consumption in human beings in a simple, safe, practical, and cost-effective manner. Oil seeds with omega 6 to 3 fatty acid ratios of 1:1 to 1: 1.25 were used to make optimized nutrient fatty acid compositions, which were stabilized with vitamin E 0.8–4 mg/ml. After cleaning, drying, and dehydrating the ingredients, oil was extracted from soybeans, flaxseed, walnut, perilla seed, rice bran, sunflower seed, evening primrose seed, high GLA safflower seed, garden cress seed, sachainchi beans, and green algae using conventional methods. After the final product formulation, 10 patients with metabolic syndrome (high blood pressure, abdominal obesity, increased fasting glucose, elevated fasting triglyceride, and low HDL) were chosen for the study. The patients were divided into two groups after their physical examination. Both groups were told to keep taking their medications and to follow a low-calorie, low-fat, low-carbohydrate diet with moderate exercise. In addition, the second group was instructed to consume 25 mL of optimized nutrient oil as cooking oil daily. When compared to the first group, the second group on optimized nutrient oil had a considerable reduction in weight, abdominal girth, blood pressure, sugar levels, CRP, SGPT, and triglyceride, as well as an increase in HDL. Furthermore, the optimized fatty acid composition could be used as food, beauty products, and a delivery vehicle for drugs and nutrients [35].

With the constant intensification of the rhythm of real-life and work, people's spirits are constantly in a state of high pressure, overload, causing a rise in blood sugar, if things continue like this, easily cause some people to develop diabetes. Which affects their standard of living. A good eating style is needed to prevent and treat the serious problems associated with diabetes. Therefore, CN107373642A disclosed a kind of diabetic special food formula including; whey separated protein powder with a precise weight of 15,30 parts by weight of plant fat powders (including 18 parts by weight of medium-chain triglycerides) 8 parts by weight sunflower oil powders, 4 parts by weight flaxseed oil), 10 parts by weight dietary fibers, 45 parts by weight carbon water compound, 2 parts by weight soybean phosphatide, 0.2 parts by weight phytosterol, 0.3 part by weight of vitamin, 0.3 parts by weight mineral matter. Following the formulation of the food product, a functional test on 30 diabetic patients was performed. Patients were told to keep taking foodstuff on daily basis. Evaluation at the end of the test discovered that the diabetic patient's symptoms had significantly improved [36].

In another patent CN106266429A, formulation of preparation of a kind of diet therapy soup including; Radix Puerariae 20–50 parts, Fructus Lycii 10–30 parts, Flaxseeds 5–30 parts, Rhizoma Dioscoreae 10–30 parts, leaf of Moringa 5–20 parts was Described. It was a powdered formulation that could reconstitute in boiling water, all ingredients of diet therapy soup worked together in harmony to treat diabetes and improve the immune system of Susceptible people by lowering blood glucose levels. Each ingredient, such as Radix Puerariae, Rhizoma Dioscoreae, Fructus Lycii, and Flaxseeds, was cleaned, pulverized, and concentrated to form a thick paste, which was then pulverized and dried to form a powder. All of the powders were combined, resulting in a uniform dry soup mix. Furthermore, this composition can be used to make beverages, food, medicine, and so on, and it is relatively suitable for diabetic patients, as it has the potential to have a blood sugar lowering function [37].

Regardless of the type of diabetes, keeping blood glucose levels within a healthy range is critical. Along with seeds, cereals have an important role in the diet of a diabetic patient. Barley is a cereal grain that has traditionally been referred to as "gluten grains." Barley is high in dietary fiber, beta-glucan, molybdenum, man- ganese, and selenium. Barley is also high in copper, vitamin B1, chromium, phosphorus, magnesium, and niacin. However, because whole grain barley is not easily palatable, it is not widely consumed as a staple food. As a result, there is a need for a technique of processing whole-grain barley that is palatable while retaining the grain's nutrients. The technique of formulation of barley-based compositions comprising barley, other cereals, pulses, spices, dry fruits, and flavoring agents was disclosed in patent application AU100571A4. The pre-mixed barley-based composition was made up of barley in the 45–75 wt percent range; wheat in the 20–40 wt percent range; and soybean in the 5–20 wt percent range. Whole green gram, broken green gram, whole Bengal gram, broken Bengal gram, whole lentil, and broken lentil were among the pulses. Coriander, cumin, carom, flaxseed, fenugreek, red chili, turmeric, asafetida, and salt were among the spices used. The steps for preparing the pre-mixed barley-based composition included cleaning the raw materials, optionally roasting the cleaned raw materials, mixing the cleaned and optionally roasted material to obtain a mixture, grinding the mixture to obtain a fine flour or powder, and packaging the flour or powder. Resultant Pre-mixed barley-based compositions could be used to make a variety of food products (flat bread, roti, chapatti, paratha, khakra, thalipeeth, cutlet, pancake, health drink) that could be consumed by both healthy people and people suffering from obesity, dyslipidemia, diabetes, and other conditions. Another advantage of this technique was powder formulation was free from chemical preservatives [38].

Flaxseed cake is a byproduct of flaxseed de-oiling that contains lignan, flaxseed gum, and cyanogenic glycoside. After consumption, cyanogenic glycosides can decompose in vivo into hydrogen cyanide, causing poisoning in the human body and death. The presence of cyanogenic glycoside in flaxseed limits the use of flaxseed and its byproducts in the food industry. To broaden the use of flaxseed in the food industry, The inventors revealed a type of red yeast rice Flaxseed cake fermentate and its application in the field of food processing technology in patent CN107136183A.The following steps were taken to prepare the product: The first step involved the detoxification of the Flaxseed cake, in which the water content of the Flaxseed cake was controlled at 170-200 g/kg and the detoxification was performed using a microwave power of 800 W for 8 minutes. The Flaxseed cake was fermented and detoxified further in the second step by activating the Monascus slant 3–4 times and accessing the seed culture medium. Then, after accessing the fermentation medium, concentration was done at 30 degrees Celcius for 3 days.

The total incubation period was 5–7 days. As cyanogenic glycoside is unable to meet food safety requirements due to the limitations of traditional Flaxseed cake detoxification mode, so, flaxseed cake is never used as raw food material, but the main advantage of this invention was that it used microwave detoxification and secondary microbial fermentation to remove the mode of poison so that cyanogenic glycoside reaches below 0.015 mg/kg. Another advantage was that the entire process was inexpensive. In the functional test, Red yeast rice Flaxseed cake fermentate was used to make biscuits. 2 L Flax seed oil was used in the preparation of the biscuits. Wheat flour, egg, white granulated sugar, edible salt, ammonium hydrogen carbon- ate, sodium bicarbonate, and water were also used. The made biscuits were found to be free of cyanogenic glycosides. In addition, Both red yeast rice and flaxseeds can help with hypotension, blood glucose control, and cholesterol reduction. Flaxseeds have antitumor, anti-aging, and anti-renal failure properties. Therefore, Red yeast rice Flaxseed cake fermentate was beneficial to diabetic patients, aged women who lack nutrition, and people suffering from obesity [39].

We all know about the benefits of flaxseed, but such ingredients should be used in conjunction with flaxseed, which can make a significant contribution to the treatment of diabetes, so such food cannot be considered anything less than a boon. Edible seaweed is high in sugar, terpene, alkaloid, and polyphenols; these active materials have a variety of benefits including hypoglycemic, decompression, antibacterial, and anti-oxidant properties. Kelp is a rich source of dietary fiber, nutrients like vegetable protein, and has a high nutritive value, is extremely beneficial for the auxiliary treatment of diabetes. In this context, patent CN108522970A disclosed a full nutrition formula food made of a type of seaweed raw material for diabetic patients. Kelp powder, Buckwheat, soybean powder, Peanut powder, Walnut powder, Flaxseed meal, Kudzu-vine root powder, and Tomato powder were used in a 3:15:8:4:1:1:8:2 ratio, along with a small amount of nutrition fortifier.

Preparation methods included polishing, mixing, homogeneous, and freeze-drying. The formula food produced could effectively control blood glucose and replace staple food to provide sufficient daily nutrition demand for diabetes patients and also cost was relatively low. Furthermore, because it is made entirely of natural food ingredients, it retains a wide range of natural sugar-reducing substances [40].

The process for preparing a mixture by mixing two types of ingredients was disclosed in Patent KR101850955B1. The type 1 ingredient was 54.5–71.5 percent by weight of flax, 8.0–11.5 percent by weight of acid, 4.5–7.4 percent by weight of onion, 1.5–2.5 percent by weight of pine needle, 1.5–2.5 percent by weight of Angelica keiskei, 1.5–2.5 percent by weight of blackberry, and type 2 ingredient was 50–54% by weight of the upper leaves and 46–50% by weight of the oak. The leaves were extracted for 2 to 3 hours at a temperature of 70 to 80 °C using hot water.

The extract was then filtered to obtain filtrate, which was concentrated until it reached 23 Brix or higher, after which the mixture of type 1 ingredients and extract concentrate of type 2 ingredients was mixed in a ratio of 1: 0.4–0.6. The final food produced could be used to prevent or treat diabetes [41].

Omega-3 fatty acids can be found in a variety of foods, including fish and flaxseed, as well as dietary supplements like fish oil. Alpha-linolenic acid (ALA), eicosapentaenoic acid (EPA), and docosahexaenoic acid (DHA) are the three main omega-3 fatty acids. Plant oils such as flaxseed, soybean, and canola oils are high in ALA. Fish and other seafood contain DHA and EPA. Omega-3 fatty acid-rich diets may help diabetics lower their insulin resistance. Omega-3 fatty acid-rich diets may lower the risk of heart attacks, strokes, macular degeneration, and certain cancers. Many types of research have been conducted to create omega-rich foods. A type of −3 sour milk powder preparation was discovered in one of these studies, and it was patented under the number CN104542974B. It was made up of 15–25 parts of flaxseed oil microcapsule powder, 70–83 parts of milk powder, and 25% parts of probiotic powder in weight proportions. Each component material was mixed in equal increments, and the resulting powder was packed in a vacuum nitrogen filling packing machine with a sterile aluminum foil bag. Omega-3 supplemented sour milk powder could be stored at room temperature and had a long shelf life.

It had numerous health benefits; it was easily soluble in water, could be dispersed into yogurt liquid, and could be reconstituted with 30–37 °C warm water to make instant drinks. It was found to be beneficial in the improvement of intestinal canal function, nutrient absorption, immunity, brain cell, intelligence, and optic nerve systematic growth, memory, and could prevent and improve chronic diseases like cardiovascular and cerebrovascular disease, cancer, and diabetes [42]. Another study, Patented in 2018 under application number CN108685006A disclosed the preparation of omega-3 solid beverage comprised of 60–96 parts by weight of linseed oil microcapsule powder, 0.02–0.1 parts by weight of probiotics, 3–10 parts by weight of compound fruit, and vegetable ferment, 3–20 parts by weight of bamboo-leaves flavones, 1–10 parts by weight of inulin. *Bifidobacteriumlongum*, *Lactobacillus paracasei*, and *Lactobacillus rhamnosus* were used as probiotics in the invention. Pawpaw, pineapple, berries, blackcurrant, melon, walnut, 10 parts black fungus, banana, grape, apple, lemon, longan, mango, radish, mushroom, cucumber, pumpkin, wild rice stem, tomato, and green cucumber were the main vegetables and fruits used in the vegetable and fruit ferment. Probiotics and linseed oil microcapsule powder were first pressed and uniformly mixed, then compound fruit and vegetable ferment was added and uniformly mixed, after that bamboo-leaves flavone were added and uniformly mixed, and finally plus inulin was uniformly mixed using the equal increments method. The resulting solid beverage was found to be effective in preventing and treating diabetes, as well as slowing diseases like kidney disease, cancer, and gastrointestinal function weakness, as well as promoting nutrient absorption [43].

US0335796A1 described the preparation of a low-calorie optimized nutrient food that contained all of the essential and non-essential nutrients in the proper proportions for optimization of body systems and, as a result, maintaining health, preventing diseases such as hypertension, heart diseases, cancer and diabetes, and delaying aging, with a calorie count of about 650. More specifically, the optimized nutrient food contained a 1:1 ratio of omega 3 to omega 6 fatty acids; high dietary fiber content; high protein with complete amino acids; and vitamins and minerals optimized according to FDA daily value recommendations, capable of providing about 650 calories per 150 g, but free of cholesterol, trans fat, preservatives, and artificial flavorants and colorants. Protein source was obtained from soya beans, goa beans, watermelon seeds, flaxseed, perilla seed, hemp seed, sachainchi seed, sunflower seed, safflower seed, almond, garden cress seed, oats, gingili seed, fenugreek, pumpkin seeds, pistachio, corn seed, rapeseed, poppy seed, sesame seed, black gram, Bengal gram, and green gram. Flaxseed, perilla seed, garden cress seed, mustard seed, canola seed, chia seed, sachainchi, hemp seed, walnut, clary sage seed were chosen as omega 3 sources, and gingili seed, sunflower seed, safflower seed, watermelon seed, corn seed, almond, cashew nut, groundnut, pumpkin seed, hemp seed, black currant seed, poppy seed, rapeseed, borage seed, were chosen as omega 6 sources. Oats bran, rice bran, fenugreek seed, soya beans, flaxseed, garden cress seed, rice bran, chicory root, aniseed, sun root, perilla seed, and safflower seed were chosen as dietary fiber sources. The functional test included 30 patients between the ages of 25 and 40 who were diabetics with fertility issues who were planning a pregnancy. Two groups of patients were created. Metformin 500 mg twice daily was given to all patients, along with dietary advice. Other standard modern medical treatment for their fertility issues was also given to them. All of the patients in the first group were instructed to cut back on fat, starch, and sugar in their diets, as well as engage in light to moderate exercise. The second group was told to eat 100 grams of optimized nutrient food every day for 2 months, along with low-calorie vegetables and fruits, and to avoid eating anything else. It was observed that both groups improved their blood sugar levels for 30 days, but the second group, which followed an optimized nutrient food diet, had significantly better blood sugar control. Both groups had improved blood sugar and HbA1C levels at the end of the 60 days. However, 80% of the first group had not achieved adequate blood sugar control before conception, necessitating the use of insulin. Due to exhaustion, the patients in the first group were unable to restrict their carbohydrate intake. Those on a nutrient-dense food diet, on the other hand, saw significant improvements in blood sugar and HbA1C. With an optimized nutrient food diet, even overweight and obese diabetic patients improved significantly, and none of them required insulin before conception. As a result, Optimized nutrient food having about 650 calorie optimal nutrition was the simple, safe, and economic answer to the prevention of lifestyle-related diseases [44].

CN110384123A disclosed the preparation method of Low-sugar high-fat mooncakes. Ketobodies are produced by a high-fat diet, which regulates glucose metabolism, reduce carbohydrate intake, block the energy source of tumor cells, reduce intracellular active oxygen ROS (reactive oxygen species), and thus including high-fat diet on daily basis can protect the human body against diabetes, Alzheimer's disease, and Parkinson's disease. In the invention, Mooncake wrapper 33 g and fillings 17 g were used to make a low-sugar, high-fat mooncake, with a fat: protein+sugar mass ratio of 1.0–10: 1–0.1. The moon cake wrapper was made of the following materials in the following mass percents: fatty microcapsule powder 26–48 percent, resistant starch 21–35 percent, sugar alcohol 4–7%, resistant dextrin 9–20 percent, polydextrose 3–6%, phosphatidase 2–6%, dried hen egg yolk 2–10%, egg liquid 11–18 percent. The material used for fillings were taken in the following mass percents: fat meat 11–22 percent, veterinary antibiotics or nut powder 19–33 percent, yolk 30–47 percent, calcium caseinate 2–6%, grease 1–5%, sugar alcohol 0.5–3%, resistant starch 3–11 percent. Antierythrite, D-sorbate, maltitol, xylitol, isomalt, lactitol, and one or more mannitol were among the sugar alcohols used. In the filling for grease, coconut oil, palm kernel oil, medium-chain triglyceride, peanut oil, rapeseed oil, olive oil, tea oil, high oleic sunflower oil, rice oil, pine-seed oil, soybean oil, corn oil, flaxseed oil, cottonseed oil, sesame oil, walnut oil, perilla herb oil, docosahexaenoic acid DHA algal oil, arachidonic acid ARA oil, fish oil, pig One or more of oil, chicken fat, and sheep oil were used. Ovum Anas Domestica yolk, egg yolk, goose egg yolk, turkey yolk, pigeon yolk, ostrich yolk, quail One or more of quail egg and emu egg were among the fillings' yolks. Filling vegetables included pumpkin, wax gourd, purple sweet potato, sweet potato, taro, Chinese chestnut, carrot, potato, water chestnut, corn, cassava, and lotus rhizome dry product; fruit included pawpaw, apple, pear, coconut, orange, lemon, strawberry, blueberry, and blackberry dry product; and nut powder included pine nut, walnut, and peanut powder. Furthermore, technical issues were resolved to make it low in sugar but still palatable [45].

US10765318B2 disclosed a regimen that included a meal plan, exercise, and specific supplements for cleansing, overcoming bodily deficiencies, and reversing Type II diabetes. In the invention, nutritional supplement preparation was described. Vitamins A, C, D, E, K, K2, B1, B2, B3, B6, B9, B12, B7, B5, Minerals Calcium, Iron, Magnesium, Zinc, Selenium, Copper, Manganese, Chromium, Molybdenum, Potassium, and stabilized fatty acid blend consisting of Borage Oil, Flax Seed Oil, and Algae Oil was included in the recommended supplement. Aside from that, the supplement contained an organic vegetable and fruit blend that included Garlic,

Jerusalem Artichoke, Cinnamon Verum, Raspberry Ketones, Rhodiola Crenulated and Glucomannanase, Turmeric, CoQ10 Ubiquinol, Black Strap Molasses, Kelp, and Probiotics like Lactobacillus, Acidophilus, L-Plantarum, Bulgaricus, Streptococcus, Thermophilus and *Enterococcus Faecium*. Furthermore, the meal plan and mild to moderate exercise were also advised in the disclosure. A case study was carried out on 20 patients who had Type 2 Diabetes and were taking diabetic medication. The study's patients all have A1c levels ranging from 6.0 to 12.5. All patients were advised to take supplements in addition to a healthy diet and exercise. It was discovered that everyone who used the invention reduced or eliminated their diabetic medications, lost weight, or had personally expressed having more energy [46].

CN111466425A revealed a method of making a low-carbohydrate flaxseed nut wafer. The wafer was made up of the following raw materials in weight order: 40–80 parts flaxseed, 5–30 parts nut grains, 0–0.06 part sweetening agent, 0–0.5 part psyllium husk powder, 15–30 parts sugar substitute powder, 2–6 parts thickening powder, 3–5 parts edible oil, 1–5 parts tackifying powder, 0–1 part yeast extract and 80–100 parts of water. Material preparation, dissolution, whipping, soaking, quantitative squeezing, trowelling and forming, and baking were all used to make the wafer. The low-carbohydrate nutritional wafer was high in protein and dietary fiber, and it was also high in Omega-3 and other minerals. It had a thin, sweet, delicious, and crispy mouthfeel and was suitable for people of all ages. It could be used as a fat-reducing meal replacement for obese people, diabetes patients, polycystic ovary syndrome patients, hyperlipidemia, and hypertension patients, and it could also be used as a leisure breakfast or nutritional meal for constipation sufferers [47].

In WO2020081417A1, a dietary supplement formulation for the treatment or prevention of digestive and immune-related disorders was disclosed. The dietary supplement could be compounded in a solid form (such as bars, wafers, or pills), a paste form, a granular form, a powder form, or a liquid form, and it could be taken orally. The ingredients of the dietary supplement include L-glutamine, mucogenic amino acid, both produced by vegan bacterial fermentation of sugar beets; lecithin derived from soy oil, oat oil, sunflower oil, safflower oil, com oil; fructooligosaccharides derived from yacon root, chicory root, Jerusalem artichoke, blue agave; beta-glucan derived from oats, barley, mushrooms, seaweed, algae, yeast cell wall; A oligosaccharide derived from the bran tissues of wheat, oats, barley, rice, millet, psyllium, flax, rye; RS-4 starch derived from oats, yacon root, chicory root, flax, acacia, com. Emulsifiers and nutricine, (that bind to and eliminates pathogenic bacteria in the digestive tract), were also included in the formulation. The dietary supplement was said to reduce systemic inflammation, which is the cause of most chronic diseases like diabetes, cardiovascular disease, and cancers, by increasing the impermeability of the gut lining and balancing the bacteria components of the microbiota [48].

An enteral or oral nutritional composition comprised of whole food components, herbs or spices blend, fat, carbohydrates, protein source, fiber source, and micronutrients was described in WO2020039277A1. Cereals, fruits, and vegetables were included in the whole food supplement. Fruits included banana, apple, guava, orange, grapes, mango, pomegranate, and chickoo; vegetables included beetroot, tomato, carrot, radish, onion, and spinach; and cereals included wheat, rice, soybean, and chickpea were utilized in the supplement. For fatty acids supplementation α-linolenic acid, docosahexaenoic acid, eicosapentaenoic acid, flaxseed, walnuts, almonds, algae, krill, were used. All of the ingredients were cooked for 10–30 minutes at 60–80 degrees Celsius, then blended and homogenized into a powder that could be reconstituted with water or juice. It was noticed that the invention of enteral/oral nutritional composition satisfied the specific nutritional needs of patients with various illnesses, such as cancer, diabetes, kidney, and liver diseases, as well as patients who had undergone surgery [49].

In KR20210012287A, meat powder and antidiabetic ingredients were mixed with whole wheat flour, rye flour, and wheat flour used to make noodles for diabetic patients. Based on a total of 100 parts by weight of nutritional noodles, the meat powder was 5 to 30 parts by weight and the antidiabetic material powder was 5 to 30 parts by weight. Broccoli, onion, eggplant, garlic, arbor, deodeok, burdock, purple pork potato, bitter gourd, flaxseed, nuts, brown rice, and barley were among the anti-diabetic ingredients. The noodles were cooked at a low temperature, which has the advantage of reducing nutrient loss by minimizing nutrient destruction.

The meat powder was made using beef, pork, chicken, lamb, and duck. The resulting noodles were low in calories, high in protein, high in vitamins and minerals, and low in cholesterol, making them an ideal diabetic food [50].

3. STATISTICS OF PUBLICATIONS

The purpose of this review is to cover the most relevant papers and patent applications that attempt to provide a clear picture of flaxseeds used in various food products and how these food products can benefit diabetics. This review included 34 kinds of research, 18 of which were paper publications and 16 of which were patent applications. These studies are classified into five major categories based on the end product. **Figure 4** depicted the percentage breakdown of the various categories covered in the review. Out of the 34 reviewed studies, 12 were bakery products, so this category took first place with 35.29 percent and contributed the most to the literature review. The remaining 64.71 percent was distributed as follows: 26.47 percent of extruded snacks and beverages secured second place with 9 researchers, 17.64 percent of nutritional supplements secured third place with 6, With 4, 11.76 percent of dairy products secured fourth place, and 8.82 percent of food mixes secured fifth place with 3 studies. Out of 16 applications, six have been taken by companies and two have been taken by research institutes, and the remaining eight applications are still not commercially available. Seven different inventors enjoy 50% of the total share with 8 patent applications out of a total of 16 patents covered in this review. One inventor, V.M Jolly captured 2 patents. These patents are not still claimed by any company, lab, or institute therefore not commercially available.

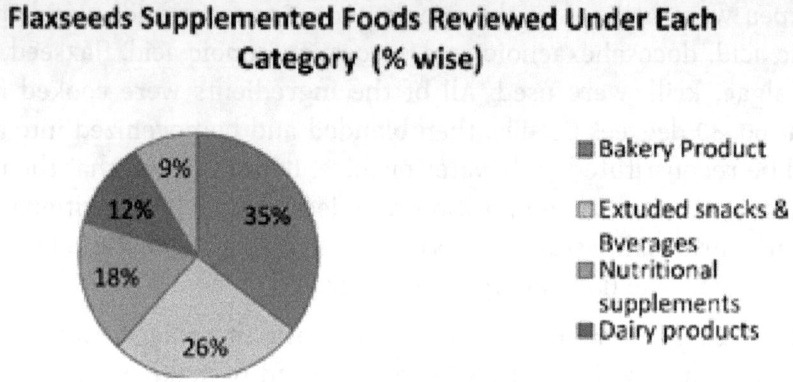

Figure 4. Types (in percentage) of food reviewed under each category.

Six different patents were captured by 4 different companies that enjoy 37.5% of the total share. Hangzhou Tianlong health products Co. Ltd. captured 3 patents and enjoys an 18.75% share of 37.5%. Shenzhen Jielikang biological Tech. Co. Ltd., Hockings consulting Inc., Guangzhou Saliai stemcell Science and Tech. Co. Ltd., each availed 6.25% of 37.5% with 3 patents. Two other research institutes Shanxi biological institutes and Zhejiang Ocean University captured a collective share of 12.5% as they claimed 1 patent each. **Figure 5** represented the distribution of different patent applications claimed by companies, institutes & inventors as assignees.

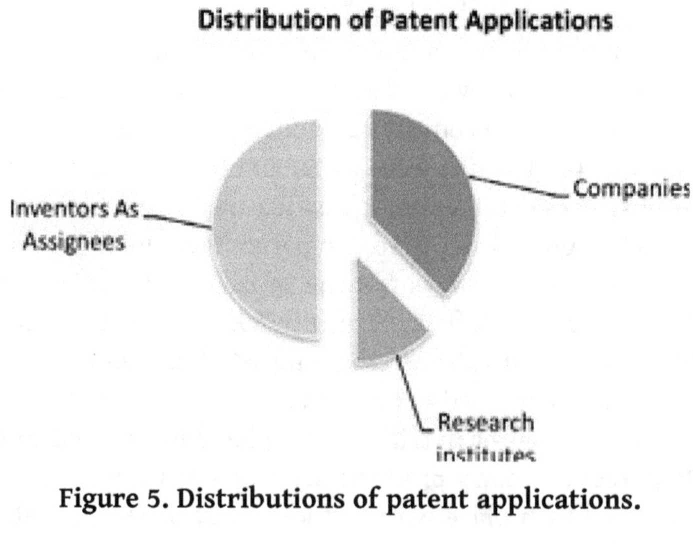

Figure 5. Distributions of patent applications.

4. CHALLENGE FACED BY FLAXSEEDS

Humans across the region are adopting healthier lifestyles, which is one of the major factors driving demand for flaxseeds, which is linked to a variety of health benefits. Even so, the flaxseeds must face some obstacles that cannot be overlooked. Chia seeds are one of the flaxseed substitutes in the market. These are a close substitute for flaxseeds due to their nutrient content and health benefits. Both flax and chia seeds are high in nutrients and have been linked to health benefits such as lower blood sugar, a healthier heart, and cancer prevention. When compared to flaxseeds, chia seeds aid in the reduction of food cravings while also being high in fiber [51]. In this context, a study was conducted in which three separate occasions, 15 healthy participants (5 Males and 10 Females having age between 23 to 27 years; BMI: 22.2 ± 0.8 kg/m2) were randomized to receive a 50 g glucose challenge alone or with either 25 g ground Salba-chia or 31.5 g flax. At fasting and 2-hours postprandially, blood glucose samples and satiety scores were collected and investigated. According to findings in comparison to glucose control, Salba-chia reduced blood glucose area under the curve more (82.5 ± 19.7 $mmol^{-1}$ P0.001) than flax which reduced the area by 60.0 ± 19.7 $mmol^{-1}$ (P = 0.014), over 120 minutes. When compared to flaxseeds, Salba-chia significantly reduced mean ratings of the urge to eat, prospective consumption, and total appetite score [52]. They also have a competitive advantage over flaxseeds in that they do not go rancid and do not require grinding due to their smaller size [53]. As a result, high competition from chia seeds has created a difficult situation for flaxseed.

5. FUTURE DIRECTIONS

The oil and fibrous plant flax (*Linum Usitatissimum* L.) has been used by for over 600 decades and it was one of the first plants to be domesticated. However, because of its high nutritional value, flaxseed is becoming more popular in the nutraceutical segment as people become more aware of its health benefits and applications in the food industry. There are some aspects of this field that have yet to be explored. An attempt has been made in this article to shed some light on those points as well.

- People today are well aware of the significant relationship between diet and health. They believe in letting the food, rather than medicine, be your medicine. Therefore, as the demand for functional foods grows, so does the amount of waste produced by them. If this waste is not properly disposed of, it can harm the environment, as well as the nutrition and economy. As a result, this aspect must also be considered. However, the presence of anti-nutritional compounds, particularly cyanogenic glycosides, which are an essential component of flax, present in the flaxseed meal left after oil extraction has limited its applications in food and feed. Many techniques for reducing cyanogenic compounds in flaxseed meals on a lab scale, such as boiling in water, microwave roasting, wet autoclaving, acid treatment, and extrusion cooking, can be used to successfully reduce or eliminate anti-nutritional components in flaxseed meal [54]. However, these methods have a significant impact on beneficial polar compounds like lignans. Cold-pressed flaxseed meal can be a good choice for removing the cyanide while retaining the beneficial nutrients, protein, fat, fiber, and lignans at the same level as untreated flaxseed meal. One research was conducted using the cold pressing method of the by-product left over after extracting oil from flaxseed as a fat-replacer in low-fat salad dressing formulation, which is an excellent waste-to-wealth strategy [55].

- Eicosapentaenoic acid (EPA) and docosahexaenoic acid (DHA) are not produced by the human; they must be consumed or formed through metabolic processes of ω-3 fatty acids, such as linolenic acid. Current eating habits have resulted in a significant reduction in daily ω-3 fatty acid intake, which is now below the recommended levels, necessitating the need to supplement food with ω-3 fatty acids. One of the current strategies for staying healthy is to incorporate Omega-3 fatty acids into new products in the medical and health food fields. However, there is a problem with ALA, it is strongly heat-sensitive, reactive towards oxygen, and metal ions, and during processing and storage, it produces rancid flavors compounds. Flaxseed oil's nature restricts its use in the food and dairy industries. Several authors have recently worked on flaxseed oil stabilization using various technologies such as ultrasonication and microen- capsulation, as well as developing emulsion and microcapsules. One significant disadvantage is that the encapsulation process could cost more than twice as much as the ω-3 ingredient. Furthermore, due to storage, the ALA content is also easily reduced [53, 56]. As a result, more work is needed to improve its stability and quantification after it has been fortified in food products.

6. CONCISE SUMMARY

This study aimed to describe the incorporation of flaxseed into various products, including baked goods, dairy items, extruded foods, and snacks, and to assess their effects on diabetes patients. Flaxseed helps promote the cultivation of health. Enhancing the nutritional value of foods by reducing salt, sugar, and saturated fat levels, while increasing omega-3 fatty acids and other bioactive components that may assist diabetic patients in managing their condition. Further study is necessary to create environmentally sustainable and cost-effective technologies for extracting bioactives from flaxseed by-products. Stability techniques for ALA in the formulation of value-added flaxseed-enriched products will serve as a promising advancement for the food sector.

REFERENCES

1. K. K. Singh, D. Mridula, J. Rehal, and P. Barnwal, "Flaxseed: A potential source of food, feed and fiber," *Crit. Rev. Food Sci. Nutr.*, vol. 51, no. 3, pp. 210-222, 2011, doi: 10.1080/10408390903537241.
2. O. N. Tolkachev and J. Zhuchenko, "Biologically active substances of flax: Medicinal and nutritional properties: (A review)," *Pharm. Chem. J.*, vol. 34, no. 7, pp. 360-367, 2004, doi: 10.1023/A: 1005217407453.
3. A. E. D. A. Bekhit *et al.*, "Flaxseed: Composition, detoxification, utilization, and opportunities," *Biocatal. Agric. Biotechnol.*, vol. 13, no. November 2017, pp. 129-152, 2018, doi: 10.1016/j. bcab.2017.11.017.
4. A. Goyal, V. Sharma, N. Upadhyay, S. Gill, and M. Sihag, "Flax and flaxseed oil: an ancient medicine & modern functional food," *J. Food Sci. Technol.*, vol. 51, no. 9, pp. 1633-1653, 2014, doi: 10.1007/s13197-013-1247-9.
5. F. Rashid, Z. Ahmed, S. Hussain, J. Y. Huang, and A. Ahmad, "*Linum usitatissimum* L. seeds: Flax gum extraction, physicochemical and functional characterization," *Carbohydr. Polym.*, vol. 215, no. June 2018, pp. 29-38, 2019, doi: 10.1016/j.carbpol.2019.03.054.
6. P. Kaur, R. Waghmare, V. Kumar, P. Rasane, S. Kaur, and Y. Gat, "Recent advances in utilization of flaxseed as potential source for value addition," *OCL - Oilseeds fats, Crop. Lipids*, vol. 25, no. 3, 2018, doi: 10.1051/ocl/2018018.
7. P. Kajla, A. Sharma, and D. R. Sood, "Flaxseed—a potential functional food source," *J. Food Sci. Technol.*, vol. 52, no. 4, pp. 1857-1871, 2015, doi: 10.1007/s13197-014-1293-y.
8. R. Bernacchia, R. Preti, and G. Vinci, "Chemical Composition and Health Benefits of Flaxseed," *Austin J. Nutr. Food Sci.*, vol. 2, no. 8, pp. 1-9, 2014.
9. C. Soukoulis, C. Gaiani, and L. Hoffmann, "Plant seed mucilage as emerging biopolymer in food industry applications," *Curr. Opin. Food Sci.*, vol. 22, pp. 28-42, 2018, doi: 10.1016/j. cofs.2018.01.004.
10. P. Z. Zimmet, D. J. Magliano, W. H. Herman, and J. E. Shaw, "Diabetes: A 21st century challenge," *Lancet Diabetes Endocrinol.*, vol. 2, no. 1, pp. 56-64, 2014, doi: 10.1016/S2213-8587(13)70112-8.
11. J. Beckman, *Global E&P*, IDF Diabetes Atlas, sixth edition, 2016.
12. M. D. Nathan, "Long-Term Complications Of Diabetes Mellitus," *N. Engl. J. Med.*, vol. 328, no. 23, pp. 1676-1685, 1993.
13. A. Alkhatib *et al.*, "Functional foods and lifestyle approaches for diabetes prevention and management," *Nutrients*, vol. 9, no. 12, pp. 1-18, 2017, doi: 10.3390/nu9121310.
14. I. Rudkowska, "Functional foods for health: Focus on diabetes," *Maturitas*, vol. 62, no. 3, pp. 263-269, 2009, doi: 10.1016/j.maturitas.2009.01.011.

15. P. Mirmiran, "Functional foods- based diet as a novel dietary approach for management of type 2 diabetes and its complications: A review," *World J. Diabetes*, vol. 5, no. 3, p. 267, 2014, doi: 10.4239/wjd.v5.i3.267.
16. U. V. Mani, I. Mani, M. Biswas, and S. N. Kumar, "An open-label study on the effect of flax seed powder (*Linum usitatissimum*) supplementation in the management of diabetes mellitus," *J. Diet. Suppl.*, vol. 8, no. 3, pp. 257-265, 2011, doi: 10.3109/19390211.2011.593615.
17. Ł. Sęczyk, M. Świeca, D. Dziki, A. Anders, and U. Gawlik-Dziki, "Antioxidant, nutritional and functional characteristics of wheat bread enriched with ground flaxseed hulls," *Food Chem.*, vol. 214, pp. 32-38, 2017, doi: 10.1016/j.foodchem.2016.07.068.
18. B. A. Kay *et al.*, "Pudding products enriched with yellow mustard mucilage, fenugreek gum or flaxseed mucilage and matched for simulated intestinal viscosity significantly reduce postprandial peak glucose and insulin in adults at risk for type 2 diabetes," *J. Funct. Foods*, vol. 37, pp. 603-611, 2017, doi: 10.1016/j. jff.2017.08.017.
19. M. A. Kurek, J. Wyrwisz, S. Karp, and A. Wierzbicka, "Effect of fiber sources on fatty acids profile, glycemic index, and phenolic compound content of in vitro digested fortified wheat bread," *J. Food Sci. Technol.*, vol. 55, no. 5, pp. 1632-1640, 2018, doi: 10.1007/ s13197-018-3061-x.
20. F. Zhu and J. Li, "Physicochemical properties of steamed bread fortified with ground linseed (*Linum usitatissimum*)," *Int. J. Food Sci. Technol.*, vol. 54, no. 5, pp. 1670-1676, 2018, doi: 10.1111/ijfs.14043.
21. F. Zhu and J. Li, "Physicochemical and sensory properties of fresh noodles fortified with ground linseed (*Linum usitatissimum*)," *Lwt*, vol. 101, pp. 847-853, 2018, doi: 10.1016/j.lwt.2018.12.003.
22. T. Wandersleben *et al.*, "Enhancement of functional and nutritional properties of bread using a mix of natural ingredients from novel varieties of flaxseed and lupine," *LWT - Food Sci. Technol.*, vol. 91, pp. 48-54, 2018, doi: 10.1016/j.lwt.2018.01.029.
23. N. Soltanian and M. Janghorbani, "A randomized trial of the effect of flaxseeds to manage constipation, weight, glycemia, and lipid in constipated patients with type 2 diabetes," *Nutr. Metab.*, vol 15, no. 36, 2018,doi: 10.1186/ s12986-018-0273-z.
24. N. Soltanian and M. Janghorbani, "Effect of flaxseed or psyllium vs. placebo on management of constipation, weight, glycemia, and lipids: A randomized trial in constipated patients with type 2 diabetes," *Clin. Nutr. ESPEN*, vol. 29, pp. 41-48, 2018, doi: 10.1016/j. clnesp.2018.11.002.
25. F. Yuksel, "Investigation of certain nutritional properties of noodle enriched with raw flaxseed," *Qual. Assur. Saf. Crop. Foods*, vol. 11, no. 2, pp. 183-189, 2019, doi: 10.3920/ QAS2018.1363.

26. N. Hasaniani, M. Rahimlou, A. Ramezani Ahmadi, A. Mehdizadeh Khalifani, and M. Alizadeh, "The Effect of Flaxseed Enriched Yogurt on the Glycemic Status and Cardiovascular Risk Factors in Patients with Type 2 Diabetes Mellitus: Randomized, Open- labeled, Controlled Study," *Clin. Nutr. Res.*, vol. 8, no. 4, p. 284, 2019, doi: 10.7762/cnr.2019.8.4.284.
27. T. M. de Oliveira Giarola, C. G. Pereira, M. E. T. Prado, L. R. de Abreu, and J. V. de Resende, "Effects of Golden Flaxseed Flour on Ice Recrystallization in Uvaia (Eugenia pyriformis Cambess.) Diet Sherbet," *Food Bioprocess Technol.*, vol. 12, no. 12, pp. 2120-2135, 2019, doi: 10.1007/s11947-019-02377-w.
28. J. L. Tobias-Espinoza *et al.*, "Effects of the addition of flaxseed and amaranth on the physicochemical and functional properties of instant-extruded products," *Foods*, vol. 8, no. 6, pp. 1-14, 2019, doi: 10.3390/foods8060183.
29. A. Wirkijowska, P. Zarzycki, A. Sobota, A. Nawrocka, A. Blicharz- Kania, and D. Andrejko, "The possibility of using by-products from the flaxseed industry for functional bread production," *Lwt*, vol. 118, p. 108860, 2020, doi: 10.1016/j.lwt.2019.108860.
30. K. Geetha, G. M. Yankanchi, S. Hulamani, and N. Hiremath, "Glycemic index of millet based food mix and its effect on pre diabetic subjects," *J. Food Sci. Technol.*, vol. 57, no. 7, pp. 2732-2738, 2020, doi: 10.1007/s13197-020-04309-5.
31. N. N. Sevostyanova, E. A. Pchelina, M. A. Vihrova, O. Y. Trezorova, and L. V. Andreeva, "Flaxseed bread for therapeutic nutrition," *IOP Conf. Ser. Earth Environ. Sci.*, vol. 613, no. 1, 2020, doi: 10.1088/1755-1315/613/1/012137.
32. A. Almehmadi, H. Lightowler, M. Chohan, and M. E. Clegg, "The effect of a split portion of flaxseed on 24-h blood glucose response," *Eur. J. Nutr.*, vol. 60, no. 3, pp. 1363-1373, 2021, doi: 10.1007/ s00394-020-02333-x.
33. M. L. L. Campidelli *et al.*, "Rheological, Physico-chemical and Sensorial Properties of Ice Cream Made with Powdered Form with Low Energetic Value and High Content of Prebiotic Fibers," *J. Culin. Sci. Technol.*, vol. 19, no. 4, pp. 331-351, 2021, doi: 10.1080/15428052.2020.1768995.
34. J. M. Vieira *et al.*, "Flaxseed gum- biopolymers interactions driving rheological behavior of oropharyngeal dysphagia-oriented products," *Food Hydrocoll.*, vol. 111, no. July 2020, p. 106257, 2021, doi: 10.1016/j.foodhyd.2020.106257.
35. V. J. Mathew, "Optimized Nutrient Fatty Acid Composition," *US0360073A*, vol. 1, 2017.
36. H. Chen, X. Ge, F. Wang, Y. Wang, J. Cheng, X. Wang, "One kind is for diabetes B patient's diet and foodstuffs and preparation method thereof," *CN107373642A*, 2017.
37. X. Li, X. Han, Y. Gao, "A kind of compositions being applicable to diabetes patient and diet therapy soup thereof and preparation method, "*CN106266429A*, 2017.

38. J. Shirodkar, "Food Products and Processes for Preparation Thereof, "*AU100571A*, vol. 4, no. 12, 2017.
39. M. Wang, P. Pan, W. Song, Z. Deng, Y. King, B. Su, Y. Cai, X. Li, X. Lu, J. Gong, "A kind of red yeast rice Flaxseed cake fermentate and its application, "*CN107136183A*, 2017.
40. C. X. Li, H. Luo, "Diabetes full nutrition formula food and preparation method thereof made of a kind of seaweeds raw material,"*CN1085 22970A*, 2018.
41. T. W. Kim, "Method of producing a food for preventing and improving diabetes,"*KR10-1850955B*, vol. 1, no. 19, pp. 1-17, 2018.
42. W. Wang, "A kind of 3 sour milk powder of omega and its preparation process,"*CN104542974B*, 2018.
43. Y. Chen, "A kind of 3 solid beverage of omega,"*CN108685006A*, 2018.
44. V. J. Mathew, P. T. Kochery, "Optimized Nutrient Food," *US0335796A*, vol. 1, 2019.
45. K. Fang, H. Zheng, Y. Wu, "Low- sugar mooncake high in fat and preparation method thereof," *CN110384123A*, vol. 29, no. 19, 2019.
46. J. M. Hockings, "Mothod for treating and reversing type 2 diabetes," *US10765318B*, vol. 2, 2020.
47. J. Lu, S. Li, X. Wu, "Low- carbohydrate flaxseed nut wafer and preparation method thereof," *CN111466425A*, vol. 31, no. 19, 2020.
48. S. Anderson, J. Hall, M. Yoho, "Human Dietary supplement and method for treating digestive system and immune-related disorders," *WO2020081417A*, vol. 1, 2020.
49. S. Salimath, "Enteral Or Oral Nutritional Composition," *WO20 20039277A*, vol. 1, no. 51, 2020.
50. J. Pyo, "Preparing method of noodle for improving diabetes, and the noodle obtained thereby," *0012287A*, 2021.
51. B. Kulczynski et al., "The chemical composition and nutritional value of chia seeds – current state of knowledge," *Nutrients*, vol. 11, no. 6, pp. 1242, 2019.
52. V. Vuksan *et al.*, "Comparison of flax (*Linum usitatissimum*) and Salba- chia (*Salvia hispanica* L.) seeds on postprandial glycemia and satiety in healthy individuals: A randomized, controlled, crossover study," *Eur. J. Clin. Nutr.*, vol. 71, no. 2, pp. 234-238, 2017, doi: 10.1038/ejcn.2016.148.
53. A. Gowda, V. Sharma, A. Goyal, A. K. Singh, and S. Arora, "Process optimization and oxidative stability of omega-3 ice cream fortified with flaxseed oil microcapsules," *J. Food Sci. Technol.*, vol. 55, no. 5, pp. 1705-1715, 2018, doi: 10.1007/s13197-018-3083-4.
54. M. Imran, F. M. Anjum, M. S. Butt, M. Siddiq, and M. A. Sheikh, "Reduction of cyanogenic compounds in flaxseed (*Linum usitatissimum* L.) meal using

thermal treatment," *Int. J. Food Prop.*, vol. 16, no. 8, pp. 1809-1818, 2013, doi: 10.1080/10942912.2011.608914.

55. Z. H. Tekin and S. Karasu, "Cold- pressed flaxseed oil by-product as a new source of fat replacers in low-fat salad dressing formulation: Steady, dynamic and 3-ITT rheological properties," *J. Food Process. Preserv.*, vol. 44, no. 9, pp. 1-13, 2020, doi: 10.1111/jfpp.14650.
56. E. Feizollahi, Z. Hadian, and Z. Honarvar, "Food Fortification with Omega-3 Fatty Acids; Micro- encapsulation as an Addition Method," *Curr. Nutr. Food Sci.*, vol. 14, no. 2, pp. 90-103, 2017, doi: 10.2174/1573401313666170728151350.

7 Chapter

UTILIZING TECHNOLOGIES FOR VEGETABLE CULTIVATION: A SOLUTION FOR FOOD AND NUTRITION INSECURITY IN GHANA

1. INTRODUCTION: AN OVERVIEW

Contemporary agriculture frequently employs advanced technologies, including robotics, temperature and moisture sensors, aerial imagery, and GPS systems. These sophisticated equipment and precision agriculture and robotic systems enable enterprises to enhance profitability, efficiency, safety, and environmental sustainability. Farmers seeking to enhance their harvests must own an effective irrigation system. Supplying plants with the correct quantity of water enhances their growth and, therefore, the agricultural yields. Historically, agricultural productivity in Ghana has been predominantly dependent on rainfall, and the application of fertilizers and pesticides has been mismanaged, partially due to insufficient extension services. Farmers had difficulties in their agricultural operations and the adoption of advanced technologies due to the significant disparity between technology development and access to pertinent information. Ghana is currently poised for a promising future due to the integration of science, technology, and innovation in agricultural output. Agriculture is evolving into a lucrative enterprise, and with appropriate scientific methodologies, Ghanaian farmers are starting to realize the rewards of their labor.

Vegetable cultivators encounter numerous obstacles, such as pests and illnesses, labor shortages, and climate change. The majority of vegetable cultivation in Ghana is conducted by smallholder farmers with rudimentary agricultural tools.

Production primarily occurs on elevated beds, ridges, or unadorned soil in wide fields. Moreover, cultivation often occurs under a rainfed system, permitting production in just one or two seasons annually, contingent upon the farm's geographical location within the country. Supplementary irrigation may be administered throughout the nursery and initial field establishing phases via watering cans, buckets, knapsack sprayers, and occasionally tiny powered pumps or various combinations of these techniques. The dependence on rainfall for vegetable cultivation, coupled with the unpredictable rainfall patterns in recent years owing to climate change, frequently leads to significant losses from inadequate yields during droughts or prolonged dry periods.

In recent decades, numerous "smart" technologies have been created for vegetable production and processing. These encompass nursery management technology, protected or screenhouse vegetable culture, grafting technology, and information technology in vegetable production. Nonetheless, vegetable cultivators face numerous hurdles when contemplating the adoption of new technology or the modification of old technology. Producers are being presented with solutions that may be incompatible with their particular production systems or may lack economic viability. This chapter showcases instances of cutting-edge technologies applicable to contemporary or imminent vegetable cultivation. It additionally examines the trends in vegetable production in Ghana and the issues related to vegetable agriculture. Furthermore, it examines the nascent technology in vegetable cultivation and delineates sophisticated methodologies in vegetable production in Ghana, focusing on genetic enhancement, soil fertility augmentation, irrigation, and pest and disease management strategies. It also illustrates the attributes of research and the engagement of farmers in innovative technology for vegetable production. It ultimately delineates the potential of vegetable production for food and nutrition security in Ghana.

2. VEGETABLE PRODUCTION TRENDS IN GHANA

In this section general vegetable production practices in Ghana are enumerated with emphasis on conventional practices, urban and peri-urban vegetable production. Although some emerging technologies such as cultivation in screen houses are becoming common in Ghana, these will be dealt with in a separate section and thus will not be the focus here.

2.1 Types of vegetables grown in Ghana

Ghana has a tropical climate and for that reason, most of the vegetables cultivated in Ghana are those that have a tropical origin. Notwithstanding, some subtropical and temperate vegetables that can thrive in tropical climates are also grown, although the yields are often lower compared to those grown in their ideal climates. Vegetables grown in Ghana can be classified as either exotic or local/traditional depending on their origin. The types of vegetables (both local and exotic) grown in a specific region in the country often depend on a number of factors including but not limited to the demand in that area and the suitability of the micro-climatic conditions for that specific vegetable crop. However, for most rural areas, the former factor is not a major consideration since in most cases the produce is transported to urban centers to be sold. In the case of those targeted for the export market, proximity to ports is a major factor in the choice of area for cultivation. Whereas some of the exotic and local vegetables are consumed by Ghanaians and expatriates, some of them are exported to generate foreign revenue for the country. In addition, some local vegetables are also exported. Major local vegetables cultivated in Ghana include tomato (*Solanum lypersicum*), pepper (*Capsicum annuum* – cayenne, chilli), scotch bonnet pepper (*Capsicum chinense*), onion (*Allium cepa*), shallot (*Allium cepa* var. *aggregatum*), garden eggs (*Solanum melongena*), and okra (*Abelmoschus esculentus*).

Other vegetables that may be important in certain areas of the country but often classified as underutilized include leafy vegetables such as local spinach (*Amaranthus* spp), Jew's mallow (*Corchorus* spp), gboma (*Solanum macrocarpon*), sweet potato leaves (*Ipomea batatas*), cocoyam leaves (*Xanthosoma sagittifolium*), roselle (*Hibiscus sabdariffa*), among others.

Exotic vegetables grown in Ghana include cabbage (*Brassica oleracea* var. *capitata*), lettuce (*Lactuca sativa*), carrot (*Daucus carrota*), cucumber (*Cucumis sativus*), spring onion (*Allium fistulosum*), garlic (*Allium sativum*), cauliflower (*Brassica oleracea* var. *botrytis*), French beans (*Phaseolus vulgaris*), beetroot (*Beta vulgaris* subsp. *vulgaris* Conditiva), radish (*Raphanus sativus*), bell/sweet pepper (*Capsicum annuum*), squash (*Cucurbita* spp), and pumpkin (*Cucurbita* spp). Asian vegetables such as yard long bean (*Vigna unguiculata* subsp. *sesquipedalis*) and Chinese cabbage (*Brassica rapa* subsp. *pekinensis*) are also cultivated mostly for the export market.

2.2 Conventional approach to vegetable production

Most vegetable production in Ghana is carried out by smallholder farmers who use basic farm implements on their farms. Production takes place predominantly on raised beds, ridges, or bare ground in open fields. Nursery establishment for those vegetables that do not need to be planted at stake is mostly on nursery beds created close to the site of cultivation. The vegetables are often cultivated on a piece of land owned or rented by these farmers as is mostly the case. Manual labor is used for most farm activities by the farmer with the help of family members or hired farm hands. For example, weeding is by use of hoe, cutlass, and/or spraying of chemicals using knapsack sprayers. However, in recent times some farmers who have larger farms are incorporating some level of mechanization such as the use of tractor-drawn tools (plough and harrow) for land preparation. Nevertheless, most of the other practices such as seeding, weeding, and harvesting are still carried out manually.

Furthermore, cultivation is often under the rainfed system which only allows for production within two seasons or even one season of the year depending on the agroecology within which the farm is located in the country. Supplementary irrigation may be provided at the nursery and early field establishment stage using either watering cans, buckets, knapsack sprayers, and sometimes small motorized pumps or various combinations of these methods. Over-reliance on rainfall for vegetable production and the erratic nature of rainfall patterns in recent years, a consequence of climate change, often results in severe economic losses due to poor yields in the event of drought or long dry spells. In addition, depending on the crop and the timing of the rains, too much rain can also cause yield losses due to the high incidence of diseases and other physiological disorders such as fruit cracking. And there is also the glut associated with seasonal cultivation, which often results in a poor price for the produce based on the economic principle of demand and supply. Most farmers incur huge losses during these periods due to the low prices or rotting of the produce as a result of buyers being overwhelmed with produce. These problems have been exacerbated due to poor knowledge of appropriate postharvest handling techniques and low to minimal processing practiced by farmers.

2.3 Urban and peri-urban vegetable farming

Urban and peri-urban vegetable farming generally refers to vegetable production in open spaces or fields mostly within and along the outskirts of major urban areas. Urban vegetable cultivation is considered to have originated with the arrival of the Europeans who created gardens around castles and forts in the country from the sixteenth century [1]. In addition, due to urbanization, there have been a rise in commercial urban and peri-urban farming to meet the needs of the ever-increasing population within the urban centers. Nevertheless, the actual number of farming sites keeps fluctuating due to the loss of some farming sites as a result of development projects and the establishment of new or previously unidentified areas [2].

Most of the farmers in these urban and peri-urban sites cultivate different crops and vegetables all year round. This is due to the ability to find ways of providing supplementary irrigation in the absence of rain with the aid of watering cans, buckets, small motorized pumps, and occasionally furrow and sprinkler systems [3]. This ensures an almost regular supply of vegetables to consumers all year long. Nevertheless, with most being small-scale farmers, production is not sufficient to prevent the shortage of particular vegetables during certain times of the year. Most of the vegetables grown by farmers involved in peri-urban and urban farming in the southern part of the country are exotic vegetables such as cabbage, lettuce, carrot, spring onions, and bell pepper. Other exotic vegetables such as cauliflower, radish, and traditional vegetables including okra, tomato, hot pepper, garden eggs, local spinach, and Jew's mallow (*ayoyo*) are also cultivated, although, these may be to a lesser extent in certain areas. In cities in the northern part of the country such as Tamale, local leafy vegetables, okra, tomatoes, and pepper are the pre-dominant vegetables (**Figure 1**) [4].

2.4 Source of water for irrigation and health implications

The proximity of urban and peri-urban farms to a near-permanent source of water allows for intensive cultivation. However, the quality of the source of water for most of these farms has been a major cause of concern for many, especially in recent times. With the exception of some major sites that are close to un-polluted riverine bodies or formal irrigation sites and those that have dug out wells or tap water for irrigation, most farmers, particularly those in the urban centers, use water from river bodies or streams that are polluted with untreated municipal wastewater.

Figure 1. A typical urban vegetable farm A typical urban vegetable farm Source [2].

Some even go to the extent of using water directly from drains for irrigation. These pose a health risk to both farmers and consumers since produce are often contaminated with fecal coliforms, helminths eggs, and heavy metals at levels higher than globally accepted [5].

2.5 Towards improving livelihoods of peri-urban farmers

A recent project funded by the Japan Social Development Fund and coordinated by the Ghana Commercial Agriculture Project, the Ghana Peri-urban Vegetables Value Chain Project (GPVVCP) is seeking to develop urban and peri-urban agriculture to improve on the livelihood of farmers as well as the safety of produce for consumers. To this end, select sites within the Greater Accra, Volta, and Ashanti regions have been targeted to pilot such a venture. The project involves setting up farmer-managed irrigation systems including the use of solar pumps and construction of washrooms and packhouses for farmers. The latter is to facilitate cleaning and value addition to produce in order to improve farmers' incomes.

2.6 Production in-home and backyard gardens

Vegetable production in-home and backyard gardens has a long history in Ghana, although much attention was given to it as a result of the Operation Feed Yourself (OFY) policy initiated by the Acheampong government in the 1970s. Many people established farms in any space they could lay their hands on, including spaces in their homes [6]. Following the end of the government, some people still maintained their farms whereas others went on to start their own gardens when they got the opportunity. The tradition and idea of the home or backyard garden have remained with many Ghanaians who still establish their own gardens when they get the opportunity. Obuobie et al. [7] estimate that in the city of Accra alone, between 50–70 ha of land distributed among 80,000 homes (about 60% of houses in the city) are used for home gardens. Nevertheless, this figure includes land used for cultivating other crops and livestock. Following concerns about the safety of vegetables produced by some commercial farmers in the urban and peri-urban centers, more people with the wherewithal in terms of space and time have started their own backyard gardens.

There are many social media groups such as the Facebook group, home gardening, Ghana where people with no farming experience can get technical knowledge about growing certain vegetables and crops in general in their backyard. Vegetables grown in the home and backyard gardens are often for home consumption, although in some cases some produce may be sold when there is excess.

2.7 Seasonal and off-season production at formal irrigation sites or banks of rivers

Attempts by the Government of Ghana and other private entities to prevent the over-reliance of agriculture on rain have led to the development of irrigation sites to enable off-season or nearly all-year-round cultivation of vegetables and other important agricultural commodities. The capacity of these irrigation schemes ranges from a few hectares to hundreds and even thousands of hectares, making use of rivers, lakes, and existing or newly created dams. A typical example of such sites is the Tono dam located in the Kassena-Nankana District and managed by the Irrigation Company of Upper Region Ltd. This dam spanning about 2.5 miles serves close to 2500 ha of land used for the cultivation of various vegetables and crops. With general oversight and development by the Ghana Irrigation Development Authority in collaboration with key stakeholders and management by Scheme Management Entities, a total of 57 schemes spread across the entire country exist to provide irrigation for the production of different vegetables and crops over a total area 16175 ha [8]. These sites may be located in rural areas or within urban centers depending on the source of water for irrigation. Invariably, those within the urban centers play a key role in urban vegetable production. More recently, the Government of Ghana under the leadership of His Excellency, Nana Addo Dankwah Akuffo-Addo rolled out a policy named 'One Village One Dam' (1V1D) which sought to create small earth dams in certain villages to serve as sources of water for agriculture (both crops and livestock). A total of 439 dams were planned, 375 of which were in various stages of completion, with 64 of them were between 90–100% completed as at October 2020 [9]. Each dam was constructed to provide irrigation water for at least 5 ha of farmland. Nevertheless, recent evidence from Sore [10] of the MyJoyOnline team shows that some of the dams were poorly constructed and almost dried up during the dry season and as such, they were only used as a source of drinking water for livestock but not for the production of vegetables or other crops.

Regardless, the prospects look good for access to irrigation by farmers for vegetable production throughout the country with the proposed development of an additional 8 irrigation sites by Ghana Irrigation Development Authority (GIDA) with funding from the Export Development and Agricultural Investment Fund (EDAIF) [8]. These new projects will cover over 12,000 ha spread between 7 regions within the country. Moreover, farmers in the Upper East Region will have an additional boost for vegetable production from the impending construction of the Pwalugu Multipurpose Irrigation Project in the Talensi District which will supply irrigation water for about 25,000 ha of land and benefit 15,000 people.

3. CHALLENGES ASSOCIATED WITH VEGETABLE PRODUCTION IN GHANA

3.1 Limited number of varieties

Vegetables continue to play important role in the diet of most Ghanaian households. These vegetables are eaten raw or cooked, serving as a major source of plant protein. There are two major sources namely local and exotic. Examples of the local vegetables are *kontomire* (cocoa yam leaves), cassava leaves, pepper, *bokorbokor*, *alefu* and many others. Examples of exotic vegetables are cabbage, carrot, lettuce, French beans, common beans (*Phaseolus vulgaris*), cowpea (*Vigna unguiculata*) among others. Compared to other food crops such as the root and tubers, and cereals, vegetable growers face the challenge of a limited number of improved varieties.

The majority of the local vegetable have not been improved or new varieties released to replace the old ones leading to reduced productivity due to low climate adaptation. The number of vegetables consumed in Ghana has decreased due to the continual selection and domestication of just few crops. People acquire a preference for specific vegetable types. As a result, they go for just those vegetables, which also inform farmers to grow only those vegetables. They reuse their stored seeds for successive planting, resulting in the extinction of other kinds [11]. Imported seeds are used by other farmers who grow exotic crops like cabbage, carrots, lettuce, and so on. Their preferences inform seed importers on which kinds to bring to market, resulting in a reduction in the number of types of those vegetables available. The exotic vegetables on the other hand are also limited due to the fact that breeding is not done in country due to the climatic requirement needed for the reproductive phase of the crop. Farmers, therefore, rely on only a few tried and tested varieties that have been introduced over the years leading to low productivity as a result of diseases and pest build-up. Most of the breeding programs have centered on the cereals, root and tubers, and the legumes to the neglect of the vegetables.

3.2 Scarcity of seeds

Another constraint is inadequate access to high-quality improved vegetable seed cultivars. Vegetable farmers in Ghana and other countries in sub-Saharan Africa are constrained by factors such as poor quality and availability of seeds. For the local vegetables, farmers rely on farmer-saved seeds from previous harvests, local markets, or from friends whiles others also buy seedlings from the local markets. These sources are normally unreliable, seeds are sometimes of poor quality (physical and genetic) leading to low productivity and high cost of production. The imported seeds also have problems with acclimatization as most of them come from the temperate regions and when they are exposed immediately to tropical conditions lose viability quickly. Improper storage condition at the retail shops also affects the quality.

In Ghana, it is difficult to find vegetable seed growers, unlike staple crops. The Grains and Legumes Development Board (GLDB) not only lacks the capacity to absorb all breeder seeds generated by research institutions, but it also lacks the capacity to create sufficient amounts of foundation seed to fulfill the demands of seed businesses and seed producers. When research institutions are unable to sell all of their breeder seeds generated in a given year, they have difficulty producing the next year. The situation has worsened due to a lack of gene banks dedicated to preserving these long-forgotten indigenous crop types. These vegetables, on the other hand, are rich in genes that are both tolerant of abiotic and resistant to biotic stimuli. Because the majority of them have become extinct, intentional efforts are undertaken to bring those kinds from outside of Ghana whenever they are required for improvement.

3.3 Absence of a functional vegetable seed system

Two main seed systems operate in Ghana namely: formal and informal (traditional) [12]. However, in certain farmer-based organizations and community seed production, a combination of these two major seed systems has been used [13]. The formal seed system, which is governed by state legislation and overseen by Research Institutes, Universities, and the Ministry of Food and Agriculture (MOFA), is controlled. By providing stakeholders with knowledge and samples to work with, these institutes reinforce the relevance of seed certification. MOFA is in charge of approximately 80% of the operations (registration of seed growers, cleaning, sorting and grading of seeds, seed inspection and certification, and packaging for sales only at the regional capitals). Bureaucracy at this level causes delays in providing services to seed firms, resulting in a loss of production. A weak distribution system exists from the stage of approved seed production to farmers, as well as annual supply failing to satisfy farmer demand. In contrast to vegetables, crops such as maize, rice sorghum, millet, peanuts, and cowpea have recently garnered increased financial focus in varietal development and release. Few crops, such as tomato, pepper, and okra, have been developed, but because of a poor seed distribution method, they have a low adoption rate [12].

3.4 Overreliance on imported seeds of common vegetables

Since only a few locally improved varieties are available for only a crop like pepper and non-existent for the other vegetables [14], vegetable growers have to rely on imported seeds for cultivation. Imported vegetable seeds are mostly available at various agro-input dealers throughout the country and mainly deal in crops such as tomato, pepper, garden eggs, and onions [15]. Since farmers in most remote communities' lack access to agro-input dealers who sell vegetable seeds, they often fall victim to poor-quality seeds that may be obtained from uncertified sources.

Variability in the timing and distribution of rainfall is also preventing farmers from purchasing certified seeds because they are frightened of accruing big debts if their crops fail completely due to bad weather [12]. In comparison, most development programs concentrate on staples like maize, cassava, and rice, with little attention paid to vegetables. The problem may also be attributed to a lack of research into our indigenous vegetables, which makes them unable to compete with alreadyimproved imported crops, making it harder for farmers to choose and reproduce them. As a result, they are unable to exhibit their entire potential. Because of these qualities, farmers often accept exotic vegetables that have been enhanced to farmers, influencing food and nutritional security in the country.

3.5 Infrastructure challenges (irrigation and processing facilities)

Improved infrastructure in the areas of irrigation, roads, transport, processing, and storage are needed for enhanced vegetable production. About 80–90% of agriculture in sub-Saharan Africa relies on rainfall for cultivation and Ghana is no exception. The major and minor planting seasons have been erratic over the years [16]. Rainfall variability over time as a result of global warming and climate change has posed a danger to vegetable productivity and food security [17]. Crops that require less water, such as cassava and maize, can thrive during these times [16]. Vegetables, on the other hand, are succulent crops that contain around 90% water [18], necessitating a continual supply of water during their growth cycle. Vegetable production is mainly under rain-fed or irrigation depending on the season and location. Water shortages can impact the output of vegetables like tomatoes, garden eggs, and notably exotic ones like cabbage, carrot, and lettuce at any stage of their development. To facilitate access to water, most vegetables in Ghana are grown near a water source. Where available, irrigation facilities offer great opportunities for vegetable growers. However, these are not available in most countries in adequate amounts to meet the needs of farmers in all communities. According to the World Bank [19] irrigated agriculture represents only 20 percent of the total cultivated land and contributes 40 percent of the total food produced worldwide. Due to this inadequacy, most vegetable growers especially in urban and peri-urban areas use other alternative water sources for the growing of vegetables. Quansah et al. [20] reported that 70% of vegetable growers in Ghana use waste drain water for irrigation, resulting in poor microbiological quality. Another challenge of vegetable production is poor road infrastructure. The perishable nature of most of the vegetables requires that they reach the markets and consumers in time to reduce post-harvest losses. The majority of the vegetable production takes place in the rural areas characterized by poor road network infrastructure making it difficult for farmers to get their produce to the market centers. Farmers in rural areas experience high post-harvest losses due to the poor nature of the road as the middle women who buy these vegetables find it difficult to reach the farmers [21, 22]. Linked to the poor road network is the high cost of transport and inappropriate means of transport.

Lack of or low mechanization equipment needed for seedbed preparation also poses a challenge in vegetable production [23]. Manual vegetable production is labor intensive. Besides, lack of handheld or small machines affects the timely undertaking of certain activities such as weeding, pests, and diseases control. Lack of or inappropriate storage conditions contribute to post-harvest losses in vegetable production. Apart from a few supermarkets that are located in urban areas, most of the vegetables are sold in the open market where they are exposed to harsh weather conditions. The humid tropical condition in Ghana predisposes these vegetables to deterioration in storage. Such storage conditions also make it easy for insects and rodents to attack [21].

Another challenge with vegetable production is glut which occurs at peak harvest due to the fact that most farmers cultivate their vegetables at the same time when the rains set in. This simultaneous harvesting of vegetables and glut are complicated with the absence of processing facilities for these vegetables. Most processing facilities in the country are for cereals, grains, legumes, cassava, and a few vegetables with extended storage life, such as pepper. Due to a lack of storage facilities for perishable food like vegetables, seed growers are forced to sell their products at a low price in order to not incur more cost for having to dispose them off.

3.6 Contamination from abuse and misuse of agrochemicals in vegetable cultivation

In Ghana, chemical pesticides are frequently used to control pests and diseases in vegetable cultivation. According to [24], pesticides are used by 87% of vegetable producers in Ghana. However, concerns about the negative impacts of this usage on public health and the environment are increasing. Agrochemical handling and storage, application, and disposal procedures were investigated in other research, and severe abuse and misuse of agrochemicals were discovered [25]. It was discovered that certain agrochemical manufacturers do not protect their products from direct or indirect sunlight, causing them to lose their effectiveness in controlling insects or pests. Most farmers do not use the prescribed dosage of agrochemicals during application and also do not wear Personal Protective Equipment (PPE). Over 80% of tomato growers in Ghana's northern region do not use personal protective equipment (PPE) [26]. Some farmers utilize agrochemicals that are outdated, prohibited, or unlicensed. The most significant problem occurs when farmers use agrochemicals too close to harvesting time, causing the preharvest period to be missed before the crop is harvested for sale. Some farmers dumped leftover pesticides in bodies of water, while others poured them onto their fields. Empty containers are dumped on the ground, while others are used as drinking containers [27].

Another challenge associated with vegetable production in Ghana is contaminants found on harvested produce. Heavy metals such as copper (Cu), zinc (Zn), lead (Pb), nickel (Ni), chromium (Cr), cadmium (Cd), and cobalt (Co) have been reported on some vegetables [28, 29]. These contaminants come from the use of waste water for irrigation, inappropriate use of chemical fertilizers, weedicides, pesticides, and insecticides application [28, 30]. In assessing the level of pesticide contamination on vegetables sold in markets in some selected regional capitals, [31] reports that 78% of the samples had chlorpyrifos residue on them. Non-adherence to harvest interval on the labels of these chemicals is a major cause of such chemical residues reported to be above the minimum recommended levels [29].

4. EMERGING TECHNOLOGIES FOR VEGETABLE CULTIVATION

4.1 Nursery management technology

A nursery is a place or an establishment where seedlings of crop plants particularly vegetables and fruit trees are raised before transplanting onto the main or permanent field [32]. Nurseries are essential in vegetable production because most of the vegetable crops cannot be directly seeded in the open field. This is because most vegetable crops require special attention during their early growth which cannot be guaranteed in the open field [33]. Secondly, most vegetable crops in nature have very small seeds that make direct seeding in the field impractical. Due to the above reasons, nurseries are set up to provide favorable environmental conditions for germination and seedling growth. Additionally, nurseries offer the opportunity to provide better care for the seedlings during their early growth [34].

4.2 Characteristics of a good nursery

A good nursery is characterized by the following;

- The soil or growth media should be well-drained;
- There should be adequate sunlight;
- Availability of clean water for irrigation and;
- The nursery should be easily accessible and secure.

4.3 Types of nurseries

There are three types of nurseries in vegetable cultivation. These are the nursery bed (seed bed), seed box (containers) and the plug trays (V-type nursery) [35]. The nursery bed is the commonest method of raising seedlings at the nursery [32]. In this method, seeds of vegetable crops are nursed on either raised or sunken beds depending on the season. Raised beds are used during the rainy season to prevent waterlogging whiles sunken beds are preferred in the dry season to conserve moisture. Vegetable nurseries can also be set up in seed boxes or containers. The advantage of using seed boxes is that they can be moved to a safer place in the event of unfavorable weather conditions. The last type of nursery is the use of plug trays. The plug trays are specially designed plastic trays for raising seedlings of vegetable crops [34, 35].

With the plug tray, each seedling occupies a cell (hole) which reduces competition and ensures uniformity among the batch seedlings. They can also be used for commercial production of seedlings of vegetable crops that are not normally nursed for example the Cucurbits. This is possible because the rooting system of the seedling is not disturbed during transplanting. Plug trays come in different sizes with the different number of cells (holes). The type of tray to use depends on the vegetable crop. Vegetable crops with small-sized seedlings should be raised in trays with smaller cells while large seedlings should be raised in trays with larger cells [34].

4.4 Sterilization of growth media

To produce healthy vegetable seedlings, the growth media (soil or soilless media) should be sterilized to kill or reduce the population disease-causing pathogens. The growth media can be sterilized using heat from fire (conventional method), heat from steam, solar radiation, and chemicals (fumigants). In seed bed nurseries, the commonest method of sterilizing the soil is through heat sterilization by burning dry grass on the beds. The solarization method is mostly practiced in greenhouse nurseries. However, it can be used to sterilize the main field. In the method, the nursery bed is moistened and covered with transparent plastic sheets which allows sunlight to pass through and heat up the moistened soil. The steam sterilization method is mostly used in sterilizing growth media for seed boxes and plug tray nurseries. Chemical sterilizers (fumigants) are also used in sterilizing the nursery bed but their use is limited due to the non-availability and high cost of the fumigants.

4.5 Types of growth media

Seeds of vegetable crops can be nursed in soil or soilless media such as cocopeat (coconut husk), decarbonized rice husk, saw dust, vermiculite, perlite, peat soil [32]. Most of these soilless media do not contain plant nutrients and therefore should be mixed with well-decomposed manure or compost to supply nutrients to the seedlings.

4.6 Good nursery management practices

Irrigation or moisture management: This is very crucial at the nursery since inadequate moisture supply may cause the death of seedlings while the excess of it will create conducive conditions for the spread of diseases. Insect pest and diseases management: The use of recommended insecticides and fungicides have been the commonest method of controlling insect pests and pathogen at the nursery. Recently, the use of colored sticky pads have proven to be effective in controlling flying insects such as whiteflies and aphids at the nursery.

Weed management: Weeds are good competitors for productive resources such as water, sunlight, and nutrients, and for this reason they must be controlled at the nursery. In the case of seed beds, manual hand weeding can be done to control weeds. Pre-emergence herbicides can be used to control weeds before sowing seeds. However, care must be taken to avoid the residual effect of the herbicides on the vegetable seeds. In the case of seed boxes and plug trays, effective sterilization will kill most of the weed seeds. Hand-picking can be done to get rid of weed seeds.

Pricking out: Overcrowded seedlings in seed beds and seed boxes should be pricked out to reduce competition for productive resources. Pricking out is not required in plug trays since each seedling occupies a cell. Hardening off: It is one of the most important agronomic practices at the nursery. It refers to a gradual process of exposing seedlings to conditions in the field where they will be transplanted. It involves a gradual withdrawal of water and exposure of the seedlings to sunlight with the aim of reducing transplanting shock and increasing percentage seedling survival.

5. PROTECTED/SCREENHOUSE VEGETABLE CULTIVATION

This refers to growing vegetable crops in protected structures such as polyhouses, greenhouses, net houses, shade houses, net tunnels [32]. Protected vegetable cultivation technology became necessary due to the challenges associated with vegetable cultivation in the open field [36]. The protective structures serve as a mechanical barrier to biotic stresses such as insect pests and disease-causing organisms and rodents [37]. The structures also protect the vegetables from harsh environmental conditions such as rainstorms, scorching sunlight, sandstorm, high winds, etc. These protected structures differ in the level of control or automation of the growing conditions such as temperature and relative humidity. The commonest protective structures in Ghana are net houses mostly without temperature controlling devices. The increasing demand for the use of protective structures for vegetable cultivation is the high yield and improved quality of the product obtained under protected cultivation compared to open field cultivation [37, 38]. Additionally, certain vegetable crops are vulnerable to high rainfall or windy conditions and therefore only perform better when grown under protected structures. Growing vegetables in protected structures can extend the growing season and ensure all- year-round production and supply of vegetables since climatic and growing conditions can be optimized in the protected structures [36]. Protected structures such as greenhouses can double as a commercial nursery where high-quality disease-free seedlings of vegetable crops could be produced and supplied to farmers to boost vegetable production. The production of true-to-type (genetically pure) seeds of cross-pollinated vegetable crops is challenging during the main cropping season. However, vegetable seed production in protected structures helps maintain genetic purity without worrying about isolation distances.

6. GRAFTING TECHNOLOGY

Grafting refers to the union between two plant parts (scion and rootstock) to produce a single plant [32]. it is an environmentally friendly method of controlling soil-borne pathogens and improving the yield of susceptible vegetable cultivars [39]. Grafting technology was first deployed in Asia by grafting watermelon scions onto squash rootstocks [40]. Grafting technology was developed to curb the indiscriminate use of insecticides and fungicides by farmers in vegetable cultivation. Farmers controlled biotics stress in vegetable cultivation by spraying pesticides which raised health concerns. However, the use of resistant grafted varieties offers the best solution to biotic and abiotic stresses [41]. The rootstock is usually an adapted cultivar that is tolerant or resistant to either biotic or abiotic stresses and develops into the rooting system of the grafted plant. The scion which develops into the shoot of the grafted plant is most susceptible to the biotic and abiotic stresses and is selected for its desirable fruit characteristics and quality. Grafting, therefore, offers the opportunity to combine the desirable traits of the two plants into a superior grafted plant with resistance to biotic and abiotic stresses and increased yield [42]. Grafting technology was first adopted in vegetable production to control soil-borne diseases such as Fusarium wilt in the Cucurbitaceae and Bacterial wilt in the Solanaceae families [43]. However, grafting technology has been deployed to produce drought tolerant, nematode resistant, low temperature tolerant, heavy metal tolerant, high salt-tolerant, and nutrient efficient vegetable varieties [44–46]. Notwithstanding the opportunities grafted technology offers, there are challenges associated with the technology. Some of the challenges include;

- Grafting requires skilled personnel and is labor intensive.
- Difficulty in the selection of compatible scion and rootstock.
- Grafting can affect the quality of vegetable fruits.
- Increased exposure of the grafted plants to disease infection.
- Grafting can cause excessive vegetative growth.
- There is an increased incidence of physiological disorders.
- Symptoms of incompatibility may show at later stages.
- Higher prices of grafted seedlings.

Several methods of grafting the scion onto the rootstock are available. However, the technique to use depends on the crop, size, and age of the seedling and the compatibility between the two plants [43]. A careful selection of the rootstock and scion is imperative in obtaining successfully grafted plants. Some of the techniques in grafting include cleft grafting, tube grafting, hole insertion grafting, and approach grafting [47, 48]. After grafting, the seedlings are allowed to heal and harden before being planted in the field. For successful healing, the grafted seedlings are kept in a dark place with high relative humidity. These conditions in the healing environment promote early healing and survival of the seedlings. Grafting technology has proven to be a reliable control method for biotic and abiotic stresses in vegetable cultivation and can be adopted as a control method to manage the challenges posed by these stresses to the vegetable sector in Ghana.

7. INFORMATION TECHNOLOGY IN VEGETABLE PRODUCTION IN GHANA

The rapid evolution of information technology (IT) is revolutionizing agriculture in a way that appreciably reduces risks associated with vegetable cultivation. Areas of IT application include diseases and pests' surveillance and management; soil water, irrigation, and fertility; weather alerts and forecasting; marketing; training and agricultural extension; transport and storage among others. Although great strides have been made there in some of these areas, others are still at the explorative phase and accuracy might not be as expected. For example, the accuracy of mobile applications for detecting many insect pest species are very low to be reliably used by an ordinary farmer. Plant symptoms, however, reliably give the type of group of pests and recommend good intervention. The identification of diseases with mobile phone and computer applications is more precise.

7.1 IT in climate smart vegetable production systems

Climate change resulting in a change in weather pattern from what has been established as the norm for centuries greatly influences agricultural productivity, especially vegetables. This is because the majority of vegetables are less hardy hence have less tolerance to drought and high temperatures. For this reason, the use of IT tools helps mitigate some of these adverse effects. There are a number of IT platforms that forecast the weather for specific locations. These platforms can forecast rainfall, relative humidity, temperature, ultraviolet index, wind speeds for about 2 weeks ahead of time. This weather information can assist in planning and determining the appropriate planting time over the stipulated period. The Ghana Meteorological Agency (GMet) also provides a yearly forecast for the onset of rain after the dry season, rainy months, intensity, and abatement for all the major ecological zones. With these information, farmers can schedule when to plant a particular vegetable under rain-fed conditions. Some radio stations and television stations also provide daily rainfall and temperature in the mornings or evenings. These help in daily farm schedules.

There are also automated systems for greenhouse vegetable production. Soil moisture, ambiance humidity, and temperature are remotely assessed and adjusted appropriately for optimal growth of crops. This efficiently mitigates against unpredictable climate variability, thereby facilitating vegetable production at all times.

Climate change also has the propensity of changing pests and diseases phenology [49] which may come at the blind side of the vegetable grower. Certain pest populations and diseases are known to surge at specific months of the year due to their accompanying rainfall and humidity patterns [50]. However, erratic rainfall patterns can disrupt these established times thereby shifting pests and diseases dynamics. For example, thrips and spider mites' outbreaks in vegetables are associated with droughts [51]. On the other hand, the surge in fungal diseases such as leaf blight and anthracnose are associated with high rains [52]. Based on these premises, deductions from weather forecasts help predict what pests and diseases to expect at any stage of vegetable cultivation. This foreknowledge allows the vegetable grower to scout and implements interventions as early as possible in order to reduce crop losses and also cut down the cost of management.

7.2 Pests and diseases identification and management tools

According to [53] estimated 20–40% of global crop production is lost to pests and crop diseases cost about $220 billion. At the local level pests and diseases has the potential to cause substantial loss and in extreme cases absolute loss, depending on the type of pest or disease, tolerance level of crop variety, and stage of infestation. Traditionally, several approaches have been used to identify and manage pests and diseases. These include experiential knowledge by the farmer as the crop is cultivated over time, acquaintance with the experienced farmer, agricultural extension, and farmer-researcher interactions. The digital space now offers a novel strategy to complement these traditional strategies. There are a number of mobile phone and portable computer applications that allow vegetable growers to identify plants, diseases, and pests as well as possible management options. A user takes and uploads a close-up photo of the pest or plant with signs of pest or disease infestation. The application matches the uploaded photo with a repository for the best match. It must be emphasized that identifications by these applications are not absolute and their accuracy is still being enhanced. For some applications, however, the user has the option of manually searching after the application has automatically filtered out unlikely candidates. When there are no search matches or the user deemed the results as inaccurate, some applications further allow the photo to be forwarded to a pool of experts. Experts then give feedback when correctly identified.

Drones or unmanned aerial vehicles (UAVs) are gradually being used in pests, diseases, and weeds' surveillance as well as water stress monitoring in agricultural production in recent times. Multispectral cameras attached to the drone scan the cropped field and based on color reflectance due to crop response to stress, such as pest or disease infestation, are able to identify these stresses. The advantage of this over traditional scouting is that infestation or stress can be detected way before they become visible to the naked eye. These IT systems facilitate early intervention and are also cost-effective [54, 55].

Another IT pest predictive and advisory IT platform is based on established degree-days of specific insect pests – rates of development of insects and many organisms are dictated by temperature, and require a fixed amount of heat to develop from one stage to another. This physiological development time is quantified into degree-day, and this is used to make predictions and recommend management actions [56]. This is well-patronized platform in many of the developed countries. The user key in some basic information such as type of crop, planting date, and location, and the system will generate pest management timelines to follow. In Africa, the degree-day model for pest management is being piloted on selected pests including the American tomato leaf miner (Tuta absoluta) in Kenya, Ghana, and Zambia under the Pest Risk Information SErvice (PRISE) (www.prise.org). The development of such IT tools to cover more vegetables and their associated pests will help rationalize the use of insecticides leading to minimizing residues in the produce.

7.3 Extension services

It is important in raising agricultural productivity, improving food security and nutrition as well as reducing poverty among the poor. Agricultural extension mediates these by extending research advancements and new knowledge in agriculture to end-users, particularly farmers. Although agricultural extension has been in existence in Ghana for over a century, its expected impact on vegetable production is still lagging. This is due to the high extension officer to farmer ratio, low funding, and poor farm accessibility. Vegetable production is, however, a sector that requires constant information exchange among the actors. It is also through a good extension system that new IT tools that enhance vegetable productivity can be introduced to farmers. Traditionally, extension practitioners physically interact with farmers and thus are restricted by time and space. This is where the e-agricultural system, which is more versatile is very helpful. E-extensions involve using the power of online networks, computer communications, and digital interactive multi-media to facilitate the dissemination of agricultural technology. There are a number of tools and techniques available for e-extension and they include personal computers/laptops, mobile phones, USSD technologies, call centers, Twitter interactive white boards, Facebook, WhatsApp, interactive voice response systems, TV, radio, SMS, etc. The current wide usage of mobile phones has made it the most preferred medium for the transfer of new agricultural technology to farmers. Through WhatsApp groups, for instance, an extension officer can connect with several farmers and other value chain actors for the exchange of information.

Scientific Animation Without Borders (SAWBO) is a digital animation platform that uses animation in addressing constraints and teaching good agricultural practices. It allows experts to design video animation to address a defined constraint, including pests and diseases. It can also be designed to target a specific audience, and thus audio could be in any dialect (https://sawbo-animations.org). SAWBO is a handy IT platform for the easy delivery of agricultural extension services to vegetable producers.

7.4 Postharvest handling technologies of vegetables

Majority of vegetables physically deteriorates quickly after harvesting due to the inherent rapid breakdown of tissues, physiological changes such as respiration, transpiration, etc. Good postharvest management practices and facilities are required to increase the shelf life of vegetables. Facilities such as harvesting and collection tools, containers and packing, rapid and cold chain transportation, and commercial storage are key to a successful vegetable postharvest value chain. Some companies are leveraging on IT solutions to address some of these postharvest management nodes such as offering cold transportation and storage for a network of farmers **(Table 1)**.

High temperatures and sometimes low humidity in the markets hasten the deterioration of vegetables. Some tropical countries have successfully used low-cost zero-energy cool chambers for temporary storage, pending sales, to prolong the shelf life of vegetables [58]. Piloting this technology in the traditional vegetable markets should be considered.

7.5 Digital marketing and use of social media to enhance the visibility of vegetable producers

Digital marketing, according to [59] is the use of the internet, mobile devices, social media, search engines, and other channels to reach consumers. It is considered a new means of marketing products and services through understanding and keeping abreast of customer behavior, which tends to be different from traditional marketing. Digital marketing is generally interactive and customers can give feedbacks. Customers are attracted via email, content marketing, search platforms, social media among others. The focus here is social media since little skills are required, cheap, and easy to use, which is more likely to be easily acceptable to the majority of Ghanaian or African farmers.

Many social media users in Ghana and many countries in sub-Saharan Africa tend to focus on the entertainment aspect of these powerful electronic tool. Statuses of millions of users on Facebook, WhatsApp, Snapchat, etc. are flooded with funny media, selfies, and all manner of commentaries. Considering that there are about 2.5 million agricultural households in Ghana with a population of 11.3 million and 55% mobile phone adoption, leveraging on social media can improve productivity along the agricultural value chain. In terms of marketing in the food chain, only the fast-food subsector has taken advantage of social media in marketing their products, especially in the urban areas. However, vegetable sales and consumption are steadily increasing because many people are actively increasing vegetable proportion in their diet because of their health benefits. This means the vegetable supply chain covering agro-input selling, labor services, transportation, and marketing needs to explore social media. This could be simple advertisements on Facebook wall, Instagram, statuses of other social media handles, and more importantly social media group platforms, which come at practically no cost. Social media group platforms such as that of the old school, church, clubs, etc. may have celebrities or influential persons. These personalities may agree to freely share advertisements on their private social media handles to catch the attention of their huge followers. The pay services by these social media outlets are also available to market vegetables.

Many consumers are skeptical about sources of vegetables consumed because of practices such as the misapplication of pesticides and the use of contaminated water. To allay fears of consumers, vegetable producers can set themselves apart by following the best practices outlined in this book and making short videos of their production practices. These videos could be shared on YouTube, Facebook, WhatsApp, etc. Agro input (seeds and fertilizer) tracking to authenticate their genuineness, effectiveness with the aid of IT is provided by some IT companies. Similar tracking of source and production information of agro produce is also available (**Table 1**). These give credibility and consumer satisfaction to the agro produce, giving it an edge over others. The national e-agriculture and other private IT platforms that vegetable growers can benefit from are listed in **Table 1**.

Table 1. Information technology agricultural platforms available to vegetable growers in Ghana.

IT platform service provider	Available services	IT services formats	Link
AcquahMeyer Drone Tech	Agricultural drone services for pests & diseases management		https:// amdronetech. com
Digital Agricultural Innovation Hub (DAIH)	4-in-1 portals/ hubs on marketing, extension, scientific article repository and technological repository	Mobile App (Agritech Advisor), web portal	http:// technologies. csirgh.com , http:// csirspace.csirgh. com, http:// kuafo.csirgh. com , http://agritech. csirgh.com
Agroseal	Commodity trading/marketing and training for smallholder farmers	Web portal	https://agroseal. com
AgroCenta	Marketing, produce transportation	IVR, SMS	https://agrocenta. com
CABI Plantwise	E-resources and MoFA extension staff field visits	Mobile app, Web portal, plant doctor services in collaboration with MoFA	https://plantwise. org
CRI TV	Dissemination of CSIR developed agro packages	Web portal, online TV channel	https://critivgh. com

Complete Farmer	Crowd farming, Marketing	IoT	https:/ completefarmer. com
Coologi-Green Tech	Cold storage		https://coologi.com
CSIR Soil Information	Soil profile data (physical, chemical, suitability, fertility) for all locations in Ghana	Web portal	www.csirsoilinfo. org
DigiExt	Remote sensing/ drone for pest & disease management, marketing, agric machinery rentals	SMS, call service	https://digiext.com
ESOKO	Vodafone farmer's club: farming advice, weather updates, market price and free calls between members (only Club's dedicated Vodafone SIM); data collection & digitization; biometric profiling; analytics	Call center, Interactive Voice Response System (IVR)	https://esoko.com
Farmerline	Input distribution, farmer training, weather updates, and market prices	SMS, Android, voice messaging	https://farmerline. co
FarmCap	Provides access to land, funds, agro inputs, weather & farm data	IoT, Web portal	https://farmcap.co
Farm Cure	Pest surveillance & management	Mobile app, web portal	https:// farmcuregh.com
Farmklass	Solar irrigation	Web portal	https://farmklass. com
Ghalani	Agro data management tool such as costs & revenue	Mobile app, web portal	https://ghalani. com
KaraAgro AI	Diseases, pests, nutrient & water deficiency diagnosis and management	Mobile app, web portal, drone services	https://karaagro. com
Khalmax Robotics	Robots for automation of farming activities, training	Web portal	https://khalmax softwaresystems com
Kwidex	Crowdfunding for agro projects	Web portal	https://kwidex.com

Ministry of Food and Agriculture E-agriculture	Up-to-date information on crop and animal production, market prices, farm management techniques and practices, and traceability in the Ghanaian agriculture	Farmer audio library/ IVR, E-field extension where officer collects field data for upload (toll-free 1848 and 30037), E-learning and resource center	https://e-agriculture. gov.gh
Okuafo Foundation	Diseases surveillance, identification, and management	Mobile app	https:// okuafofoundation org
QualiTrace	Agro input & produce tracing and anti-counterfeiting	Mobile app	https:/ qualitracegh. com
Techshelta	Greenhouse automation & management, marketing, training	Web portal, IoT	https://techshelta. com
TrusteeFarm	Agro produce traceability, marketing	Blockchain, QR/ USSD codes)	https:/ trusteefarm. com
Trotro Tractor	Connect farmers to tractor operators	SMS, Web portal	https:/ trotrotractor. com

8.ADVANCEMENTOFTECHNOLOGIESINVEGETABLEPRODUCTION IN GHANA

This section gives a description of advancement technologies in vegetable production in Ghana based on genetic improvement, soil fertility improvement, Irrigation as well as pest and diseases management practices. It also shows the characteristics of research and farmers' involvement in advanced technologies in vegetable production.

8.1 Genetic improvement

Crop genetic improvement target is by using diversity collections to identify new genes and gene combinations that can be introduced into crops, in a natural way using conventional breeding techniques without using Genetic Modification. It aims at increasing the frequency of desirable genes and decreasing the frequency of undesirable genes in a human perspective. Vegetables are a key component of every meal in Ghanaian and significantly contribute to improving the livelihood of its growers. In spite of the significance of vegetables in Ghana, past breeding programs have been disorganized and few developments of new varieties that meet the needs of consumers as well as environmental stresses have been done. Also, there has been advancement in a few vegetables such as tomatoes, garden eggs, Okra, and peppers. Generally, there are advances in areas of vegetable production constraints in Ghana, which mainly focused on the collection, evaluation, and screening of germplasm against biotic and abiotic stresses as well as morphological and agronomic characterization. The established variability and the outcomes of the evaluations against the various biotic and abiotic stresses have not been fully utilized in the development of new varieties.

8.2 Soil fertility improvement

Soil fertility constraints to crop production are now recognized as the major obstacles to food security and agro-ecosystem sustainability in Sub-Saharan Africa of which Ghana cannot be excluded [60]. There is therefore the need for improvement in soil organic matter for sustainable cropping. Fallowing is the oldest method of restoring soil fertility and productivity in traditional cropping systems. The longer the fallow period, the higher the level of fertility generated. However, population pressure and land fragmentation have led to a shortening of the fallow periods in several parts of the country [61]. According to [62], farmers fallowed farmlands for 1–3 years and only about 12.3% of the farmers maintain some fallow with the fallow duration skewed towards the lower limit of 1–2 years. However, due to the rapid growth of population, there is a scarcity of cultivated land for this practice. Currently, the use of both organic and inorganic fertilizers is generally adopted by vegetable farmers' during production. However, conservation agriculture has been one of the techniques introduced to help mitigate depletion of soil nutrients, conserve water, minimize soil erosion, reduce land degradation and boost farm productivity [63]. Fundamentally, Conservation agriculture includes minimum tillage, crop rotation, and organic soil cover [64]. Minimum tillage decreases the risk of land degradation and retains the soil structure by reducing the intensity of soil disturbance and this has been adapted by many vegetable growers in Ghana. [62] Indicated that the adoption of sustainable agricultural technologies remains the route through which developing nations could combat poverty and attain food security.

8.3 Irrigation

Ghanaian farmers traditionally rely on rainfed cultivation to grow their crops. However, between the 1960s and the year 1980, growth of irrigation started and a total of approximately 19,000 ha of irrigated land had been developed and in 2007 the area in irrigation had expanded to 33,800 ha. Irrigation systems observed in Ghana are mainly initiated and developed by the Ghanaian government or various nongovernmental organizations (NGOs), and emerging systems, which are initiated and developed by private entrepreneurs and farmers. Ghana Irrigation

Development Authority (GIDA), controlled by the Ministry of Food and Agriculture manages irritation and is the only public organization connected directly to irrigation development and management in the country. GIDA has developed 22 irrigation project districts of varying sizes covering a total area of 8,800 ha across the country. Since its establishment, GIDA has developed and managed public irrigation systems utilizing government subsidies, and public funds to cover staff costs. As part of structural adjustment, GIDA's budget was dramatically reduced and that resulted in difficulties in managing all the 22 public irrigation districts sustainably [65]. However, in order to maintain irrigation activities, a "Participatory Irrigation Management (PIM)" was introduced, whereby beneficiary farmers and others could manage the irrigation facilities during their vegetable and other crop production. Generally, the watering cans, buckets, motorized pumps with hosepipe, surface, and sprinkler irrigation methods, are mostly being used by farmers in the vegetable production areas in the country [7, 66]

8.4 Organic farming technology

The Ghana Government accepts the premise that national food security can only be realized through the wide-scale introduction of modern commercial agriculture and that such modernization will spearhead the development of the economy [67]. Health concerns with respect to vegetable production have meant that organic farming is gaining popularity in many countries, including Ghana. The factors influencing the adoption of organic vegetable technology in the country are based on extension contacts and farmers' engagement in training workshops, formation of farmer-based organizations (FBO), and farming experience over the past years. Organic vegetable farming can be used as a strategy to improve food production and reduce poverty in the country.

8.5 Pest management

The advent of synthetic insecticides revolutionized how insect pests were managed through a leap from suppression to "kill them all". Increasing reports of insect pests building resistance to these insecticides prompted the adoption and promotion of all available management measures through careful planning. These complementary options include natural enemies, microbials, botanicals, cultural practices, and synthetic insecticides. Similarly, the management of diseases solely based on synthetic chemicals has shifted to the adoption of complementary approaches.

Thus, current pests and diseases management strategies involve the adoption of a multifaceted approach; integrated management. The high vulnerability of vegetables to pests and diseases makes this aspect of the production very critical. For this reason, developing new complementary pests and diseases management technologies is crucial for the sustainable yield of vegetables. Some of these technologies accessible to the Ghanaian vegetable farmer are discussed below:

Microbial pesticides: They contain one or more microorganisms (e.g., fungus, bacterium, virus, nematode, or protozoan) as active ingredients. They may be target-specific or broad-spectrum and tend to be environmentally friendly since they are naturally occurring. Common microbial pesticide formulations available for vegetables include *Bacillus thuringiensis, Metarhizium anisopliae, Beauveria bassiana.* Plant-incorporated protectants which involve incorporating genetic material into a plant (i.e. Genetically Modified Organism) to offer insecticidal property are not yet approved for use in Ghana.

Botanical pesticides: A number of plants-derived products have exhibited insecticidal activity against a number of pests of vegetables. The most popular are the neem products with Azadirachtin as the active ingredients. [68] Provided an exhaustive list of plants with insecticidal properties, including those suitable for vegetables.

Natural enemies: Pests are normally perceived by many as inimical organisms that just increase in number without any natural check. On the contrary, pests are not completely immune from the adverse effects of other animals. There are a good number of animals, predators and parasitoids, that attack and kill pests in every vegetable production system. Predators seize and kill pests while parasitoids normally inoculate their immature forms into different developmental stages (eggs, larvae, pupae, or adult) of pests. All these help suppress pest populations below the economic threshold to maintain a balanced production system. Although naturally, pest populations may surge up irrespective of the natural enemies, anthropogenic disturbances such as the use of synthetic insecticides, monocropping, etc. are the main sources of pest outbreaks. Strategies to boost natural enemy populations and their attendant services include intercropping, mixed cropping, crop rotation, and leaving strips of flowering or natural vegetation along boundaries of vegetable fields.

Pheromone lures: This employs the use of pests' own chemical communication against them. The most commonly communication chemicals used in agriculture are sex pheromones. Although a wide array has been developed, particularly for monitoring and decision making, only a few are available in Ghana. The common pheromone lures for vegetables in Ghana, mostly used at the research level are that of a tomato leaf miner (*Tuta absoluta*) and diamondback moth for tomato and cabbage respectively. Nevertheless, there is a great prospect for mass adoption.

Sticky traps: These are color, mainly yellow, flexible plastic cards with the sticky surface to trap flight-capable insect pests. They are relatively costly since they are scarce locally. However, homemade sticky traps can easily be created by cutting yellow plastic strips (10 × 25cm) smeared with grease or petroleum jelly. Traps are hung on a pole just above the height of the crop. Traps can be reused by scrapping off trapped insects when the surface is almost filled up. A number of the above strategies can be combined to suppress pests in vegetables, and the inclusion of synthetic insecticides is always considered as the last resort.

8.6 Diseases management

Many vegetables are prone to diseases and their impact on production can be dire. Causative agents of diseases range from fungi, bacteria, and viruses. Understanding of disease triangle which consists of the host, the pathogen, and the environmental condition is important in managing these problems. A break in the link of the triangle truncates the perpetuation of the disease [69]. Thus, effective management of diseases in vegetables involves the use of disease-free seeds and varieties that are tolerant/resistant to the major diseases that are prevalent in the production enclave. Although general climatic factors are beyond the control of the vegetable grower, a micro climate that reduces the survival of the causative pathogen can be provided. An example is using well-drained soil and avoiding weed growth on the cropped field. In situations where a vector is involved in the transmission of the disease, it is important to manage the pest. A typical example is white flies, which transmit viral diseases.

Other intervention, which has also become popular among growers is the use of pesticides such as fungicides, bactericides, etc. These may be applied as preventative or curative depending on the active ingredient and thus the time of application is very crucial.

9. CONTRIBUTION/PROSPECTS OF VEGETABLE PRODUCTION TO FOOD AND NUTRITION SECURITY IN GHANA

In Ghana, the promotion of food security has been centered on starch-based staple crops such as maize, rice, cassava, yam, sweet potato, plantain, etc. which are mainly dietary sources of energy. However, in recent times, the introduction and addition of the term "nutrition security" has reshaped the concept and created awareness on the dietary importance of vegetables [70]. Intake of a diverse range of food categories in the right proportions is required for a healthy living. Food and nutrition security can, therefore, never be achieved without realizing and acknowledging the roles of vegetables in our diets. Their rich source of water, protein, minerals, vitamins, and other important phytochemicals champions the need for dietary diversity for good health, protecting us from heart disease, stroke, cancer, and diabetes [71]. Thus, vegetable intake is one of the affordable approaches to achieve both food and nutrition security.

In Ghana, the vegetable sector has recently recorded a spike in production [72] and consumption [73] of both indigenous and exotic vegetable crops. This typically shows the roles of vegetable production in improving the livelihoods of people involved in its value chain. It creates important economic opportunities for smallholder farmers particularly in the rural communities as well as fragmented areas in urban and peri-urban and thus, offering opportunities for poverty reduction [19]. It is worth noting that, income generated from the vegetable sales helps these farmers to purchase other foodstuffs which are equally important in achieving a healthy diet. Farmers can explore and adopt a mixed cropping system, thereby, integrating vegetable crops into the cultivation of the major staple crops or shifting to solely vegetable production to maintain economic or financial stability [70]. Vegetable production may offer a profitable business opportunity. The recent introduction of vegetable production under controlled environments such as greenhouses has improved the availability of vegetables as a result of year-round production, as well as the quality. Although the initial cost of production is high, farmers who go into greenhouse production with good management practices and associated high yields increase their returns per unit area of production compared with those into open-field production [74]. This system has offered an alternative attractive source of employments for the youths who previously disregarded farming. For instance, graduates with engineering backgrounds can explore areas such as design, construction, and installation of greenhouses and accessories. Others with pre-requisite training in greenhouse vegetable production can take managerial roles. In addition, vegetable production offers huge cash income for market-oriented vegetable farmers particularly women who are the largest group of beneficiaries. There is also a huge opportunity in the export of vegetables. The greenhouse cultivation offers this opportunity to farmers to enable them to meet the strict export requirements of vegetables and leverage to maximize the profit margin of the farmers as well as foreign exchange for the economic growth of the country. Production of vegetables under greenhouse conditions limits and/or avoids the uncontrolled use of agro- chemicals which pose serious threats to food safety and security [75]. The vegetable production under controlled conditions has contributed to introducing the new scope of research and teaching in the tertiary levels. For instance, some universities have already revised their course contents or curriculum and mounted an entire course and program at undergraduate and postgraduate levels. This is the surest way of promoting sustainable vegetable production and the indirect approach of ensuring sustainable food and nutrition security. Thus, students who successfully attain training in diverse areas in vegetable production become the main actors in the vegetable value chains such as the vegetable farmers, managers, exporters, policymakers, and researchers, etc.

Moreover, vegetable production could serve as a feedstock for various industrial productions. Unfortunately, the opportunities associated with vegetable production have not been utilized adequately. There are so many recipes of vegetable juices that can be developed and commercialized in Ghana as done in developed countries. Improved processing and packaging with excellent hygienic conditions will ease in preparation and consumption of vegetables. Vegetables such as cabbage, lettuce, carrots, spring onions can be chopped and packaged and made available in groceries. Industrialized vegetable production, processing, and marketing could offer huge employment opportunities to the youth. In addition, this will eventually promote vegetable availability, accessibility, and utilization thereby contributing to food and nutrition security.

In Ghana, there are underutilized vegetable species such as Turkey berries (*Solanum torvum*) which are being used for the preparation of local delicacies as well as medicinal purposes, especially in rural and peri-urban areas. Such vegetables may be rich in essential secondary metabolites which could be harnessed to promote daily consumption thereby addressing or improving food and nutrition security [76]. However, there is little, or no research works on these vegetables which are growing distinct whiles others are scattered in various localities and highly exposed/ endangered to destruction by mechanical weeding, herbicide applications, and pest infestations. Therefore, considering their nutritional and medicinal properties, these underutilized vegetables could be scientifically explored and promoted to improve the livelihoods of the local dwellers using them as food or medicines [76, 77]. In addition, advanced research could lead to the commercialization and industrialization of these vegetable crops contributing to various economic benefits.

10. CONCISE SUMMARY

The chapter examined cutting-edge technologies in vegetable cultivation, current trends, and the issues related to production. The discussion focused on the varieties of vegetables and the innovative technology for vegetable cultivation in Ghana. It delineates sophisticated technology in vegetable cultivation in Ghana, focusing on genetic enhancement, soil fertility augmentation, irrigation, and pest and disease management strategies. Production is predominantly characterized by human labor for the majority of agricultural activities; although, recently, some larger-scale farmers are integrating a degree of mechanization. Cultivation typically occurs on a rainfed system, permitting output in either one or two seasons annually, contingent upon the farm's geographical location within the country. The dependence on rainfall for vegetable cultivation, coupled with the unpredictable rainfall patterns in recent years attributed to climate change, leads to suboptimal yields. Limited output is unable to avert the scarcity of some vegetables during certain seasonal periods. The mitigation of agriculture's reliance on rainfall has resulted in the establishment of irrigation facilities to facilitate off-season or practically year-round vegetable cultivation.

Challenges in vegetable cultivation in Ghana encompass a restricted variety selection resulting from the demise of indigenous cultivars. The absence of a functional vegetable seed system, in contrast to other crops, impairs the availability of quality seeds. The dependence on imported seeds for common vegetables is due to insufficient research on indigenous varieties. Particularly infrastructure challenges.

Poor irrigation and processing facilities, along with inadequate roads, impede sector growth. The perishable characteristics of most vegetables necessitate their timely delivery to markets and customers to minimize post-harvest losses. Contamination resulting from the overuse and misuse of agrochemicals in vegetable agriculture presents an additional obstacle to vegetable production in Ghana. Innovative technologies for vegetable growing encompass nursery management techniques, such as the sterilization of growth media and grafting methods, as well as the utilization of protected environments or screenhouses. The application of information technology in vegetable cultivation in Ghana is another aspect. The swift advancement of information technology in climate-smart vegetable production systems increasingly incorporates drones, or unmanned aerial vehicles, for the observation of pests, illnesses, and weeds. Digital marketing and the utilization of social media to augment the exposure of vegetable growers and postharvest management solutions for vegetables are gaining traction. This chapter clearly indicates that numerous agro-techniques exist that could revolutionize Ghana's vegetable economy. Nonetheless, appropriate transformational mechanisms and a conducive atmosphere must be established to facilitate the adoption of these technologies. Technology serves as a catalyst for swift advancement; so, enhanced agricultural technology must be effectively utilized. This would not only guarantee food security but will significantly enhance the livelihoods of smallholder farmers who have adopted vegetable production as their primary source of sustenance..

REFERENCES

1. La S. Vegetable gardening in Accra. The Ghana Farmer. 1963;**1**(6):228-230
2. Danso G, Drechsel P, Obuobie E, Forkuor G, Kranjac-Berisavljevic G. Urban vegetable farming sites, crops and cropping practices. In: Drechsel P, Keraita B, editors. Irrigated Urban Vegetable Production in Ghana: Characteristics, Benefits and Risk Mitigation. 2nd ed. Colombo, Sri Lanka: International Water Management Institute (IWMI); 2014. pp. 7-27. DOI: 10.5337/2014.219
3. Keraita B, Cofie O. Irrigation and soil fertility management. In: Drechsel P, Keraita B, editors. Irrigated Urban Vegetable Production in Ghana: Characteristics, Benefits and Risk Mitigation. 2nd ed. Colombo, Sri Lanka: International Water Management Institute (IWMI); 2014. pp. 74-88. DOI: 10.5337/2014.219
4. Bellwood-Howard I, Chimsi E, Abdul-Ganiyu S, van Veenhuizen R, Amoah P. Urban and Peri-urban agriculture in Tamale: A policy narrative. 2015. Available from: http://www.urbanfoodplus.org/fileadmin/user_ upload/

Pictures/Publications/Policy_ narrative_UPA_in_Tamale_2015_11.pdf

5. Amoah P, Lente I, Asem-Hiablie S, Abaidoo RC. Quality of vegetables in Ghanaian urban farms and markets. In: Drechsel P, Keraita B, editors. Irrigated Urban Vegetable Production in Ghana: Characteristics, Benefits and Risk Mitigation. 2nd ed. Colombo, Sri Lanka: International Water Management Institute (IWMI); 2014. pp. 74-88. DOI: 10.5337/2014.219
6. Asomani-Boateng R. Urban cultivation in Accra: an examination of the nature, practices, problems, potentials and urban planning implications. Habitat International. 2002;**26**:591-607
7. Obuobie E, Keraita B, Danso G, Amoah P, Olufunke OC, Raschid-Sally L, et al. Irrigated Urban Vegetable Production in Ghana: Characteristics, Benefits and Risks. Accra, Ghana: IWMI-RUAF-CPWF; 2006
8. Ministry of Food and Agriculture. Ghana Irrigation Development Authority (GIDA). 2021. Available from: http://www.e-agriculture.gov.gh/index.php/about-mofa/subvented- organisations/ghana-irrigation-development- authority
9. Communications Bureau. President Akufo-Addo Inspects Completed Namoligo 1-Village-1-Dam Project. The Presidency, Republic of Ghana. 2020. Available from: https://presidency.gov. gh/index.php/briefing-room/news-style-2/1740-president-akufo-addo-inspects-completed-namoligo-1-village-1-dam-project
10. Sore A. 'One Village, One Dam': Government yet to improve on poorly constructed dams. MyJoyOnline. 2021. Available from: https://www. myjoyonline.com/one-village-one-dam-government-yet- to-improve-on-poorly-constructed-dams/
11. Bortey HM, Banful B, Olympio NS. Quality of farmer-saved tomato seeds and its effect on fruit yield in Ghana. Ghana Journal of Horticulture. 2011;**9**:25-33
12. Etwire PM, Atokple IDK, Buah SSJ, Abdulai AL, Karikari AS, Asungre P. Analysis of the seed system in Ghana. International Journal of Advance Agricultural Research. 2013;**1**:7-13
13. Sperling L, Cooper HD. Understanding seed systems in seed security. In: Improving the Effectiveness and Sustainability of Seed Relief. In: Proceedings of a Stakeholders Workshop in Rome. Rome: Food and Agricultural Organization; 2003. p. 40
14. Anonymous. Catalogue of Crop Varieties Released and Registered in Ghana. Accra, Ghana: Directorate of Crop Services, Ministry of Food and Agriculture; 2019. p. 81
15. Krausova M, Banful AB. Overview of the Agricultural Input Sector in Ghana. Development Strategy and Governance Division. IFPRI Discussion Paper

01024;2010

16. Padi M. Food crop farming and the climate in Southern Ghana. Acta Scientific Agriculture. 2018;**2**(7):23-26
17. Nyadzi E, Werners SE, Biesbroek R, Ludwig F. Techniques and skills of indigenous weather and seasonal climate forecast in Northern Ghana. Climate Development. 2021;**13**:551-562
18. Solankey SS, Singh P, Shukla P, Kumar R. Vulnerability of vegetable crops to climate change. National seminar on "Recent Advances in Agriculture for Sustainable Rural Development (RAASRD-2019)", at VKSCoA, Dumraon (BAU, Sabour), Bihar; 2019
19. World Bank. Project information document (PID) – Concept stage. Report No.: 105933; 2016
20. Quansah JK, Escalante CL, Kunadu PH, Saalia FK, Chen J. Pre- and post- harvest practices of urban leafy green vegetables farmers in Accra, Ghana and thei association with microbial quality of vegetables produced. Agriculture. 2020;**10**(18):1-14
21. Sambo BE, Ahmed A, Omotugba JT, Adedapo JO, Shekwolo G. The challenge of infrastructure development on preservation of perishable horticultural produce for sustained urban food nutrition in Metropolitan Kaduna, Nigeria. International Journal of Horticulture and Forestry Research. 2016;**1**(1):13-26
22. Yeboah S. Influence of condition of road transport infrastructure on rural agricultural development in the Jaman South District. MSc. Thesis submitted to the School of Graduate Studies, Kwame Nkrumah University of Science and Technology in partial fulfilment of the requirements for the degree of MSc in Development Policy and Planning; 2015
23. Nyo AK. Inadequate infrastructure: The bane behind food loss and food security in the savannah zone of Ghana. Journal of Developments in Sustainable Agriculture. 2016;**11**:43-47
24. Okoffo ED, Mensah M, Fosu-Mensah BY. Pesticides exposure and the use of personal protective equipment by cocoa farmers in Ghana. Environmental Systems Research. 2016;**5**(17):1-15
25. Aidoo AK, Arthur S, Bolfrey-Arku G, Mochiah MB. Pesticides abuse and health implications in Ghana: A review. International Journal of Environment, Agriculture and Biotechnology. 2019;**4**(3):874-883
26. Dari L, Addo A, Dzisi KA. Pesticide use in the production of Tomato (*Solanum lycopersicum* L.) in some areas of Northern Ghana. African Journal of Agricultural Research – Academic Journals. 2016;**11**(5):352-355
27. Afari-Sefa V, Asare-Bediako E, Kenyon L, Micah JA. Pesticide use practices and perceptions of vegetable farmers in the cocoa belts of the Ashanti and Western Regions of Ghana. Advances in Crop Science and Technology.

2015;**3**(174):1-10

28. Drechsel P, Keraita B, (Eds.). Irrigated Urban Vegetable Production in Ghana: Characteristics, Benefits and Risk Mitigation. 2nd ed. Colombo, Sri Lanka: International Water Management Institute (IWMI); 2014. p. 247
29. Abaidoo R, Keraita B, Drechsel P, Dissanayake P, Maxwell A. Soil and crop contamination through wastewater irrigation and options for risk reduction in developing countries. In: Dion P, editor. Soil Biology and Agriculture in the Tropics. Heidelberg: Springer Verlag; 2009. pp. 275-297. DOI: 10.1007/978-3-642-05076-3_13
30. Amoah P, Drechsel P, Abaidoo RC. Irrigated urban vegetable production in Ghana: Sources of pathogen contamination and health risk elimination. Irrigation and Drainage. 2005;**54**(1):49-61
31. Amoah P, Drechsel P, Abaidoo RC, Ntow WJ. Pesticide and pathogen contamination of vegetables in Ghana's urban markets. Archives of Environmental Contamination and Toxicology. 2006;**50**(1):1-6
32. Ratha KP, Rajwant KK, Tewari JC, Roy MM. Plant Nursery Management: Principles and Practices. Jodhpur: Central Arid Zone Research Institute; 2014. p. 40
33. Singh RR, Meena LK, Singh P. High tech nursery management in horticultural crops: A way for enhancing income. International Journal of Current Microbiology and Applied Sciences. 2017;**6**(6):3162-3172
34. Pandiyaraj P, Kumar YR, Vijayakumar S, Arindam S. Modern nursery raising systems in vegetables. International Journal of Agricultural Sciences. 2017;**9**(52):4889-4892
35. Singh B, Kumar M, Yadav HL. Plug-tray nursery raising technology for vegetables. Indian Horticulture. 2005;**49**(4):10-12
36. Chakraborty H, Sethi LN. Prospects of protected cultivation of vegetable crops in North Eastern Hilly Region. International Journal of Basic and Applied Biology. 2015;**2**(5):284-289
37. Sanwal SK, Patel KK, Yadav DS. Vegetable production under protected conditions in NEH Region: Problems and Prospects. Himalayan Ecology. 2004;**12**(2):1-12
38. Ganesan M. Effect of poly-greenhouse on plant microclimate and fruit yield of tomato. IE (I). J.-AG. 2004;**80**:12-16
39. Lee JM, Oda M. Grafting of herbaceous vegetables and ornamental crops. Horticultural Reviews. 2003;**28**:61-124
40. Kubota C, McClure MA, Kokalis-Burelle N, Bausher MG, Rosskopf EN. Vegetable grafting: History, use, and current technology status in North America. Horticultural Science. 2008;**43**(6):1664-1669
41. Rivard CL, Louws FJ. Grafting to manage soil-borne diseases in heirloom

tomato production. Horticultural Science. 2008;**43**:2008-2011

42. Kumar P, Rana S, Sharma P, Negi V. Vegetable grafting: A boon to vegetable growers to combat biotic and abiotic stresses. Himachal Journal of Agricultural Research. 2015;**41**(1):1-5
43. Lee JM, Kubota C, Tsao SJ, Bie Z, Hoyos Echevarria P, Morra L, et al. Current status of vegetable grafting: Diffusion, grafting techniques, automation. Scientia Horticulturae. 2010;**127**(2010):93-105
44. Albacete A, Andujar C, Pérez-Alfocea F, Lozano J, Asins M. Rootstock mediated variation in tomato vegetative growth under low potassium or phosphorous supplies. Acta Horticulturae. 2015;**1086**:147-152
45. Al-Razaq AHA. Grafting techniques in vegetable crops: A review. Plant Archives. 2019;**19**(1):49-91
46. Li Y, Tian X, Wei M, Shi Q, Yang F, Wang X. Mechanisms of tolerance differences in cucumber seedlings grafted on rootstocks with different tolerance to low temperature and weak light stresses. Turkish Journal of Botany. 2015;**39**:606-614
47. Lee JM. Advances in vegetable grafting. Chronica Horticulturae. 2003;**43**(2):13-19
48. Oda M. New grafting method for fruit-bearing vegetables in Japan. Japan Agricultural Research Quarterly. 1999;**29**:187-194
49. Collier RH. Pest and disease prediction models. Encyclopedia of Applied Plant Sciences. 2017;**31**:120-123
50. Doody A. Pests and diseases and climate change: Is there a connection?. Climate Adaptation and Mitigation, Environmental Health Biodiversity. 2020. Available from: https://www. cimmyt.org/news/pests-and-diseases-and-climate-change- is-there-a-connection/
51. Litskas DV, Migeon A, Navajas M, Tixier M-S, Stavrinides CM. Impacts of climate change on tomato, a notorious pest and its natural enemy: Small scale agriculture at higher risk. Environmental Research Letters. 2019;**14**(8):084041
52. Del R, Simpson MB. Climate Change in Mali: Expected Impacts on Pests and Diseases Afflicting Selected Crops. 2014. Available from: https://www. climatelinks.org/sites/default/files/ assset/document/Pests_and_Diseases_Crops_Sahel.pdf
53. FAO. New Standards to Curb the Global Spread of Plant Pests and Diseases. 2019. Available from: https:// www.fao.org/news/story/en/ item/1187738/icode/
54. Huang Y, Reddy KN, Fletcher RS, Pennington D. UAV low-altitude remote sensing for precision weed management. Weed Technology. 2018;**32**:1-5.

DOI: 10.1017/wet.2017.89

55. Lost Filho HF, Heldens BW, Kong Z, de Lange SE. Drones: Innovative technology for use in precision pest management. Journal of Economic Management. 2020;**113**(1):1-25

56. University of California Agriculture and Natural Resources. How to Manage Pests: Degree-Days. 2016. Available from: https://www.prise.org

57. Palmer T, Darabian N. Farmers' club; A mobile application service by Vodafone Ghana. GSMA mAgri Programme. 2017. Available from: www. gsma.com/mobilefordevelopment/ programmes/magri

58. Mangal JL, Dhyani A. Postharvest technology of fruits and vegetables. In: Dhyani A, Nautiyal S, Mangal JL, Gaur A, editors. Innovative Production Systems in Horticulture, Souvenir. Greater Noida: Institute of Horticulture Technology; 2013. pp. 36-40

59. Barone A. Digital marketing. Investopedia. 2020. Available from: https://www.investopedia.com/terms/d/ digital-marketing.asp

60. Sanchez PA, Jama B, Niang AI, Palm CA. Soil fertility, small farm intensification and the environment in Africa. In: Lee DR, Barrett CB, editors. Tradeoffs or Synergies. Wallingford, UK: CAB International; 2000.pp. 327-346

61. Quansah C, Drechsel P, Yirenkyi BB, Asante-Mensah S. Farmers' perceptions and management of soil organic matter-a case study from West Africa. Nutrient Cycling in Agroecosystems. 2001;**61**(1):205-213. DOI: 10.1023/A:1013337421594

62. Simtowe F, Kassie M, Diagne A, Silim S, Muange E, Asfaw S, et al. Determinants of agricultural technology adoption: the case of pigeon pea varieties in Tanzania. Quarterly Journal of International Agriculture.2011;**50**(4):325-345

63. Halbrendt J, Kimura AH, Gray SA, Radovich T, Reed B, Tamang BB. Implications of conservation agriculture for men's and women's workloads among marginalized farmers in the Central Middle Hills of Nepal. Mountain Research and Development. 2014;**34**(3):214-222

64. FAO. Towards the Future We Want: End Hunger and Make the Transitionto Sustainable Agricultural and Food Systems. 2012. Available from: http://www.fao.org/docre p/015/an894 e/ an894 e00.pdf

65. Miyoshi T, Nagayo N. A study of the effectiveness and problems of JICA's technical cooperation from a capacity development perspective: Case study of support for the advancement of Ghana's irrigated agriculture. In: Case Study Report on Capacity Development. Tokyo, Japan: Institute for International Cooperation, Japan International Cooperation Agency; 2006

66. Keraita B, Drechsel P, Huibers F, Raschid-Sally L. Wastewater use in informal irrigation in Urban and Peri-urban areas of Kumasi, Ghana. Urban Ag-

riculture Magazine. 2002;**8**:11-13

67. World Development Report. Agricultural for development. 2008. DOI: 10.1596/978-0-8213-6807-7
68. Bosu PP, Paintsil S, Cobbinah JR. Hand of Plants with Insecticidal Properties in Ghana. Kumasi: CSIR-Forestry Research Institute; 2017
69. Scholthof KB. The disease triangle: Pathogens, the environment and society. Nature Reviews Microbiology. 2007;**5**:152-156. DOI: 10.1038/nrmicro1596
70. Schreinemachers P, Simmons EB, Wopereis MC. Tapping the economic and nutritional power of vegetables. Global Food Security. 2018;**16**:36-45
71. Vincente AR, Manganaris GA, Ortiz CM, Sozzi GO, Crisosto CH. Nutritional quality of fruits and vegetables. In: Postharvest Handling. Academic Press; 2014. pp. 69-122
72. Abdulai J, Nimoh F, DarkoKoomson S, Kassoh KFS. Performance of vegetable production and marketing in peri-urban. Kumasi, Ghana. Journal of Agricultural Science. 2017;**9**(3):202-218
73. Azupogo F, Seidu JA, Issaka YB. Higher vegetable intake and vegetable variety is associated with a better self-reported health-related quality of life (HR-QoL) in a cross-sectional survey of rural northern Ghanaian women in fertile age. BMC Public Health. 2018;**18**(1):1-13
74. FAO. Unlocking the potential of protected agriculture in the countries of the Gulf Cooperation Council – Saving water and improving nutrition. Cairo. 2021;**2021**:216. DOI: 10.4060/cb4070en
75. Gil R, Bojacá CR, Schrevens E. Does optimized agrochemicals management help to reduce the environmental impact in tomato production? A comparative analysis between greenhouse and open field systems. International Symposium on Advanced Technologies and Management for Innovative Greenhouses: GreenSys. 2019;**1296**:1145-1152
76. Chacha JS, Laswai HS. Micronutrients potential of underutilized vegetables and their role in fighting hidden hunger. International Journal of Food Science. 2020;**2020**:1-5
77. Arowosegbe S. Preliminary domestication and cultivation efforts on some medicinally importance wild vegetables in Ado-Ekiti, Nigeria. Bio-Science Research Bulletin. 2013;**29**(2):101-107

8 Chapter

THE IMPACT OF MOSSES ON THE HYDRAULIC PROPERTIES OF VEGETABLE PLANTS IN CIREBON, WEST JAVA, INDONESIA

1. INTRODUCTION: AN OVERVIEW

Indonesia is among the several countries globally characterized by rich diversity. Indonesia is situated between the continents of Asia and Australia. This advantageous geographical position contributes to Indonesia's status as a global center of diversity and its recognition as a megadiverse nation. Each location cultivates distinct varieties of moss. Putri et al. [2] conducted research on mosses in forest regions, Cut Raihan et al. [1] in waterfalls, Jumiati [3] in watersheds, Zahara [4] in the Leuser ecosystem, and Sujatmiko and Vitara [5] in the UGM campus vicinity.

Moss is highly beneficial for the conservation of nature in forests, mountains, waterfalls, campus settings, and surrounding habitats. However, moss can be harmful if it proliferates alongside the plants we are producing. Moss proliferates in hydroponic systems.

Moss is frequently observed in hydroponic gardening systems. Without prevention and care, the proliferation of moss will become increasingly vigorous. Indeed, all the essential elements required for moss to thrive and develop are present in a hydroponic system, including nutrient-rich water, oxygen, and adequate sunlight.

Horticulture serves as a source of human nourishment, providing essential vitamins, minerals, and fiber. Horticulture encompasses a variety of vegetables, including mustard greens, spinach, and chaise. Vegetables are utilized to enhance the daily diet. Cultivating horticulture on soil necessitates extensive land and a substantial labor force, with outcomes frequently compromised by pests and diseases. Consequently, an innovation in vegetable cultivation utilizing hydroponics was introduced.

Hydroponic technology offers numerous advantages over conventional agriculture, including environmental sustainability, hygienically produced products, accelerated plant growth, consistent vegetable quality, enhanced yield, and improved health due to the absence of soil contamination from heavy metals and plastic waste.

2. THE HYDRAULIC MEDIA OF VEGETABLE PLANTS

Increasing vegetable production needs to be supported by various efforts, one of which is extensification with increasing use of non-agricultural land and agricultural intensification [6].

Hydroponics is a way of cultivating plants by utilizing water without using soil as a planting medium, so hydroponics is very important in meeting plant nutrients. In the hydroponic method, the function of the soil replacement medium is to support the plant and the most important thing is water which functions to dissolve nutrients that will be absorbed by plant roots.

Hydroponics requires less water and is more efficient than plants with soil media (conventional), so it is very good if applied to areas that have a limited water supply. This system does not require a large area of land, it is enough to use the yard of the house or narrow places in urban areas.

Rockwool or often also called mineral wool is an inorganic material made by blowing air or steam into melted rock. The result of this method is a type of fiber that has cavities with diameters between 6 and 10 micrometers. Rockwool can hold water and air in good amounts to support the development of plant roots [7].

Another advantage of planting with the hydroponic method is:

1. Eco-friendly.
2. Safe from pesticides.
3. More innovative.
4. Vegetables are fresher.
5. Harvest can be arranged.
6. Plants grow faster.
7. Free from nuisance plants/weeds.
8. Easy to harvest results.

3. BRYOPHYTA ON THE HYDRAULIC MEDIA

Moss (Bryophyta) is a group of lower plants that grow widely on land. Moss is a small plant that grows attached to the substrate (stones, trees, wood, and soil). The life of moss is influenced by environmental factors such as temperature, humidity and light [3].

Moss has green leaves and contains chlorophyll so that with the help of sunlight it can carry out the process of photosynthesis [4]. Leaf moss (musci) is an expan- seof leaf moss consists of one leaf moss plant that grows tightly and densely in groups to strengthen and support each other (**Figure 1**). This moss is not attached to its substrate but has rhizoids attached to its body. The body is generally erect, in the form of a thallus, leaves like scales that are dense, dense, and flattened or piled up. Moss forms a thallus in the form of a pillow or thick, and on the forest, the soil is often a layer resembling a velvet [8]. Leaf moss (moss) is the most famous moss plant. Leaf mosses consist of dense groups of mosses, which support each other.

Each plant incorporated in the bed is attached to the substrate by elongated cells or cellular filaments called rhizoids. Unfortunately, in this paper, the identification of mosses has not been carried out.

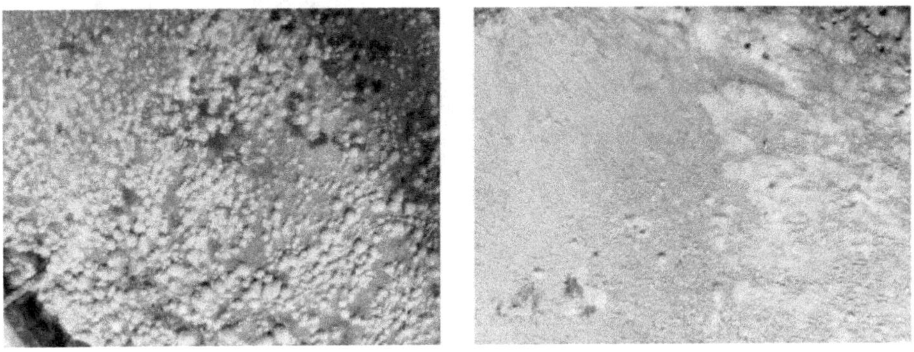

Figure 1. Moss on Hydrophonic Media in Cirebon, West Java, Indonesia.

4. THE EFFECT OF MOSSES ON THE HYDRAULIC MEDIA

In conditions with minimal moss, vegetable growth looks very optimal, with indications of wider leaves and dense stem conditions. On the other hand, in conditions where there is quite a lot of moss growth, the growth of kale plants looks not optimal, with indications of thinner leaves and poor stem conditions. Solid (**Figure 2**). The appearance of moss in a hydroponic system will be very difficult to prevent. Farmers cannot avoid the factors that lead to the appearance of moss, even at the beginning of growing seedlings.

The reason moss is not good for hydroponic plants

1. Growing moss needs nutrients, so moss will consume the nutrients available to plants. Thus the plants we plant will competefor nutrients with moss.

Figure 2. Moss in the pipe of Hidrophonic instrument.

Moss can reduce the level of acidity in water. Plants absorb nutrients well in a hydroponic system at a water ph level (water acidity) of 5.8 to 6.5. So in the presence of moss, the acidity of the water will be reduced. This will interfere with the plant's ability to absorb nutrients.

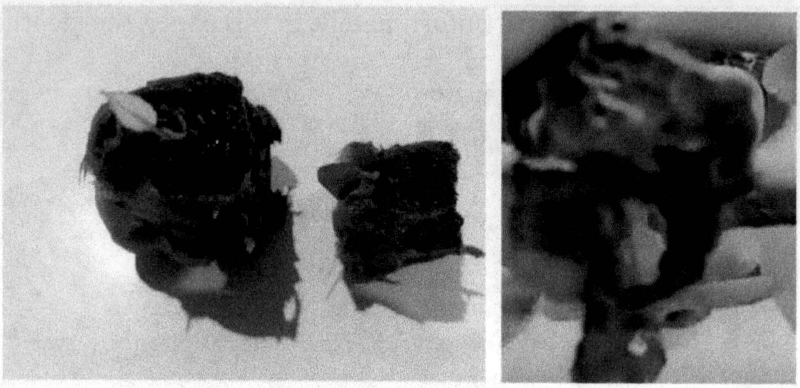

Figure 3. The death of the vegetable that grows together with the moss.

2. Lichens will consume oxygen in nutrient solutions at night and produce toxic products. This will interfere with plant growth.
3. Moss can produce an unpleasant odor. The vegetables we grow will smell less fresh (**Figure 3**)

5. OVERCOMING MOSS IN HYDROPONIC PLANTS

What must be done is to minimize or suppress the growth of moss. Things that can be done include cleaning gutters after each harvest. Cleaning the nursery gutters every time the seedlings move to the gutters (if the seedlings are growing using gutters). Drain and provide shade or cover on the reservoir. Using reservoirs and hoses that do not penetrate the sun. Minimize the gaps where sunlight enters the gutter.

Here we share tips that we do in our garden:

1. Make sure to clean the gully/bucket that we use until it's completely clean at EVERY HARVEST. Because even if there is only a small amount of algae left in the installation, it will spur the growth of more moss in the next planting and so on.
2. Make sure to periodically clean the installation pipelines (inlet and outlet installation networks) to avoid moss interference from inside. Usually, we do it every 6 months to 1 year (depending on the form of each system).
3. At the time of seeding, make sure that the sides of the rock wool are exposed to the sun as little as possible (but for plants you MUST get full sun).
4. We do this by making the Rockwool arrangement neatly lined up in our seedling installation. In this way, there are at least some sides of the Rockwool that are not disturbed by moss.

5. Make sure the root growth is faster than the growth of algae so that the roots are not covered by moss and disturbed the absorption process.
6. If it is already covered with moss, you can spray the roots with water with low pressure and be very careful. The moss will fall off by itself.
7. Close the reservoir. The nutrient liquid in the reservoir, which is full of nutrients, will be the right stimulus for the growth of moss on the walls of the reservoir.
8. Use a black hose. Just like the reservoir, the nutrient liquid flowing down the hose will stimulate the growth of moss on the hose wall. By using a black hose, the sun's rays cannot penetrate.

Another part that likes to be overgrown with moss is in the paragon pipes around the plant roots. The sun's rays penetrate the gaps between the net pot and the PVC pipe, stimulating the growth of moss. For this one, the growth of moss can be overcome by doing regular cleaning.

To overcome the problem of moss growth in Cirebon, West Java, we use H_2O_2. The benefit of this H_2O_2 is an anti-microbial that is safe from the recommendations of the FDA (United States Food and Drug Administration).

Some of the benefits of H_2O_2 for hydroponic plants:

1. prevent root rot.
2. Plant pest control.
3. Care for seeds before planting.
4. Sprayed on the stems and leaves to eradicate the fungus.
5. Treating wounds or damaged parts of plants.

H_2O_2can supply extra oxygen to plant roots, so the absorption of nutrients to the roots will be maximized, when plants get enough nutrients, growth will be faster, healthier, and also stronger against pests. In water for hydroponics, you can use the dose of 2 ml H_2O_2/liter water, first while looking at the condition of the plants because if there are too many hydroponic plants, they can wither.

The second way to deal with moss we use an ultraviolet lamp (**Figure 4**).

Figure 4. Ultraviolet light.

The benefit of UV light, the nutrients needed by hydroponic plants are maintained and get enough light for a full 24 hours so that plants grow more optimally. The rapid growth of 24-hour plants will beat the growth of moss so that moss disturbances can be overcome using an ultraviolet lamp. With a stable irrigation system and lighting with UV lamps, organic vegetables can be harvested in just 30–35 days, faster than the normal 45 days. In terms of weight, it can reach 200–250 grams for each plant stem. Certainly heavier than ordinary hydroponics which usually only reaches 150 grams per plant stem. In terms of quality, plants that use a hydroponic system with a UV light have brighter leaves, bright white roots. This is an indicator that the plant is growing healthily. (**Figure 5**)

Figure 5. Vegetable grows healthy without moss.

6. CONCISE SUMMARY

The conclusion of this research indicates that the proliferation of moss significantly influenced plant development. The growth of kale appeared most ideal with less moss, exhibiting broader leaves and a robust stem condition. Conversely, the spinach affected by moss exhibited suboptimal growth compared to our kale, characterized by narrower leaves and reduced density.

REFERENCES

1. Cut Raihan, Nurasiah, Zahara, N. 2018. Keanekaragaman Tumbuhan Lumut (Bryophyta) Di Air Terjun Peucari Bueng Jantho Kabupaten Aceh Besar (Diversity of Moss Plants (Bryophyta) in Peucari Bueng Jantho Waterfall, Aceh Besar District).. Prosiding Seminar Nasional Biotik 2018. 439-451.
2. Putri, S.E., Prayogo, H., Wulandar, R.S., 2019. Inventarisasi Jenis-Jenis Lumut Di Kawasan Hutan Adat Bukit Benuah Kabupaten Kubu Raya. *(Inventory of Moss Types in the Bukit Benuah Customary Forest Area, Kubu Raya Regency).* Jurnal Hutan Lestari Vol 7, No 5: 1035-1047.
3. Jumiati, Endang, T., Pramesthi, D.I.A., 2020. Inventarisasi Jenis-jenis Lumut (Bryophyta) di Daerah Aliran Sungai Kabura-Burana Kecamatan Batauga

Kabupaten Buton Selatan. *(Inventory of Moss Species (Bryophyta) in the Kabura-Burana River Basin, Batauga District, South Buton Regency).*Jurnal Biologi Tropis, 20 (2): 161 – 172

4. Zahara, M. 2019. Jenis-Jenis Tumbuhan Lumut (Bryophyta) Di Stasiun Penelitian Soraya Kawasan Ekosistem Leuser Sebagai Referensi Mata Kuliah Botani Tumbuhan Rendah. (*Types of Moss Plants (Bryophyta) At Soraya Research Station Leuser Ecosystem Area as a Reference for Low Plant Botany Courses).* Skripsi Fakultas Tarbiyah dan Keguruan. Universitas Islam Negeri Ar-Raniry. Banda Aceh.
5. Sujatmiko, H. Vitara, P.E., 2021. Tumbuhan Lumut di Kampus UGM. *(Moss Plants at UGM Campus).* Gadjah Mada University Press.
6. Kamalia, S., Dewanti, P., Soedradjat, R., 2017. Teknologi Hidroponik Sistem Sumbu Pada Produksi Selada Lollo Rossa (Lactuca sativa, L) dengan Penambahan CaCl2 sebagai Nutrisi Hidroponik. *Hydroponic Technology of Wick System on Lollo Rossa Production (Lactuca sativa L.) with Addition of Cacl2 as a Hydroponic Nutrition.* Jurnal Agroteknologi, Vol. 11 No.01 (2017): 96-104.
7. Ummu, N.A. 2019. Kajian Usahatani Sayuran Hidroponik Kota Makassar (Studi Kasus CV. Akar Hidroponik Kelurahan Masale Kecamatan Panakkukang Kota Makassar).*(Research on Hydroponic Vegetable Farming in Makassar City (Case Study of CV. Hydroponic Roots, Masale Village, Panakkukang District, Makassar City)).*Skripsi. Jurusan Agribisnis. Fakultas Pertanian. Universitas Muhammadiyah Makasar.
8. Lukitasari, M. 2018. Mengenal Tumbuhan Lumut (Bryophyta) Deskripsi, Klasifikasi, Potensi, dan Cara Mempelajarinya. *(Getting to Know Moss Plants (Bryophyta) Description, Classification, Potential, and How to Study It).* CV AE Media Grafika. Magetan. West Java Indonesia.

9 Chapter

CULTIVATION OF ORGANIC VEGETABLES

1. INTRODUCTION: AN OVERVIEW

Recently, organic vegetable cultivation has gained significance due to heightened awareness among the expanding population regarding food safety and quality. This can be ascribed to the reality that traditional agricultural methods, characterized by excessive and indiscriminate application of chemical fertilizers and pesticides, are no longer safe as they directly infiltrate the food chain. The conventionally grown crops present numerous health dangers due to elevated pesticide residues, heavy metals, and genetically engineered organisms. In cereals and pulses with a moisture content of approximately eight percent, pesticide residue is minimal and tends to diminish over time during storage. Conversely, vegetables and fruits, which have moisture levels exceeding ninety percent, exhibit maximum pesticide residue, necessitating consumption within a reasonable timeframe due to their perishable nature. Organic gardening not only safeguards the environment but also exerts a significant socio-economic influence on a nation. The genuine impact is fundamentally rooted in sustainable food production, which significantly contributes to a nation's environmental stability and economic resilience. The International Federation of Organic Agriculture and Movements (IFOAM) has proposed principles to improve biological cycles within agricultural systems, augment soil fertility, minimize pollution, avoid the use of chemical fertilizers and pesticides, preserve genetic diversity, assess the socio-economic implications of food production, and generate high-quality food in adequate quantities. The National Organic Program established by the USDA under the Organic Food Production Act (OFPA, 1990) has delineated criteria for the organic production of crops to be deemed acceptable as organic.

Organically cultivated veggies account for over 20% of the overall vegetable share in many European nations. Conversely, the market share of organic vegetables is minimal in tropical and subtropical nations, resulting in their exportation. In response to the growing awareness of organically cultivated veggies, nations such as Brazil, Argentina, and China have adopted organic farming practices.

2. IMPORTANCE OF ORGANIC FARMING

Vegetable crops when grown organically produce lesser yield. However due to its nutritive quality and storage attributes they are valued well than conventionally grown vegetables. Organoleptic studies have shown that vegetables like tomato and potato tastes better when grown organically. Likewise, the fruits had a better taste, flavor, texture and juicier when compared to conventionally grown ones. Similarly, organically grown okra and carrots were found to possess better quality attributes like taste, flavor and sugar content than those grown conventionally.

Excessive nitrate intake is posing a serious threat to human health. Leafy vegetables, in particular, accumulate more nitrates followed by root vegetables and potato. Studies have confirmed that organically produced vegetables like potato, carrot, cabbage beetroot, celery, leak, parsley and lettuce contain lesser levels of nitrates and higher levels of vitamin C content when compared to conventionally grown vegetable crops. Similarly, it has been found in studies that organically grown vegetables accumulate higher content of total sugars, minerals like phosphorous and magnesium and phenolic compounds in vegetables like carrot, potato, spinach, brinjal, lettuce and cabbage. Organically grown vegetables like sweet pepper and brinjal exhibited higher levels of phenolic compounds, peroxidase and capsidol activity offering resistance to diseases.

3. STATUS OF ORGANIC VEGETABLE PRODUCTION

During the past few decades, there is increasing concern on the food safety which in turn is drawing attention of the farmers, researchers and policy makers on the alternate production systems like organic cultivation. Organic farming is estimated to be growing at a rate of 30% a year worldwide in response to the market needs [2]. The demand for certified organic produce, particularly vegetables is presently exceeding the supply, thus fetching premium market price. Globally, the agricultural production system is undergoing a rapid transformation since there is a rise in demand for healthier and environmentally safer food. A larger proportion of growers are now shifting to organic production practices to meet the increasing consumer demands.

Organic cultivation has a significant role to play in maintaining the soil fertility by boosting the microbial flora of the soil. This can substantially lead to increase in yield, plant composition as well as nutritional quality. Organic treatments resulted in higher carrot root production compared to conventional treatments [3]. The yield of cabbage and tomato grown under organic practices yielded better than that grown under conventional system [4]. Similar results were obtained in cucumber [5]. Better results in terms of fruit yield in vegetables could be attributed to the fact that organic amendments of soil changed the soil dynamics as well as the plant composition and nutritional quality. Organic inputs can proportionately increase the microflora which in turn facilitates production of substances such as citrate and lactate which combine with soil minerals to increase the availability of mineral nutrition to the plant roots [6].

Besides increasing the soil fertility status, the nutritional quality of organically grown vegetables has been found to be appreciable. Higher levels of iron and magnesium were recorded in vegetable crops like carrot, beetroot, lettuce, kale, leek, turnip, onion, celery and tomato when grown organically [7]. Vitamin C holds an important place in the daily recommended diet due to its higher antioxidant properties. Hence, the focus of many research experiments has been on the vitamin C content in organically grown vegetables.

Experiments on tomato, celery and kale have shown higher vitamin C content when they are grown under organic practices than when grown under traditional systems of cultivation [7]. On an average, the vitamin C content of organically grown vegetables is 27% more when compared to the conventionally grown vegetables [6]. Likewise, the vitamin C content in organically grown leafy vegetables was found to be higher compared to that grown with chemical inputs [8]. Plant system responds to chemical fertilizer by increasing the production of proteins rather than carbohydrates since chemical fertilizers are rich sources of nitrogen. However, vitamin C production is the outcome of carbohydrate production and hence more quantum of vitamin C is expected to be produced with increased quantities of organic manures which have lesser nitrogen content and thereby lesser protein production and more of carbohydrate production.

4. FEASIBILITY OF ORGANIC PRACTICES

Organic farming implies application of composted plant residues and animal manures and growing legume crops as cover crop to meet the nutrient requirement of the crop. Soil fertility status is also enhanced by adopting practices like crop rotation, sequential cropping and also by minimizing the plowing activities. These techniques have a profound impact in maximizing the carbon sequestration in lands where organic practices are followed. On the contrary, in lands where conventional farming practices are adopted increased tillage operations lead to depletion in the organic matter accompanied by greater loss in mineral composition of the soil. The annual sequestration rate has been found to increase substantially by upto 3.2tons of CO_2/ha/year by organic farming practices [9] which has a direct implication in reducing the green house gases.

With the unpredictable climatic condition in the present day, organic farming which can increase soil water retention capacity can go a long way in fighting the drought situation. Increased carbon retention in the soil helps in withstanding climatic challenges as well as soil erosion.

Organic farming requires 28–32% lesser energy compared to the traditional farming practices as this cuts down the cost on fertilizers, pesticides and machinery [10].

Organic agriculture can potentially lower the green house gas emission [11]. The estimated quantum of green house gas emission due application of chemical fertilizers is 1000 million tonnes annually. Organic farming can by and large reduce the Green House Gas emission by sequestering carbon into soil.

5. SUSTAINABILITY OF ORGANIC FARMING

The global human population is projected to expand to 9.3 billion by 2050. Hence, sustainable production needs to be addressed to meet the increasing food requirements of the human population. Sustainable agriculture provides a potential solution to enable agricultural systems to feed a growing population within the changing environmental conditions. For successful organic farming maintaining soil health by addition of organic residues is imperative. However, it may not be feasible because of continuous sourcing of the crop residues/organic matter which has become scanty in recent times. Soil health is not only maintaining the carbon content in the soil but also maintaining a balance between carbon and nitrogen which is the most important factor that determines the nutrient availability besides the population dynamics of the microflora in the soil. Unlike conventional agriculture which involves usage of harmful chemical fertilizers and pesticides, organic farming sustains, maintains and enhances the quality of ecosystem. Though the plant absorbs nutrient in the simpler forms of nutrients similar to inorganic fertilizers, the source of nutrients is important factor to be considered in any organic farming practices. In the global market, vegetables grown organically fetch higher price because of its quality consideration and long term storability compared to inorganic green vegetables.

6. ORGANIC CERTIFICATION

The body, CODEX Alimentarius of the Food and Agriculture Organization of the United Nations aims to protect the health of the consumers and ensure fair prices in the international market. Internationally, organic certification is underway to facilitate international trade between countries. One such body is International Federation of Organic Agriculture Movement (IFOAM).

USDA organic certification confirms that the farm complies with USDA organic regulations. Farms are certified by State or Private entities which have been accredited by USDA. Any farm that produces over $5000 annually through organic sales needs to be certified.

Land utilized for organic vegetable production must not have used chemical fertilizers, pesticides, Genetically Modified Organisms (GMO's) for atleast 3 years prior to growing for organic purpose. Whole farm or part of a farm can be certified as organic. It is particularly important for vegetable growers to document the last date of application of prohibited chemicals, particularly when crops like, Lettuce or Spinach is grown.

7. ORGANIC CERTIFICATION IN INDIA

Organic farming in India is showing steady increase. Farmers involved in organic farming are of three categories *viz.*, farmers of low input zones, traditionally doing organic farming, farmers who have shifted to organic farming as a result of realizing the ill effects of conventional farming and farmers who have ventured into organic practices to fetch premium price for their produce. In India, the National Programme on Organic Production (NPOP) has provided the regulatory framework while National Project on Organic Farming (NPOF) is involved in providing support for expanding the area under certified organic production [12].

8. REGULATORY MECHANISM

The National Programme on Organic Production (NPOP) offers regulatory mechanism to acts for domestic and export markets. NPOP under Foreign Trade Development and Regulation Act (FTDR) caters the export requirement and has been acclaimed by European Union, Sweden and USDA. Hence, any organic product certified by NPOP can be exported to Europe, Sweden and USA. Similarly, to meet the domestic demands, NPOP notified under Agriculture Produce Grading, Marketing and Certification Act (APGMC) comes to play lead role. Regulatory body of NPOP under FTDR is Agricultural and Processed Foods Export Development Authority (APEDA) under Ministry of Commerce and NPOP under APGMC Act is Agricultural Marketing Advisor (AMA) under Ministry of Agriculture. Accreditation of Certification and Inspection Agencies is being granted by a common National Accreditation Body (NAB) [12].

9. NEED FOR CERTIFICATION

Certification ensures quality produce. "Certified Organic" serves as a product assurance to consumers. Certification necessarily aims to regulate and facilitate the sale of organic vegetables to consumers.

10. CERTIFICATION PROCESS

In order to ensure certification by any agency, the producer has to satisfy the following criteria:

- Adherence of the organic standards as prescribed by the certification authority.
- The production practices and farm facilities have to comply with the norms and standards.
- Detailed documentation of the entire farming procedures adopted and farm facilities is required.
- Periodical inspection by the authorities concerned.

11. PRODUCT LABELING

Organic legislation defines three levels of organic labelling in many countries. Products are given a "100% organic" label when the products are made entirely with certified organic ingredients. Products with the label "Organic" indicate 95% organic ingredients being used. The third label "made with organic ingredients" shows that a minimum of 70% of organic ingredients has been used.

12. PROSPECTS AND CONSTRAINTS OF ORGANIC VEGETABLE PRODUCTION

There is immense scope for organic production of vegetable crops in India since the agricultural sector has enormous organic resources like crop residues, livestock and other bio-products from agro industries. Organic farming is growing at a rapid pace among Indian farmers and entrepreneurs, particularly in rainfed and hilly areas where fertilizer consumption is less than 25 kg/ha/year [13].

Market development, particularly domestic sector continues to be one of the biggest challenges in organic farming. Lack of infrastructural facilities for post production practices also poses a challenge as it sets a constraint in meeting the organic standards. Similarly the cost involved in certification process and the extensive documentation procedure is a major setback.

13. CONCISE SUMMARY

Organic agriculture is expanding in numerous countries that achieve self-sufficiency in vegetable production. Conversely, in economically disadvantaged and emerging regions, prioritizing the adequate cultivation of vegetables to nourish the population is essential.

The barrier to implementing organic farming methods in vegetable crops is the elevated input cost. Nonetheless, in light of environmental concerns and the long-term sustainability of organic farming, the justification for the additional prices must be communicated to farmers and entrepreneurs.

REFERENCES

1. International Federation of Organic Agriculture Movements (IFOAM), 1998. The IFOAM basic standards for organic production and processing. General assembly, Argentina, November, IFOAM, Germany. Organic Food Production Act of 1990 (U.S.C)s. 2013.
2. Ashley, R., Bishop, A and Dennis, J. (2007). Intensive organic vegetable production integrated development. A report for the Rural Industries Research and Development Corporation. RIRDC Publication No. 04/121 RIRDC Project No. DAT-37A, 46pp.

3. Bruno, R.L.A., Viana, J. S, Silva, V.F.G.B, Moura, M.F. (2007). Production and quality of seeds and roots of carrot cultivated under organic and mineral fertilization. Hortic Bras. 25: 170-174.
4. Ma, C.H., Chen, J. H, Yang, R. Y, Palada, M. C, Ous, C, Lin, Y. H, Chen, L.H. (2009). Monitoring soil and vegetable quality under six fertilization strategies in organic and conventional farming systems. In: The 9th international conference of the east and Southeast Asia Federation of Soil Science Societies. pp. 373-374.
5. Mahmoud E, El-Kader NA, Robin P, Akkal-Cor fi ni N, El-Rahman LA (2009) Effects of different organic and inorganic fertilizers on cucumber yield and some soil properties. World J Agric Sci 5(4):408-414
6. Worthington, V. (2001). Nutritional quality of organic versus conventional fruits, vegetables and grains. J. Altern. Complement Med., 7(2): 161-173.
7. Laison, D. (2010). Nutritional quality and safety of organic food. A review. Agron. Sustain. Dev. 30: 33-41.
8. Xu, H.L., Wang, R. Y, Mridha, M.A.U, Goyal, S. (2003). Yield and quality of leafy vegetables grown with organic fertilizations. Acta Hortic., 627: 25-33.
9. Smith, P., Martino, D., Cai, Z., Gwary, D., Janzen, H.H., Kumar, P., Mc Cael, B, ogle, S., O Mara, F., Rice, C., Scholes, B. and Sirotenko, O. (2007). Agriculture. In climate change 2007: Mitigation, contribution of working group III to the fourth assessment report of the Intergovermental panel on climate change (8.Metz, O.R. Davidson, P.R. Bosch, R. Dave, L.a. Mayer (eds)) Cambridge university press, United Kingdom and new Yark, NY, USA.
10. Kimble, J.M., Rice, C.W., Reed, D., Mooney, S., Follet, R. F, and Lal, R. (2007). Soil carbon management, economics, Environmental and Social Benefits. CRC press, Taylor and Francic Group.
11. Niggli, U., Fliebach, A., Hepperly, P. and Scialabba, N. (2009). Low green house gas agriculture: Mitigation and adaptation potential of sustainable farming systems. Rev. 2. Rome, FAO, April.
12. Hazra, P and M.G. Som. 2015. Vegetable science. ISBN 978-93-272- 5451-8. Kalyani publishers, India
13. Mitra, S. and Devi, H. (2016). Organic horticulture in India. Acta Hort. 975: 303-307.

10 Chapter

ORGANIC VEGETABLE CULTIVATION: AN EFFECTIVE METHOD FOR ENSURING SUSTAINABILITY AND PROFITABILITY

1. INTRODUCTION: AN OVERVIEW

The global population has consistently risen since the conclusion of the Black Death in 1350, when it was believed to exceed 370 million individuals[1]. From 1955 to 1975, the global population saw its most rapid growth, exceeding 1.8% annually, with a peak of 2.1% between 1965 and 1970. The United Nations Department of Economic and Social Affairs forecasts a population of 9 to 10 billion by 2050, with an 80% confidence range ranging from 10 to 12 billion by the century's conclusion[2]. Other demographers predict that in the latter half of the twenty-first century, the global population will commence a decline[3]. In 2012, a common estimate for the planet's sustainable population was 8 billion individuals. As of March 2020, the global population stood at 7.8 billion, and with standard population growth forecasts, Earth is projected to experience overpopulation by 2050 or before. To accommodate a larger population, scientists must increase agricultural production.

This will significantly affect agricultural markets in an unprecedented manner. Farmers globally must enhance crop yield by either increasing the agricultural land designated for cultivation or by augmenting productivity on current farmland through fertilizers and irrigation, while also adopting advanced technologies such as precision farming. Various factors influencing agricultural productivity, including illnesses, diminished microbial communities, pesticide application, suboptimal or excessive pH levels, salinity, and restricted nutrient availability, adversely impact crop output [4–6]. Moreover, the application of fertilizers, insecticides, and other synthetic substances may effectively increase food production for the needy, despite the severe repercussions associated with these conventional approaches. The overall expenditure on pesticide purchasing in Pakistan amounts to 8,138 million rupees. The import bill of Rs 8.139 billion for 2003 was estimated to total Rs 19,612 billion [7]. In many instances, green revolution strategies, including the augmented application of synthetic agrochemicals such as fertilizers and pesticides, the adoption of mineral-responsive, high-yield crop genotypes, and enhanced utilization of irrigation capabilities, have elevated production levels. Alterations in soil responses, development of mineral deficiencies, detriment to soil flora and fauna, decline in earthworm activity, reduction in soil humus and organic matter, atmospheric changes, diminished soil productivity, ventilation, and water retention capacity are all repercussions of the improper management and continuous application of these high-energy inputs [8]. The research indicated that the excessive usage and persistent use of chemical fertilizers pose significant health and environmental concerns, prompting a call for farmers in wealthy countries to transition their existing farms to organic practices.

Organic farming is a food production method that underscores the importance of biodiversity, soil variety, and biological cycles. It is a production approach aimed at preventing or substantially reducing the usage of synthetic fertilizers [9]. The active decomposition of organic fertilizers, including diverse agricultural residues and animal excrement, can be facilitated through the utilization of legumes and biological nitrogen fixation. Organic agriculture systems are generally seen as more sustainable than conventional and organically-managed farms, encompassing over 44 million hectares globally, with expectations for additional expansion. Enhanced sustainable agricultural output can be attained through the application of organic matter and the enhancement of land health via the development of advantageous physical soil characteristics [10]. It emphasizes the enhancement of natural soil fertility, enabling plants to assimilate essential nutrients consistently from the minerals generated and removed in alignment with their requirements. Effective disease and weed management is accomplished by establishing environmental stability within the system and employing bio-pesticides alongside other cultural techniques such as crop rotation, integrated planting, and agricultural practices. Organic farmers utilize all waste and manure on the farm; yet, the exportation of produce results in a continuous depletion of minerals. Organic farming enhances the soil's water retention capacity. Organic agricultural products typically exhibit superior size, flavor, and quality. The subterranean water in the region designated for organic cultivation is devoid of hazardous compounds. Regulates the carbon-to-nitrogen ratio in the soil, enhancing its fertility and production efficiency. Currently, the primary sources of organic farming include farmyard manure, compost, and biochar, which can enhance the productivity, quality, and profitability of numerous cereal and non-cereal crops, as well as vegetables.

Vegetable cultivation is a vital source of revenue for farmers globally, encompassing organic agriculture. Vegetable crops constitute approximately 7% of total acreage, with this percentage typically being higher in affluent nations [11]. Vegetables are highly adaptable and beneficial to farmers in utilizing organic nutrient sources. Organic inputs such as vermicompost, charcoal, and farmyard manure may enhance vegetable cultivation. Rekha [12] discovered that the application of vermicompost increased the quantity of branches and fruits in Capsicum annum (Linn.) Hepper. Vermicomposting positively impacts agricultural productivity [13]. A substantial yield of brinjal, a crop grown with vermicompost, and a notable rise in production of sweet pepper were observed [14]. Yadav and Vijayakumari [15] reported an elevation in the quality of numerous vegetables following the incorporation of organic sources.

1.1 Importance of organic farming

Organic agriculture is an environmentally friendly production option available. It is necessary for guaranteeing food supply, relieving impoverishment, and protecting dynamic mineral deposits on which current and future generations will be completely reliant for their persistence and security. Over 90% of developing countries, particularly in Asia, would be in towns, which will adopt the green revolution approach to an ever-greater level [16]. Organic farming contributes to the protection of the environment and aids to consolidate environmental problems such as soil management and organic farming by creating a crop cycle to supplement the soil with a natural nutrient reservoir. Because of its friendly approach, it helps to reduce pollution of the earth, water, and air. Thus, it serves as a natural means for conservation of the environment and maintainable development [17]. Organic farmers must go along with the procedures established by regional organic farming organizations and are not permitted to grow genetically modified (GM) crops [18].

1.2 Sustainability of agriculture

Agricultural researchers and experts are well aware of the importance of sustainable agriculture and the necessity to put it into practice, i.e., developing appropriate ways to assess the farming system's sustainability [19]. Organic agriculture raises concerns about the negative consequences of cropping and agricultural systems such as water pollution from nitrates and pesticides and emissions of gasses from inputs of nitrogen; it is conventional agriculture that raises the most issues.

But addressing the negative repercussions of productivism is not enough to ensure long-term viability. Other variables outside of the traditional system can contribute to a lack of long-term viability. As a result, the long-term viability of organic agricultural systems must be recognized. In reality, the long-term viability of organic farming is assessed using the same set of indicators to compare conventional, integrated, and organic farming systems [20].

1.2.1 Emerging approaches in organic farming

Organic amendments are being used as an alternative to inorganic fertilizers; currently, these amendments are an emerging approach [21]. Organic farming methods accord with the four basic principles that reveal their essence: health, ecology, fairness, and care. Various other approaches include, crop rotation, cover crop, green manures, animal manures, and integrated pest and weed management. Among the researchers, organic amendments such as biochar and compost have growing interests. Many studies have been done on exploring their role in the enhancement of plant nutrition, quality, yield of crops, soil fertility protection, and ensuring the sustainability of the environment [22, 23]. There are different factors that are responsible for the special effects when they are added into the soil, for example, properties of feedstock, processing methods, rate of application, type of soil, species of crop, and environmental conditions [24, 25].

1.3 Manures

Different types of manure and certain manure-derived compost, which contain larger levels of nutrients, are applied to soils to increase vegetable output and meet the rising demand for their consumption [26]. By providing necessary nutrients through substrate and decomposition to generate organic matter, the farmyard manure plays a critical role in the productivity of a variety of agricultural systems. By adding farmyard manure [27], soil microbial activity is enhanced, which may increase the rate of organic matter breakdown. Organic matter significantly enhances soil physical properties such as soil hydraulic conductivity, soil porosity, and soil water-holding capacity, all of which are important components of soil quality [5, 28]. It was noticed that the incorporation of organic manures (farmyard manure, poultry manure) to the soil resulted in remarkable improvement of physiological attributes in various vegetables [29]. When biochar and poultry manure were applied to the soil alone or in combination, they improved the physical properties of soil significantly as compared to control. They decreased soil bulk density and improved soil moisture content and soil porosity [30]. Application of manure improved the properties of soil that increased cucumber yield. Higher rates of manure application resulted in a higher yield of cucumber [31]. Miaha et al. [32], also observed similar results and said that yield of cucumber was increased by increasing higher rates of organic fertilizer. Njoku et al. [33], observed that plots that were amended with manure produced a higher cucumber yield. Another study said that the production of vegetable "cauliflower" was increased when organic manure was application and time of application was also a considerable factor in increasing crop production. Thus, while considering the productivity of the crop and the economic return of the vegetable crop, the application of organic manure as well as certain other aspects such as application timing may be significant for better as well as higher quality production of cauliflower [34].

1.4 Biochar

Biochar is porous in nature, rich in carbon contents, and is an alkaline solid product. It is prepared by pyrolysis of waste biomass [35]. Biochar has the ability to increase soil organic carbon, organic matter, and/or soil humus contents. It improves the nutrient and water holding capacity of soil [36]. It has a high cation exchange and adsorption capacity. Biochar has ability to delay fertilizer release in soil and it improves the rate of utilization of fertilizer nutrients. As the structure of biochar is porous with higher water and nutrients adsorption ability, it provides suitable habitat to the soil microorganisms thus promotes activities and propagation of beneficial soil microorganisms. Application of PAD (peanut-shell biochar-based amendment) at the optimal concentrations i.e., 1.5–3% enhanced yield of vegetable. This was mainly due to improved soil qualities and increased contents of available nutrients. Another study was carried out and results showed that application of rice husk biochar and rice straw biochar increased the yield of spring onion by 22% and 35% as compared to raw rice husk that is a current practice followed by farmers practice [37]. Nobil, [38] indicated that incorporation of biochar having low density and higher porosity leads to the higher production of basil and lettuce biomass. It is mostly related to biochar beneficial effects on the availability of water. Another study said that the addition of biochar increased potassium availability and its uptake in soil and it is primarily responsible for higher root growth in ginger [39]. Biochar application to tomatoes in the field resulted in taller tomatoes plants; it increased root growth and biomass [40].

1.5 Vermicompost

In near future "sustainable agriculture" can be ensured through the system of organic farming that includes different biological processes like the application of compost and vermicompost. Many forms of organic material can be used to prepare vermicompost, it includes manure of animals, wastes of manufacturing industries like paper waste, sugar waste of cane or the cotton residues, kitchen waste, agricultural wastes, and the municipal wastes having an organic origin [41]. Higher concentrations of vermicast and the vermitea improves the heath of the plant, provide protection, improve growth, and also provide optimum production of eggplant [42]. According to [43], the application of vermicompost effectively reduced the continuous cropping obstacles in the soil and improved crop growth, its productivity and quality through improving the soil physical, chemical, and biological properties alone and combined application of biochar and the vermicompost improved properties of soil, quality, and yield of cucumber. According to [44], the effect of vermi & parthenium compost on the growth of brinjal seedlings was observed. Growth parameters i.e. the height of plants, the number of plant leaves, area of the leaf, length of root, the number of flowers per plant, and the number of fruits per plant were increased in the parthenium compost as compared to vermicompost and control. [32], said that more nutrients were available to the okra plants when they were mixed as compared to alone organic materials. Growth and yield of vegetable crop okra was highest when it received integrated nutrient management treatment with the lowest rates of vermi-compost. But when Vermicompost was mixed with farmyard manure, it gave better results as compared to their individual application. Similar results were found that said combined application of inorganic fertilizer, organic fertilizer, and bio-fertilizers improved okra fruit quality and health of the soil. It was also said that it can produce more yield with better growth of okra vegetative [45].

2. PROFITABILITY OF ORGANIC FARMING

Organic agriculture is significantly more profitable (22–35% greater net present values) and had higher benefit/cost ratios (20–24%) than conventional agriculture under actual conditions with price premiums. When organic premiums are removed, net present values (27–23%) and benefit/cost ratios (8–7%) of organic agriculture were significantly higher. Even though premiums were 29–32%, the breakeven premiums required for organic earnings to meet conventional profits were just 5–7%, despite organic yields being 10–18% lower. While total expenses were similar, labor costs were substantially higher (7–13%) with organic agricultural techniques. Organic farming has space to expand: by 2050, it may account for 10–20% of farmland, up from 1% now [46]. Organic farming has been demonstrated to provide abundant and inexpensive food while also safeguarding the environment, assisting farm finances, and adding to the well-being of farmers and farm employees, according to research. Organic and alternatively produced foods are becoming increasingly popular in grocery shops and farmers' markets [47].

3. CONCISE SUMMARY

The terms organic and organic agriculture are widely recognized nowadays. Organic farming is an agricultural practice that conforms to the tenets of sustainable development. It is a form of agricultural production management that eschews the use of pesticides, chemical fertilizers, industrial manufactured products, and genetically modified organisms. Organic agriculture fosters sustainable development in society (health, employment, etc.), the environment (methane emissions, water resources, etc.), and the economy (wealth generation, etc.). To encourage the adoption of organic and innovative farming systems, it is necessary to provide incentives for appropriate markets, modify agricultural legislation, and realign publicly funded agricultural research. Reduced yields are less problematic if society begins to appreciate the other three attributes of organic and alternative agriculture systems: enhanced economic, social, and environmental sustainability.

REFERENCES

1. Biraben J-N. Essai sur l'évolution du nombre des hommes. Population. 1979;**34**(1):13-25. Jean-Noël. Original paper in French. London, New York: Oxford University Press; 4. pp. 1-13

2. Cave D, Bubola E, Sang-Hun C. Long Slide Looms for World Population, With Sweeping Ramifications. London: The New York Times; 2021. ISSN 0362-4331 [Accessed: 23 May 2021]
3. World Population Prospects. UN.org. 2019. Archived from the original on 11 December 2020. [Accessed: 19 January 2021]
4. Kumar SS, Gondal AH, Hayat F, Mahale AG, Farooq Q , Umer H. Weed and disease eradication in crops through genetically modified microbes and soil microorganisms: A promising treatment. Journal of Agriculture and Allied Fields. 2021;**3**(1):37-51
5. Gondal AH, Zafar H, Yousaf H, Farooq Q , Imran B, Ch MDT, et al. Impacts of tillage technologies on soil, plant, environment and its management: A short communication. Indian Journal of Pure and Applied Biosciences. 2021;**9**(3):76-83
6. Zainab DE, Gondal AH, Zafar A, Sohail S, Zafar H, Ijaz AB, Farooq Q, et al. A Brief study of quinoa role and its adaptation towards salinity and drought stress. Current Research in Agriculture Farming. 2021;**2**(3):27-40
7. Khooharo AA, Memon RA, Mallah MU. An empirical analysis of pesticide marketing in Pakistan. Pakistan Economic and Social Review. 2008;**46**:57-74
8. Behera KK, Alam A, Vats S, Sharma HP, Sharma V. Organic farming history and techniques. In: Agroecology and Strategies for Climate Change. Dordrecht: Springer; 2012. pp. 287-328
9. Howard A. An Agricultural Testaments. London, New York: Oxford University Press; 1940
10. Patiram LK. Effect of integrated use of inorganic, biological and organic manures on rice productivity and soil fertility in ultisols of Mizoram. Journal of the Indian Society of Soil Science. 2006;**54**:213-220
11. Li D, Wang X. Nitric oxide emission from a typical vegetable field in Pearl River delta China. Atmospheric Environment. 2007;**41**:9498-9505
12. Rekha GS, Kaleena PK, Elumalai D, Srikumaran MP, Maheswari VN. Effects of vermicompost and plant growth enhancers on the exo-morphological features of Capsicum annum (Linn.) Hepper. International Journal of Recycling of Organic Waste in Agriculture. 2018;**7**(1):83-88
13. Singh KN, Prasad B, Sinha SK. Effect of integrated nutrient management on a typic haplaquant on yield and nutrient availability in a ricewheat cropping system. Australian Journal of Agricultural Research. 2001;**52**:855-858
14. Kalembasa S, Deska J. The possibility of utilizing vermin-compost in the cultivation of radish and paprika. Annals of Agricultural University of Poznan. 1998;**27**:131-136
15. Yadav RH, Vijayakumari B. Impact of vermicompost on biochemical characters of Chilli (*Capsicum annum*). Journal of Ecotoxicology and Environmental Monitoring. 2004; **14**:51-56

16. Gryn, Naomi. "Dame Miriam Rothschild." Shalvi/Hyman Encyclopedia of Jewish Women. 31 December 1999. Jewish Women's Archive. (Viewed on November 4, 2021) https://jwa.org/encyclopedia/article/ rothschild-dame-miriam
17. Skoufogianni E, Solomou A, Molla A, Martinos K. Organic farming as an essential tool of the multifunctional agriculture. In: Organic Farming: A Promising Way of Food Production. Vol. BoD–Books on Demand; 2016. p. 29
18. Haslberger A. Genetically modified and organic crops in developing. A review of options for food security. Advances in Biotechnology. 2010; **28**:160-168
19. Van Calker KJ, Berentsen PBM, Romero C, Giesen GWJ, Huirne R. Development and application of multi-attribute sustainability function for Dutch dairy. Ecological Economics. 2006;**57**:640-658
20. Pacini C, Wossink A, Giesen G, Vazzana C, Huirne R. Evaluation of sustainability of organic, integrated and conventional farming systems: A farm and field-scale analysis. Agriculture, Ecosystems and Environment. 2003; **95**:273-288
21. Gondal AH, Farooq Q , Hussain I, Toor MD. Role of microbes in plant growth and food preservation. Agrinula: Jurnal Agroteknologi Dan Perkebunan. 2021;**4**(2):106-121
22. Nair VD, Nair PKR, Dari B, Freitas AM, Chatterjee N, Pinheiro FM. Biochar in the agroecosystem-climate- change-sustainability nexus. Frontiers in Plant Science. 2017;**8**:1-9
23. Gondal HA, Bakar Ijaz SS, Bisma Imran CH, Taufiq Nawaz S, Haider IAA, Dua-e-Zainab BR, et al. Alleviation of zinc deficiency from humans through plants by organic sources: A powerful tonic. International Journal of Applied Research. 2021;**7**(4):240-243
24. Libutti A, Cammerino ARB, Francavilla M, Massimo M. Soil amendment with biochar affects water drainage and nutrient losses by leaching: Experimental evidence under field- grown conditions. Agronomy. 2019;**9**:758
25. Husnain A, Farooq Q, Sohail S, Shasang S, Toor D, Zafar A, et al. Adaptability of soil pH through innovative microbial approach. Current Research in Agricultural Sciences. 2021;**8**(2):71-79
26. Candela L, Fabregat S, Josa A, Suriol J, Vigués N, Mas J. Assessment of soil and groundwater impacts by treated urban wastewater reuse. A case study: Application in a golf course (Girona, Spain). Science of the Total Environment. 2007;**374**:26-35
27. Ghimire R, Lamichhane S, Acharya BS, Bista P, Sainju UM. Tillage, crop residue, and nutrient management effects on soil organic carbon in rice-based cropping systems: A review. Journal of Integrative Agriculture. 2017; **16**(1):1-15

28. Shepherd MA, Harrison R, Webb J. Managing soil organic matter— Implications for soil structure on organic farms. Soil Use and Management. 2002;**18**:284-292
29. Kaur A. Impact of various organic manures on yield, yield attributes and economics of cabbage (*Brassica oleracea* var. capitata L.). Journal of Pharmacognosy and Phytochemistry. 2020;**9**(2):1439-1442
30. Adekiya AO, Agbede TM, Aboyeji CM, Dunsin O, Simeon VT. Effects of biochar and poultry manure on soil characteristics and the yield of radish. Scientia Horticulturae. 2019; **243**:457-463
31. Njoku C, Okoro BC. Effect of manure on physico-chemical properties of soil and the yield of cucumber (*Cucumis sativus*) in Abakaliki. Indian Journal of Ecology. 2019;**46**(1):60-64
32. Miaha R, Methelaa NJ, Ruhib RA. Effect of integrated nutrient management on growth and yield of okra. Tropical Agrobiodiversity (TRAB). 2020;**1**(2):72-76
33. Njoku C, Inyang ED, Agwu JO. Soil physical properties and yield of cucumberas influenced by biochar, wood ash and rice husk dust applicationin Abakaliki Southeastern Nigeria. IOSR Journal of Agriculture and Veterinary Science. 2017;**10**(8):21-25
34. Akhter S, Mostarin T, Khatun K, Haq ME, Soniya IA, Monira S, et al. Effect of organic manure and its application timing on the growth and yield of cauliflower (*Brassica oleracea* var. *botrytis*). Asian Journal of Agricultural and Horticultural Research. 2019;**4**(3):1-14
35. Ramlow EJ, Foster SJ, Del Grosso MF. Cotrufo broadcast woody biochar provides limited benefits to deficit irrigation maize in Colorado Agric. Ecosystems and Environment. 2019;**269**:71-81
36. Qi LP, Pokharel SX, Chang P, Zhou H, Niu X, He Z, et al. Gao Biochar application increased methane emission, soil carbon storage and net ecosystem carbon budget in a 2-year vegetable-rice rotation. Agriculture, Ecosystem and Environment. 2020;**292**:106-831
37. Hemowng S, Sangrit C, Phunthupan P, Butnan S, Vityakon P. Rice-derived biochars enhance the yield of spring onion (*Allium cepa* l. var. *aggregatum*), while reducing pesticide contamination in soil and plant. Applied Ecology and Environmental Research. 2021;**19**(1):349-358
38. Nobile C, Denier J, Houben D. Linking biochar properties to biomass of basil, lettuce and pansy cultivated in growing media. Scientia Horticulturae. 2020;**261**:109001
39. Farrar MB, Wallace HM, Xu CY, Joseph S, Dunn PK, Nguyen TTN, et al. Biochar co-applied with organic amendments increased soil-plant potassium and root biomass but not crop yield. Journal of Soils and Sediments. 2021;**21**(2):784-798

40. Vaccari FP, Maienza A, Miglietta F, Baronti S, Di Lonardo S, Giagnoni L, et al. Biochar stimulates plant growth but not fruit yield of processing tomato in a fertile soil. Agriculture, Ecosystems and Environment. 2015;**207**:163-170
41. Alves WL, Passom AA. Compost and vermicompost of urban solid waste in Licaniatomentosa (Benth) seedlings production to arborization. Pesquisa Agropecuária Brasileira. 1997;**32**(10): 1053-1058
42. El-Goud A, Amal K. Efficiency response of vermicompost and vermitea levels on growth and yield of eggplant (*Solanum melongena*, L.). Alexandria Science Exchange Journal. 2020; **41**(JANUARY-MARCH):69-75
43. Liu C, Wang H, Li P, Xian Q , Tang X. Biochar's impact on dissolved organic matter (DOM) export from a cropland soil during natural rainfalls. Science of the Total Environment. 2019;**650**:1988-2019
44. Chitale RD, Mali BS. Effect of vermicompost and Parthenium compost on growth and yield of Brinjal (*Solanum melongena* L. cv. Panchganga). 2020;**26**(1):117-120
45. Kumar V, Saikia J, Nath DJ. Effect of integrated nutrient managementon growth, yield, and quality of okra (*Abelmoschus esculentus* L. Moench) cv. Arka Anamika. IJCS. 2017;**5**(5): 2001-2003
46. FIBL and IFOAM. The World of Organic Agriculture 2011. Statistics and Emerging Trends. BioFech; 2011
47. Francis C, Van Wart J. History of organic farming and certification. In: Francis C, editor. Organic Farming: The Ecological System. Raleigh, NC: ASA; 2009. pp. 3-17

11 Chapter

BENEFITS OF CULTIVATING VEGETABLE CROPS IN CONTEMPORARY GREENHOUSES

1. INTRODUCTION: AN OVERVIEW

Vegetable production is an applied science topic that examines the biology and technology of cultivating vegetable crops in open fields and greenhouses. The primary objective is to achieve high-yield vegetable crops with superior quality edible sections that are safe for human consumption, while also ensuring environmental preservation. Vegetables are categorized as annual, biennial, or perennial herbaceous plants, which seldom form a woody stem by the conclusion of their vegetative phase, primarily in the lower portion of the stem. The edible portions of vegetables are predominantly composed of approximately 95% water and can be utilized either fresh and uncooked or in processed forms. After harvesting, the consumable portions can be preserved for a limited duration (ranging from several days to a maximum of 9 months, contingent upon the type of storage facility and its equipment). The edible components of vegetables include roots and tubers, stems and stalks, sprouts, bulbs, leaves, petioles, immature inflorescences, fruits (both mature and immature), and seeds (both mature and immature). Given the UN DESA study indicating a projected global population increase to around 8.5 billion by 2030 and 9.7 billion by 2050, alongside diminishing labor-intensive climate changes, it is imperative to explore innovative solutions in food production overall. From this perspective, it is intriguing to examine the contemporary benefits of cultivating crops in diverse protected environments, particularly within modern greenhouses. Primary plant production constitutes a segment of both fresh and processed food production, necessitating innovative approaches to existing cultivation methods or the adoption of new technology. It is crucial, particularly in greenhouses regardless of natural light, to implement innovative techniques, software, and various advanced automations in plant cultivation and post-harvest handling, alongside comprehensive equipment that ensures optimal climatic conditions for successful plant development. In summary, it entails the advancement and practical use of the latest crop production methods in contemporary greenhouses, yielding significant commercial benefits. This approach would facilitate access to nutritious, high-quality, and safe food while ensuring robust environmental protection and economic outcomes. Vegetables are utilized in the pharmaceutical sector due to their medicinal ingredients. Generally, veggies are low in calories, fat, and protein. Simultaneously, they serve as a crucial source of essential vitamins, minerals, and trace elements for human nutrition. The average recommended vegetable intake for human consumption is approximately 400 grams per day. The USDA has published intriguing dietary advice for vegetable consumption on their website (USDA Dietary advice for Americans). A comprehensive account of daily vegetable servings for individuals residing in various regions of the world exists. Growing vegetable crops in contemporary greenhouses offers various advantages over old methods and standard greenhouses. [3]. In modern greenhouses, particularly in the glass ones, it is possible to control the necessary climate conditions entirely. One of the advantages is a possibility of growing vegetables in periods when it is impossible to do in open field. This enables all-year round and/or off-season production, which is increasingly in demand in markets all over the world. This particularly goes for vegetables crops typical of warm season (tomatoes, cucumbers, peppers), but also for those of cool season (lettuce, spinach, radishes, broccoli). In Serbia, winter production of vegetables, as an off-season production, generally is not economical as the renewable sources of energy (geothermal, solar, wind, biomass, etc.) are not still in use in their full capacity. However, the significance also lies in a possibility of arranging for early and late greenhouse production of warm climate vegetable varieties. In early spring and late autumn, the warm climate vegetable varieties cannot be grown in open-field environment, although there is a high demand for them, as well as with high prices. During the cold periods, greenhouses can also yield the cool-season varieties (lettuce and spinach), which are also in high demand on the market. By combining the warm and cool-season vegetable crop varieties, the modern greenhouse can be used for almost all-year round and saved in the deep winter, when it is freezing cold, and the energy prices in Serbia are rather high. The winter period, when no vegetables are grown in the greenhouse, usually can be used for cleaning, disinfecting, and preparing it for the new season and new crops. In most of the North-West European and in some East European countries, a winter break in the greenhouse vegetable crop growing lasts from the end of November until mid-January. The USDA organization has developed a software program called a Virtual Grower that helps growers calculate the heating costs of their greenhouse. Users can use the software to predict heating and energy use specific to their location, greenhouse design, crop produced, and the management preferences. The software program, as well as a short video, can be downloaded for free from the following Web site: https://ag.umass.edu greenhouse-floriculture/ fact-sheets/- virtual-grower-link-to-usda-software.

Due to specific climatic conditions, in southern parts of Europe, a great number of warm-season vegetable crop varieties can be grown all-year round in both the greenhouses and open field. That means that the vegetable crop growers from South Europe are in a much better position when compared with those from the cooler regions of our continent. However, even in those countries with warm and humid or warm and dry climate, there are periods that are not so favorable for the openfield vegetable crop growing or in simple greenhouses (e.g., plastic tunnel greenhouse). Therefore, a well-equipped modern greenhouse is the only wise choice in vegetable crop growing. Hemming et al. [4] describe a method, based on a number of mathematical models, which helps in predicting an appropriate greenhouse model for the given climatic conditions. The local climatic conditions (light intensity, temperature, humidity, and wind velocity) are the input data needed for making a greenhouse and vegetable crop growing technology model in a given area. The climatic conditions and crop growth within such greenhouse can be easily simulated [5], and the obtained data can be used as initial information for making greenhouse an economical model. In this way, the most suited greenhouse for a specific area can be designed. The world population is constantly increasing, and it is necessary to find out possibilities for intensifying agriculture in general and horticulture in particular. In this way, it is important to increase the production per area unit, economise on energy (fossil fuels) and water consumption, and use less chemicals (pesticides in particular) and fertilizers, thus improving the quality of all the edible vegetable parts, as well as overall food safety. For that reason, it is necessary to expand cultivated land areas under greenhouses (modern glass and plastic ones) and to intensify growing various vegetable crops under such protected conditions. Furthermore, the recognised competent and expert institutions in the field of agriculture should, among other things, organize and arrange for experimentation and testing of new technologies, their further development in horticulture through devising cultivation strategies and skills, and finally, provide education through participation in such projects ("learning by doing"). Therefore, it is important to come up with a good strategy for organizing the entire production, experimentation, and application in practice. It is necessary to do a feasibility study, either a limited or a comprehensive one. Concerning these "preventive" efforts, other factors must be taken into account: A meticulously chosen site suitable for formulating the crop cultivation plan, a practical greenhouse design, and established project specifications based on the vegetable cultivation approach. Furthermore, it is imperative that the early projects are systematically arranged and appropriately aligned with the specific potentials of the respective site or region. In horticulture and vegetable crop cultivation, operational support and production monitoring of a specific crop are crucial preventive measures. Consequently, proactive management and production planning must be instituted, individualized support offered, and guidance through education and employment provided by well-trained consultants, alongside oversight and production monitoring by foreign experts from nations with advanced horticultural practices[6]..

2. GREENHOUSE SITE SELECTION IN VEGETABLE CROP GROWING

A modern greenhouse (a glass or plastic one) is actually a mechanical barrier between the outdoor climatic conditions and the indoor area with the controlled climatic conditions that are optimal for growing the selected vegetable crops. In this way, the vegetable crops are protected from the extreme temperatures, wind, snow, rain, hail, birds, and insects [7]. The work efficiency and productivity in a greenhouse depend on its type and on the type and quality of the indoor fixtures and installations [2]. When selecting a location for building a modern greenhouse, the following factors are to be taken into account:

2.1 Site microclimate

At different geographical latitudes and longitudes, the climatic conditions may differ substantially, and certain factors may even pose limitations for vegetable crop growing—insufficient daylight intensity, frequent fogs, extremely bad weather conditions, too much shade from trees, or high mountains, etc. Such outdoor climatic conditions are unfavorable for growing most of the vegetable crop varieties, which means that it would require substantial investments in indoor installations if one would opt for building a greenhouse in such a location. One of the most important factors is the wind rose chart. One should gather all the data from a relevant hydro-meteorological service for a larger number of years (up to 30, and if the data can be obtained, it would be best to get it for many years, even up to a 100, so as to know what the microclimatic extremes could expected in a given area).

2.2 Water availability and quality

Water source and supply for greenhouse crop irrigation and other necessities are often overlooked. Enough quantities of good-quality water are highly important for the vegetable crops grown in a modern greenhouse, as vegetables are very sensitive to water deficiency. Before starting with any type of vegetable crop production, either in open field or in a greenhouse, it is crucial to send samples of a potential water supply to an irrigation water testing laboratory for analysis—quality test, along with both chemical and biological ones. A water testing laboratory should be a recognized and accredited one (ISO 17025). Water chemical testing is very important for modern greenhouse crop irrigation, due to an application of modern technologies and modern irrigation systems and vegetable crop feeding. Particularly significant factor is the concentration of nitrates and certain microelements, which in high doses may be toxic to plants (e.g. iron (Fe) compounds). In such cases, there are particular technological ways of decreasing the unfavorable concentrations of certain elements and compounds. For greenhouse irrigation and other necessities, the main sources of water are groundwater from wells, surface water (rivers and ponds), rain, and municipal water. Irrigation water microbiological analysis is important for two reasons:

- Presence of pathogenic microorganisms that may affect the crops,
- Presence of pathogenic microorganisms that may affect people.

2.3 Greenhouse site physical requirements

The topography of the site affects where a growing structure is built. (Topography refers to the shape of the land, e.g., hilly, steep, rocky, flat.) The surface of the ground should be level. A 0 to 5 percent slope is recommended. Placing a growing structure on a flat surface is efficient because it facilitates easy adjustments to various mechanical controls in the greenhouse, which is economical. On steep terrains, it is recommended to build several separate greenhouses with axes parallel to contour lines [8]. Provisions must be made for the evacuation of rainfall water, and greenhouses should not be situated in hollow lands prone to landslides.

2.4 Space for expansion of the nursery

Additional space of land, which is larger than the grower's current needs, is acquired. Such area should be counted on the predicted figured out area, in order to accommodate service buildings, storage, access drives, a parking lot, etc. It means that extra space could be allotted to cover unforeseen needs.

2.5 Availability of labor force

Present and future labor needs should be estimated correctly, in order to provide necessary labor on time. Availability of labor supply has been a perennial problem in the horticulture industry. It is necessary to determine if available labor is skilled to perform routine, harvest-time, and post-harvest-time duties.

2.6 Infrastructure

Proximity to transport networks (e.g., roads, railway), access to communication systems (e.g., telephone, Internet), and availability of energy (e.g., gas, electricity) must all be considered when selecting a site for the greenhouse. Greenhouses also need convenient access to materials for growing plants (growing media, fertilizers, pesticides, etc.).

2.7 Legal considerations

When selecting and purchasing a site, there are various legal considerations Permits, licenses, and zoning regulations govern where a greenhouse may be built and often even dictate what type of building materials may be used. Such considerations may also involve relevant mandates from the Occupational Safety and Health Administration that ensures employee safety. Selecting an appropriate site also involves how the greenhouse operation affects its neighbours (schools, hospitals, parks, farms or ecological areas, etc.).

2.8 Greenhouse orientation

General recommendation is to place the greenhouse in north-west orientation [2] so that the growing plants would be exposed to optimal total light and heating, which is of a particular consideration for winter season vegetable crops. When the greenhouse is in an east-west orientation, it means that the longer sides of the greenhouse should face north and south, with the shorter ends facing east and west. A greenhouse could be built on various terrains and at different slope grades. Also, climate regions may vary, like a desert, a Monsoon region, or a cold one, but in that case, appropriate parameters should be taken into account. Before making a final decision, a feasibility study should be done in two versions: a short one and an extensive one.

3. TYPES OF GREENHOUSES

The aim here is to give a detailed description of the types of greenhouses in vegetable crop growing industry. Also, there will be parameters that should be considered when selecting a type of the greenhouse to be built, as well as the vegetable crops to be grown. Modern greenhouses are structures where the predetermined climatic and other conditions necessary for the vegetable crops production can be closely monitored and controlled automatically. In case of some sudden changes, one can respond immediately and re-establish the climatic conditions. For such elaborate greenhouse structures, there are special expert teams involved in order to come up with the best solution for an investor. Expert teams from various fields provide various software programs [9], which, by way of special sensors distributed around the greenhouse, collect information on the certain crop growing conditions and process them so that they are then compared with the predetermined values necessary for the given crop production. If any change in the conditions occurs, it is displayed on special computers built just for horticulture (different from the conventional ones). Crop growers get instructed in any new software, so that they can manage the greenhouse easily. Anyway, when a computer and the software are purchased from a renowned company, they provide excellent additional services. Also, there numerous mobile phone apps that help monitor any changes in the greenhouse vegetable crop growing.

3.1 Greenhouse type selection

The greenhouse type selection is of crucial importance in the investment profitability. It should mitigate the investment with regard to the time and the production means, as well as increase the crop yield. There are also special needs and abilities of the investor to be considered, as well as climatic conditions of the given area. Furthermore, when selecting a type of the greenhouse [8], their other important factors are to be considered before making a final decision.

Defining the grower's or investor's requirements

- Type of crops and the desired final produce quality,
- The necessary time frame for growing the desired crop(s),
- The climatic and food requirements for the crops,
- Investment planning,

The greenhouse selected crop production automatization and control level.

Climatic data:

The area climatic data should include the following:

- Minimal and maximal temperature values,
- Relative air humidity,
- Duration and intensity of daylight,
- Wind and gusts speed and direction,
- Storms, precipitation (rain and snow in particular),
- Topography (land shape),
- Types of soil or availability (obtain/use) of organic or hydroponic substrates,
- Sources and availability of good-quality water with regard to its chemical and biological properties,
- Wind rose (meteorology).

Structural solutions:

There are a number of considerations when opting for a greenhouse:

- Height, width, length,
- Number of arches or ridges,
- Number of benches,
- Ventilation (roof, lateral, or combined),
- Types of covering (plastic or glass),
- Quality irrigation and plant-feeding systems,
- Heating system (water or air),
- Hydroponic substrates,

- Fertigation and climatic control systems.

Planning:

- Defining a greenhouse structure,
- Greenhouse directional orientation, depending on the sunlight orientation and wind strength,
- Terrain sloping grade and the precipitation drainage system,
- Automatization,
- Indoor climatic conditions control (temperature and humidity),
- Irrigation and fertigation systems,
- Climatic conditions and fertigation command controls and their connection to the main power source,
- Labor.

Feasibility study:

A greenhouse feasibility study makes it possible to have an overall view of the existing abilities and a desire for modernization or for building a new, modern greenhouse. It can provide a framework within the given climatic conditions and predict the vegetable crops probable yield and quality, as well as a suitable profit. If the feasibility study shows that the greenhouse project is sustainable, then the next step is to prepare and draw a detailed business plan. Both the feasibility study and the business plan should be left to experts who shall, in the best of ways, introduce the growers or investors to the world of the greenhouse vegetable crop production.

Modern greenhouses include the following:

- Modern plastic-covered greenhouses and
- Modern glass-covered greenhouses.

The two types of greenhouses have similar characteristics with regard to its structure and indoor installations. Protection against external environment, vegetable early ripening (its edible parts), a shorter vegetation period, simplicity in controlling the climatic conditions optimal for the crop(s) growth and development, protection against diseases (conventional protection and/or biological) are just some of the numerous important aspects and necessities in the greenhouse vegetable crop growing.

Factors to consider in opting for the greenhouse vegetable crop growing are as follows:

- Longer vegetation period of certain vegetables,

- Insufficient length of a continuous frost-free period,
- High demand for the off-season vegetable produce,
- Global climate change,
- Greenhouse and open-field vegetable crop growing combination,
- A need to supply various markets, local and foreign, with the vegetable produce all-year round.

Considering the aforesaid, most of the vegetable varieties (tomatoes, peppers, eggplants, cucumbers, watermelons, cabbage, lettuce, etc.) are grown as follows:

The modern greenhouses are tall (4.5 m and more) made up of plastic or glass structures, equipped with various indoor installations that provide favorable conditions for the selected vegetable crops, as well as complete control in growing crops for the duration of the entire vegetation period (control of the crops feed, irrigation, timely implementation of agrotechnical measures, disease monitoring, and preventive actions). A general recommendation is to build the greenhouse on a flat terrain (or on a low-gradient slope, 1–2%). On the other hand, if necessary and financially sound, such structures may be built on a higher-gradient slope, but with the help of proper engineering technology that can provide a flat production area. This means that the land parcel selected for the modern greenhouse has to be leveled so that the foundation of the structure can be built properly to last for a number of years (15– 20 or more). In fact, when considering investing in the modern greenhouse and its long-lasting quality usage, there are three important factors to be taken into account and never economized on the following:

1. **Terrain leveling** of the area designated for building the modern greenhouse, as well as leveling the production area and building the foundation. The modern greenhouse should last for quite a long time (10–20 years or more) and there is folk proverb that says "no good roof without a good foundation", or better still, no good greenhouse.
2. **Greenhouse frame** should be strong and sturdy, well-built, and the best material for that is galvanized steel. If possible, the galvanized steel frame should be perforated and painted in white so as to increase indoor lighting (particularly in overcast weather during the vegetation period). As the frame casts a large shadow during the day, white paint and perforation help increase the light intensity. One should always bear in mind that modern greenhouses are mostly used for growing warm season vegetable varieties that need lots of light and its higher intensity during daytime (tomatoes, peppers, eggplants, melons, and watermelons up to 1 kg). The greenhouse roof frame is usually built of aluminium construction of various profiles.
3. **A boiler room**, well built, with at least two boilers (or more, depending on the greenhouse total area and volume). One type of fuel or even different

types may be used for boilers. It is very important that best-quality boilers are used and that the fuel is readily available on the market. So, in case one boiler breaks downs, the other is ready to be used. It is of utmost importance for the early spring and late autumn vegetable crop production. The fuel used in boilers may be oil derivatives, coal, gas, wood, pellets, geothermal sources, biomass, solar energy. The boilers heat the water that goes to the greenhouse, pumped through a system of pipes or tubes. Those then heat the indoor space quite evenly. The hot water pipe system can be arranged in several levels along the indoor space sides. Depending on the climatic conditions during the cold season and the crop vegetation period, the heating levels are decided upon.

The modern greenhouse total area depends on various factors. It usually covers 1 ha and goes up to dozens of hectares. The volume of such structures is huge, which makes it difficult for the air to circulate and refresh the space. So, the ventilation can be improved by the greenhouse vents on the roof and/or on the sidewalls, as well as by the use of fans, usually fixed along the ceiling. The optimal number of fans per area/volume unit, as well as the area and type of the vents is determined upon certain greenhouse parameters, the vegetable crop climatic needs, as well as the outdoor climatic conditions of the particular region where the greenhouse is located.

During the building of a greenhouse, one of the most important things of the construction process is building the foundation of the greenhouses. Among several types of foundations, the right type of foundation can be determined by the style of greenhouse and building codes. It is important to emphasize that the foundation is the complete system on which the greenhouse structure sits. One of the most important components of the foundation is the footing. Footers are typically poured concrete and their exact depth is determined by local building codes and the location's frost line to avoid structural damage, which can occur with the freezing of surface-level soil. Generally, the climatic conditions of a region are a crucial factor in selecting a type of the greenhouse and all its structural elements and indoor installations. It is advisable to collect all the data from a relevant hydrometeorological service for a larger number of years, of minimum 10 or up to 30 (or even more, if possible) in order to see what the extreme climatic conditions are, which helps define the greenhouse foundation and frame types, as well as the type and quality of the covering material to be used for the construction. Thus, a team of experts of various professions in charge of the greenhouse design and construction can easily propose the best of solutions for a particular greenhouse where particular vegetable crops are to be grown and in a particular region. The modern greenhouse has specific structural elements, as well as indoor installations.

4. SOME MAJOR GREENHOUSE STRUCTURAL ELEMENTS

1. **Foundation/footing** is one of the most important elements in the mod-

ern greenhouse construction for building the foundation. There are several different ways of doing it, but the particular foundation or footer type depends on the selected greenhouse style and the local building codes [2]. Actually, the foundation is the base for the entire greenhouse and should ensure its strength and stability for a number of years, even in conditions of slight terrain movement. Here, the most important factor is the soil type or structure, the land where the footer is to be placed. Footers are typically poured concrete and their exact depth is determined by local building codes and the location's frost line to avoid structural damage, which can occur with the freezing of surface-level soil.

2. **The modern greenhouse frame**—The best material for the frame is galvanized steel, while the roof frame should be made of aluminium. The greenhouse walls and roof frames depend on the greenhouse style. Many side posts and columns supporting the structure cast a lot of shadows inside the greenhouse, which affects the intensity of the photosynthesis process. So, the entire frame should be perforated and painted in white, so as to increase the indoor light intensity, particularly in the plant's vegetation period.
3. **Roof and side walls**—The roof and side wall surfaces are usually made of various types of glass or different plastic materials.
4. **Roof and side wall vents:**

   (a) Modern glass greenhouses are usually built only with roof vents,

   (b) Modern plastic greenhouses may be built with only roof vents, but may also have both roof and side vents.
5. **Roof and side wall vent insect screens**—Insect screens prevent or at least curb insect and birds' intrusion into the greenhouse. Insects may cause or transmit numerous diseases, and pathogenic microorganisms, while birds, besides transmitting disease, most often may destroy the crops physically.
6. **Doors**—The greenhouse doors, both indoor and entry ones, their size, and number depend on the size and type of the greenhouse. It is advisable to have as few entry ones as possible.

5. SOME MAJOR GREENHOUSE INDOOR INSTALLATIONS

- **Heating system installation** usually consists of a boiler room, a fuel tank, and a hot water pipe system that heats the greenhouse. Besides hot water heating system, the greenhouse can also be heated with hot air. There should be at least two boilers in the boiler room. The fuel may be the same for both or different. One boiler always serves as backup in case the other breaks down. The fuel used in boilers may be oil derivatives, coal, gas, wood, pellets, geothermal sources, biomass, solar energy.

- Cooling system
- Plumbing
- Electrical installations
- Lighting system
- Shading and thermoregulation systems
- Plant tie support system
- A system of devices that control temperature, humidity, direct sunlight intensity, or diffuse light, as well as additional lighting intensity
- Carbon dioxide supplementation system
- Benches and tables
- Hydroponic plant growing installation system
- Crop biological protection system
- A small weather stations
- Special computers and software programs that monitor and control the greenhouse climatic conditions; in a case of change, the preset parameters can be easily reset
- A power transformer substation

The modern greenhouse vegetable crop production can have various purposes:

- From seeds (or from the purchased seedlings) until the entire fruit is formed, ripened and ready to be picked
- Seedling or plug production for greenhouse or open-field plant growing (important for early warm season crops)
- From seed to seed (vegetable seed production)

The twenty-first century greenhouse has a major role in extended vegetable crops growing period, providing a great variety of produce in different times of the year [10]. Advantages of greenhouse vegetable crops growing are really numerous.

- Protection from outdoor weather conditions,
- Early plant and fruit ripening,
- Off-season vegetable production (with ever so high market demand, particularly in the North-West Europe regions),
- Control of climatic conditions favourable for the crops growth and development,
- Possibilities of controlled irrigation—the drip irrigation system,

- Simpler ways of biological and conventional protection of crops against pathogenic microorganisms and pests,
- Easier tasks for greenhouses workers,
- Environment protection and many other factors that will be discussed further on.

Generally speaking, the most important aspect, the twenty-first century vegetable crop greenhouse highlight is dominant computerisation of all the processes and the initial steps in robotic application in performing and implementing certain agro- technical measures, thus replacing human labor that is rapidly decreasing, which is actually a contradiction to an ever-increasing global population. Estimates predict that by 2015 there will be about 10 billion people on this planet of ours, and they will all have to be provided with basic living conditions, such as potable water and food. To that end, many scientific facilities and institutions, universities, and big companies involved in agricultural sector, horticulture in particular, all do research in creating possibilities for finding solutions to numerous challenges in the food and water production industry for the years to come.

6. GLASS GREENHOUSES

Glasshouses are the best protected indoor area for growing vegetables in regions where climatic conditions are unfavourable for their growth and development in either open field or plastic-glazed greenhouses. They are the best environment for controlling the climatic conditions and the mineral feed of the crops during their vegetative period. At the same time, also, the warm season crop vegetative period could be prolonged when compared with growing such crops in open field. Furthermore, the modern glass greenhouses are perfect with regard to applying modern technologies and scientific developments in growing crops. Of course, the economic factor is of utmost importance for commercial crop production. So, both short and extensive financial plans have to be made. Glass greenhouses are built mostly in cold climates, but as they provide the environment where the climatic conditions can be completely controlled, they are also a good choice for growing vegetables even in warm climate regions. Modern vegetable crops growing in glass greenhouses are very intensive from the perspective of capital investments, application of the latest production technologies, know-how, labor, energy, substrate, hydroponics. The greatest issue in modern greenhouses is the high cost of energy. There have been, however, some significant shifts toward reducing the cost of heating. The Venlo greenhouse type [8] is probably the most-built type worldwide and the most economical one among the professional vegetable crop growers. The design of the Venlo greenhouse is suitable for all sorts of crops and under various climatic conditions. This type of greenhouses was developed in the

Netherlands and was named after the region and the place of Venlo where it was first made (south- east the Netherlands, toward the German border). The whole region has one of the largest concentrations of greenhouse horticulture.

The Venlo greenhouse is the galvanized steel frame structure, as the best and most durable material that can deal with the weight of the entire greenhouse and the roof frame. The roof frame can be made of aluminium or galvanized steel of various profiles. The roof frame also carries the windows/openings that in the Venlo glasshouse can be opened from both sides and make 20–30% of the total glass surface area. As the frame structure consists of a great number of columns of various sizes and dimensions, there is actually much shade inside the greenhouse that affects the crops. For that reason, it is advisable for the entire frame structure to be perforated and painted in white so as to increase the indoor lighting. The crops that are mostly grown in greenhouses are as follows: tomatoes, peppers, eggplants, cucumber, melons and small watermelons (up 1 kg), lettuce, spinach, grapes, raspberries, blackberries, strawberries, blueberries, and, of course, seedlings of vegetables and flowers and various edible berry fruit. Often grown in greenhouses are also broccoli, cauliflower, leek, bananas, ginger. Actually, all horticultural varieties can be grown in greenhouses, but it would be best to check first how cost-effective it could be, that is, to do a feasibility study. If a feasibility study has to be done as a guaranty for the bank loan or a subsidy, then such a study should be detailed and done by a recognized firm or a consultant. Since greenhouses provide the environment for almost complete control of the selected crop production, the grown crops are safe for human consumption. That means they meet the best standards with regard to both safety and quality. Furthermore, there is a minimal environmental pollution. Other benefits are that the early crops come to market sooner and also, high yields are achieved. For instance, in hydroponic tomato production (on the rockwool growing substrate), the yield can be even up to 100 kg/m2, or 1000 t/ha per annum. The fact is that the yield (or the production rate) in the modern glass greenhouse can be 10 times as much as in the open field. High yields are achieved through intensive growing methods and by prolonging the production season.

Besides the vegetable quality and the environment pollution reduction, the yield should be high also because of a return of the investment and payments of additional costs. The majority of the glass greenhouses manufacturers put on offer glasshouses and the horticultural production technology, which may provide optimal production under the given market conditions, so that the additional costs would pay out through the earnings from the additional production. Consumers in most of the developed countries expect to find good-quality vegetable produce grown in glass greenhouses on their respective markets. In most cases, it is the improved quality of the vegetable produce that gives a motive for starting the glasshouse vegetable production. For different types of produce and different markets, there are also different parameters for food quality. The quality parameters for the fruit and vegetables required on the market may include the following: taste, flavor, aroma, size, color, stage of ripeness, firmness or softness—texture, shelf life, the absence of any spots or defects, minimum or no waste, no signs of pests or disease, with or without leaves, and many more. One of the most important standards in the primary agricultural production, which is also a quality guarantee of both the primary agricultural production and the produce itself, is the GLOBALG.A. P. standard/certificate. According to the construction and engineering characteristics, appearance and purpose, there are several types of glass greenhouses that could be single-span, multi-span, lean-to, gabled.

6.1 Single-span greenhouses

- Single-span glass greenhouses occupy a single area that is not partitioned in any way.
- They are easy to air and ventilate, but are hard to heat.

6.2 Lean-to greenhouses

- Usually of small dimensions/area (e.g., 80–200 sq. m)
- Its north side is usually higher than its south side, and with its north side, it leans to the south wall of another structure (but necessarily)
- Usually used for growing vegetable seedlings and vegetables in containers or pots arranged on shelves or benches.
- Width, length, and height (from the ridge board to the eave) can be at will,
- Lean-to-glass greenhouse vertical cross section (front): about 80–200 sq. m, 2.0–5.5 m (or higher)

6.3 Gabled glass greenhouse

- Gabled roof
- Area: 500–1000sq.m (or more)

6.4 Multi-span glass greenhouses

- In structural terms, the multi-span greenhouse consists of a number of even span greenhouses connected along the length of the house.
- At the attaching points, there is only the frame without glass glazing. If the area has to be partitioned, it can be done with both glass panes and plastic film that can be easily put up or down (whatever is necessary).
- The total area of the multi-span glasshouses can be from 1 ha to several ha (10 ha–15 ha or more) (**Figure 1**).

The ratio between the glass (cooling) surface and the multi-span greenhouse total volume is such that can cut the heating costs, but are hard to ventilate through vents (windows). In such cases, it is necessary to install a special ventilation system (including fans as well) that allows good indoor air exchange and circulation (**Figure 2**).

Figure 1. The 21st century multi-span greenhouses with a clearly defined road network and the meticulously kept weedless lawns (The Netherlands).

Figure 2. Multi-span greenhouses, ancillary structures, and a weather station on one of the production complex ancillary buildings (Serbia).

A modern greenhouse weather station usually consists of instruments that in regular intervals measure some of the weather parameters that are significant for the greenhouse crop production. Such instruments may include an anemometer, a wind vane, a thermometer, a hygrometer, a barometer, then sensors for measuring rainfall, solar radiation (direct and diffuse solar radiation). The obtained data are used locally for arranging the greenhouse production. Also, the data can be sent, in a special computer format, to the World Meteorological Organization (WMO) to help build a model for weather forecasting all over the world. WMO is a specialized UN agency and is dedicated to international collaboration and cooperation between the countries in terms of climate conditions control on this planet of ours, as well as making a model for the horticultural production monitoring and control.

7. GLASS GREENHOUSES BASIC STRUCTURAL ELEMENTS

- **Foundation**—consists of a footer (20 cm–50 cm high); built of concrete or concrete slabs.
- **Supporting frame**—consists of columns (galvanized steel), trusses, beams, purlins and braces, making a grid-like frame, painted in white, and usually of small profiles so as not to cast shade.
- **Roof and side glass glazing**—consists of glass panels 3 mm–5 mm thick with (usually) aluminum frames; glass fixed with silicone sealant.
- **Roof and side vents/windows**—make about 20–30% of the total glass area; in modern greenhouses they are mostly on the roof.
- **Doors**—can be single or double doors; the number depends on the greenhouse size and type.

The photos below show the greenhouse galvanized, painted in white steel frame and the roof frame, vent insect, and bird screens (preventing penetration of insects that may cause or carry diseases to crops and birds that can damage the plants). Also, it could be seen a glass outer surface cleaning machine (**Figures 3–6**) [2].

Figure 3. Greenhouse cleaning, washing and disinfection (Kucura, Serbia).

Figure 4. Glass greenhouse-galvanized steel frame (the Netherlands).

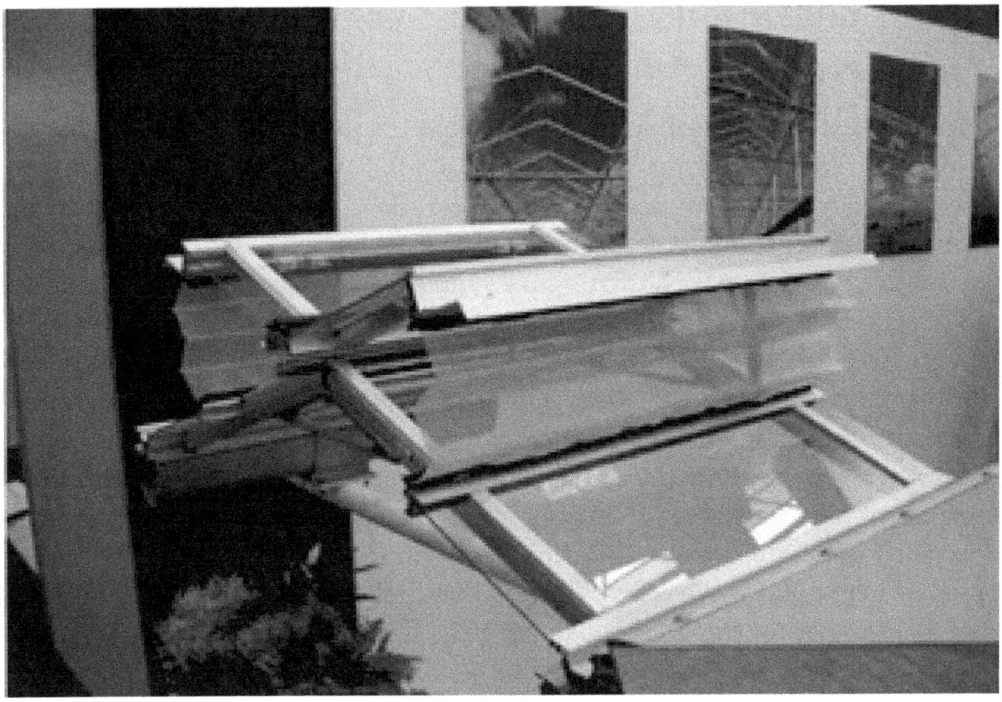

Figure 5. Roof and side vent insect and bird screen (crop protection against insects and birds).

Figure 6. Greenhouse glass roof washer.

8. GLASS GREENHOUSE ESSENTIAL SYSTEMS

8.1 Heating system

One of the essential greenhouse installations is the heating system, for the commercial reasons and for creating the suitable indoor climatic conditions. However, it may be the most expensive of the greenhouse installations and a major concern in the vegetable crop production. The system consists of a boiler room with suitable boilers—at least two (**Figure 7**) and an array of pipes which carry, say, the warm water, which is pumped through the pipes, heating the environment. The total area and length of the heat radiating pipes depend on the greenhouse size, water input temperature, both indoor and outdoor air temperatures, and so on. On the other hand, it is the best way to heat the greenhouse indoors as the heat is released evenly, so the air is heated evenly, without major oscillations in temperature. The warm water carrying pipes may be arranged on several levels, particularly along the glass walls. At moderately cold weather, only the first level of pipes is in use. But if the winters are very cold, then the second and third levels may be put in use. In this way, the upper zone is also heated, the zone under the roof, which allows roof glass defrost or snow melt and its discharge.

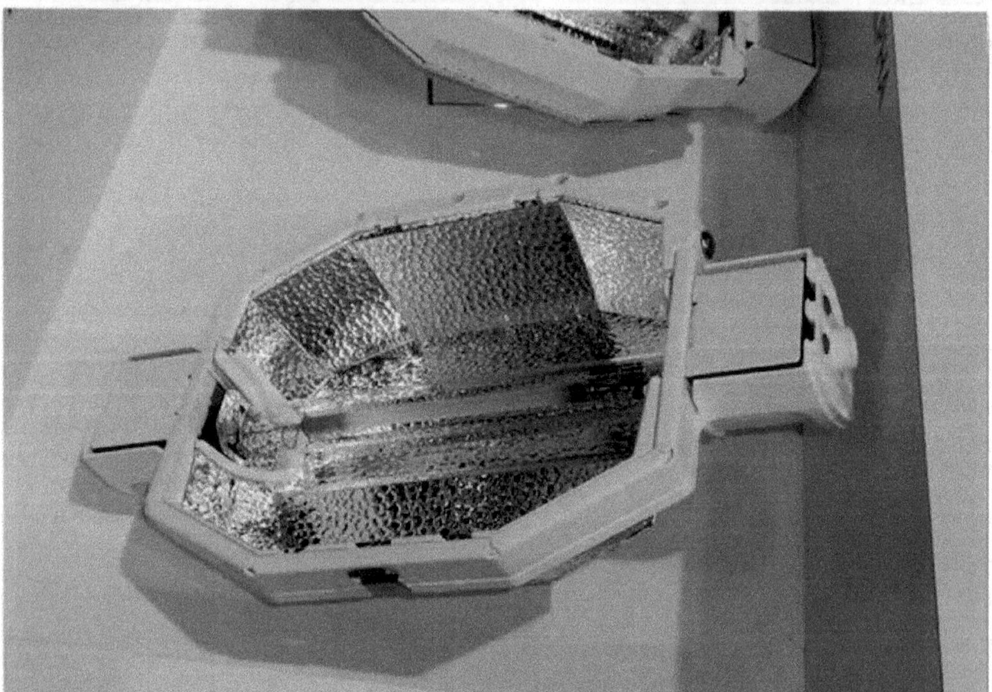

Figure 7. Supplemental lighting lamps (the Netherlands).

The heating system has to provide heat that may be lost in various ways:

- Heat loss through the glass glazing (conduction),
- Heat losses of the greenhouse can be occurred through ventilation and infiltration (fans and air leaks). Such heat transfer includes the movement of air and the movement of water vapor. When water in the greenhouse evaporates, it absorbs energy, and when water vapor condenses back to a liquid, it releases energy. It means, if water vapor condenses on the surface of an object, it releases energy to the outside environment.
- Another type of heat loss is air infiltration, which depends on the age, condition, and type of greenhouse. Older greenhouses or those in poor condition could have invisible cracks around doors, or perforations in covering material through which large amounts of cold air may enter.
- Radiation heat transfer occurs between two bodies without direct contact or the need for a medium such as air. There are two types of radiation that affect greenhouses: solar, or shorter wavelengths and longer wavelengths, or infrared (IR) radiation. Shorter wavelengths from the sun pass through the glazing and heat the greenhouse plants, soil, and structures. The amount of radiant heat loss depends on the type of glazing, ambient temperature, and amount of cloud cover. Greenhouse radiation cooling is most noticeable on clear winter nights.

Nowadays, many types of heating systems are available for the greenhouses. And selecting the proper heating system is important. Some heating systems cost less to buy, or can use less expensive fuels. Others may have a higher initial cost, but they are more efficient, easier, and safer to operate. The most common and least expensive is the unit heater system. In this system, warm air is blown from unit heaters that have self-contained fireboxes. Heaters are evenly located throughout the greenhouse, depending on its volume. These heaters use electricity, which increases its consumption, although the heaters themselves are not expensive. The best and the safest heating system is a central boiler—a boiler room, and a system of pumps and pipes that distribute hot water or steam. The best system is where the water is heated in the boilers and is then distributed around the greenhouse with a system of pumps and pipes. The warm pipe surface evenly radiates heat and also evenly heats the glass or plastic greenhouse indoors. Types of fuel for the greenhouse heating system may be solid, liquid or gaseous, then electricity (generated using wind power), and for some time now, the solar power has had an important role in heating various structures, including greenhouses. Types of fuel that are mostly used are fuel oil, coal, wood, biomass (if there is a steady source of biomass), natural and LP gas, geothermal sources. Electricity can also be generated using wind power harnessed by wind turbines erected in windy places and the generator converts the wind power into electricity. And, of course, there is the solar energy, where solar panels convert it to electricity. The cost and availability of these sources will vary somewhat from one area to another or from one country to another. Convenience, investment, and operating costs are all further considerations so as to select a type of heating that is most suitable for modern, successful, and economical crop production. Savings in labor force (or where it is not readily available) could justify a more expensive heating system with automatic controls (**Figures 8–12**).

Figure 8. A boiler in the greenhouse boiler room (the Netherlands).

Due to a rapid exhaustion of natural resources that are used for fuel and so in order to be environment friendly, the renewable energy sources are increasingly in use in modern crop growing industry. The renewable energy sources are solar energy, wind and water power, and geothermal springs.

Figure 9. A geothermal spring (Serbia).

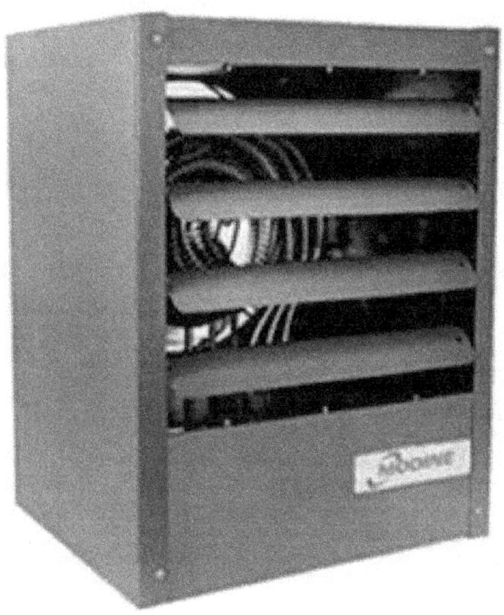

Figure 10. A greenhouse unit heater (the Netherlands).

Figure 11. Solar panels that convert solar energy to electricity (the Netherlands).

Figure 12. A wind turbine that converts solar energy to electricity (the Netherlands).

8.2 Plumbing and irrigation systems

The greenhouse plumbing and irrigation system is designed according to the requirements for the particular vegetable crops that are to be grown indoors, a selected irrigation system, and the chemical and biological quality of the water coming from an available source. Selecting a modern greenhouse irrigation system is an important step as it has to meet all the requirements of the modern vegetable crop production. It also has to be an efficient system as the supply of potable water is in decrease, which goes for the agricultural water, as well. So, the main goal is to achieve high-yield and good-quality vegetables with a highly controlled use of water. The water supply coming from an available source should be of good quality in chemical and microbiological terms. The chemical properties of the irrigation water are very important because the source itself (even the large rainwater collection tanks and ponds near the greenhouse) may contain undesirable elements or compounds, particularly the microelements that even in small quantities can be toxic to the crops or may make compounds that the crops cannot use. Also, the irrigation water may contain compounds (e.g., calcium carbonate) that can plug the plumbing system, particularly if it is a drip irrigation system.

There is also a microbiological property of water. The water may contain a number of pathogenic microorganisms that may be harmful to the crops, causing diseases in plants. Moreover, such water may contain microorganisms that are harmful to people, as they may be found in the vegetables we eat (although never in concentrations harmful to people). However, such microorganisms may be found on the surface of vegetables, thus causing some of the infectious diseases in people (due to poor hygiene after harvesting or if consuming unwashed vegetables at our homes). The most common pathogenic bacteria are *Escherichia coli, Listeria sp*., and others.

Figure 13. Rainwater pond for the greenhouse irrigation (the Netherlands).

A recommended irrigation system for greenhouses is the drip system. Due to all the above-mentioned reasons, the greenhouse irrigation system has to be selected wisely, has to be a good-quality one, and well maintained afterward. The quality of water is perhaps the most serious factor when selecting the irrigation system. If the water is of a poor quality, the system will not operate properly and will be hard to maintain. A water analysis may provide the investor/grower with a much clearer picture of the potential problems on a given location, with regard to the water supply and its use in the drip system. Also, a water test may be of help in considering all the possible solutions to the irrigation problems. The water analysis should include tests to pH values, dissolved solid matter, manganese, iron, hydrogen sulfide, carbonates, and bicarbonates. The amount and the size of the particles of certain compounds should be determined, as it determines the size of the emitter filters. The drip irrigation system may provide favorable grounds for the development of bacteria, which results in the formation of sludge deposits and anaerobic environment. Certain bacteria may cause a build-up of manganese, sulfur, and iron compounds deposits, which may plug the drip system. Furthermore, there are certain algae that may enter the irrigation system from the water supply and build up larger particle deposits. It often occurs when the irrigation water has a high biological activity and high iron and hydrogen sulfide content. The soluble iron compounds (Fe^{2+}) are the primary source of energy for certain bacteria, thus leading to iron deposits (**Figure 13**).

For all the stated reasons, it is important to use the high-quality water, as well as the high-quality system filters, or the poor-quality water may cause a lot of trouble which does not allow the modern vegetable crop production. So, regardless of the quality of water, the greenhouse irrigation water filtration has to be regularly maintained. The system has to be cleaned frequently and the water chemically treated, if necessary. There are some steps to be taken in order to plan further actions: regular basic water quality checkup, water flow and water pressure monitoring, and watching for the color of the water itself. In many countries, the main irrigation water source is rainwater collected in large ponds (wherefrom it cannot leak to the ground) and regularly tested at accredited laboratories where chemical and microbiological content is checked and then, certain measures are undertaken to make it suitable for its safe use in vegetable production (**Figure 14**).

Figure 14. Tanks for water, nutrient solution, and nitric acid used for adjusting the nutrient solution pH.

Drip irrigation system basic components:

- A pump station consisting of a motor and a centrifugal well pump and a turbine. The pump is used to distribute water evenly to each and every individual plant, with even flow and pressure values.
- An array of pipes and emitters consisting of a main duct and side pipe network. After filtration, the main line brings water to the greenhouse and on, through the sub-main line to the lateral ones.
- Computerized measuring that the correct amount of water has been used and delivered to the greenhouse. This is to record advances in irrigation efficiency for the better management of changes or improvements of the irrigation system itself. Meticulous and precise water measuring in the crops determines the correct volume of water to be used.
- Opaque tanks to prevent chemical and microbiological changes in the nutrient solution and water under the influence of sunlight.
- Filters, as already mentioned, have an important role in preventing the drip system plugging and blockage. There are various types of filters, depending on a source of blockage, which can be determined in water testing.
- A correct choice and placement of valves in the irrigation system is crucial. The valves control the water pressure, flow, and distribution through the system, particularly under various conditions that may occur while the irrigation system is in operation.
- Water flow gauges to measure the flow between the water source and the greenhouse (**Figure 15**).

Figure 15. Drip irrigation water/nutrient solution pumps (The Netherlands), Drip irrigation system—drippers around the plants on the white gutters (the Netherlands).

8.3 Electrical installation

Electrical installation may well be the most important one in the glass greenhouse as it is necessary for various production operations. A crucial one is starting a motor that supplies power to different devices (opening and closing the vents/windows, shading, switching on/off the fans, additional lighting, etc.). Every glass greenhouse for professional and commercial production has to have its own power transformer substation so as to provide a reliable energy supply in a case of a power cut, which would otherwise impede the production.

8.4 Greenhouse supplemental lighting

Supplemental lighting is necessary in seedling production, as well as in growing vegetable crops that have a longer vegetative period and high demands for light (tomatoes, peppers, cucumber). In greenhouses, it is usually needed during long winter months and periods of overcast. The supplemental lighting prolongs a day, compensates for a natural light limiting effect in winter, and enhances the amount of the available light. Supplemental lighting should not be confused with photoperiodic lighting, which is applied to create long days, thus controlling the plant growth and development processes.

Figure 16. Block-greenhouse with supplemental lighting.

Supplemental lighting is high-intensity light (6000–7000 lx or more), whereas photoperiodic lighting is of a much lower intensity (70 lx–100 lx). For now, in the vegetable crop production practice, the high-intensity discharge (HID) lamps and LED lightning are in use, which provide wavelength light from 300 to 700 nm (which is needed for the photosynthesis process). The HID lamps, besides needed light wavelength, emit high heat (up to 50%), so they must be placed at about 02 m or more above the crops. Otherwise, such lamps could overheat the greenhouses (what change the climate in the greenhouses) and could damage the plants, too. The LED lightning are placed at about 40 cm or more above the crops or in between the crop's rows in a greenhouse, as they almost do not produce heat and could not damage the crops. There are studies about the ways how the plants use the incoming photosynthetic active radiation (PAR), which is based on a principle of an exponential increase of absorbed photosynthetic active radiation, with the increase of the leaf area index (LAI) [11–16] (*Figure 16*).

8.5 High-intensity discharge (HID) lamps

The high-intensity discharge lamps include two types of lamps:

- Metal halide (MH)

Metal halides (MH) lights are commonly used during vegetative plant growth but are less popular than HPS lamps for flowering and fruiting. If MH lamps are used in the flowering or fruiting stages, they are often of a higher rated power to provide more red-light output.

- High-pressure sodium (HPS)

High-pressure sodium (HPS) lamps produce light mainly in the yellow and red end of the light spectrum, which makes these lighting systems a great fit for late-phase (flowering and fruiting) plant growth. They can also deliver greater light intensities than lamps used exclusively for vegetative growth (for instance, MH lamps).

8.6 LED lightning and modern urban vegetable growing light-emitting diodes (LEDs)

Light-emitting diodes (LEDs) represent a technology for the indoor vegetable production, which has technical advantages over traditional lighting sources, as well as a significantly positive impact on the plant photosynthesis process and on the crop yield. LED lightning has been tested for horticultural applications, both in greenhouses and in special chambers with a total control of climatic and other conditions necessary for the crop's growth and development. It is used in growing leafy vegetables and herbs, as well as in the production of tomato, pepper, cucumber, and some more vegetables in greenhouses. Depending on the vegetable varieties and their edible parts (vegetative part, fruit, immature flower heads), LEDs could be designed to emit light recipes for each phenophase of the crop's growth and development. One of the most important features of LEDs for vegetable growing is that the LED lightning almost does not produce the heat, which could damage the plants. There is a combination of additional lightning and its quality and supplemental carbon dioxide empowerment vegetable growing in the greenhouses. It is vital that both light and carbon dioxide are provided in sufficient amounts within the greenhouse, or otherwise, a lack of either may pose a limiting factor for the photosynthetic process and consequently for the crop's productivity. Controlling climate conditions and nutrition in protected area, without or with daylight, gives opportunity to produce vegetables, which is safe for humans and environment. It is called ““City Farming“ or “Urban Agriculture.” The most important factors are the LED lighting, carbon dioxide concentration, air humidity, and conditions that keep the leaf stomata open in order to uptake carbon dioxide [14–19] (**Figure 17**).

Figure 17. LED lighting (The Netherlands), Growth chambers without natural daylight, providing optimal conditions for vegetable crops growing under LED lighting (the Netherlands).

8.7 Light quality in supplemental lighting

The suitable light quality in the greenhouse per vegetable species refers to the wavelengths (colors) that are efficient in triggering photosynthesis and other growing processes in vegetable crops. Successful growing vegetables in greenhouses depend on the visible spectrum wavelengths of about 390 nm to 760 nm, which is only a small portion of the sunlight (radiation) spectrum. In general, the visible light consists of the following: violet (380 nm to 430 nm), blue (430 nm to 500 nm), green (500 nm to 570 nm), yellow (570 nm to 590 nm), orange (590 nm to 630 nm), and red light (630 nm to 760 nm). The most important visible light range mainly corresponds to the PAR from about 400 nm to 700 nm. Those wavelengths have the right amount of energy for the biochemical processes, and their participation in the available light is important for the quality of light during vegetable growing. About half of the incoming sunlight energy participates in the photosynthetic processes. The rest of the energy is from the sunlight short wavelength spectrum (UV—ultraviolet radiation) and sunlight long wavelength spectrum (IR—infrared radiation).

Blue section of the spectrum is known as a cool light, which encourages vegetative and leaf growth through strong root growth and higher intensity of photosynthesis. **Red section of the spectrum** induces stem growth, tuber and bulb formation, flowering, and fruit production, and chlorophyll formation. **Far-red light** may cause plants to stretch (elongate) and may trigger flowering in some long-day plants.

Figure 18. LED lighting between the rows of cucumber plants growing on rockwool in a modern glass greenhouse (the Netherlands).

The plants are exposed more to the far-red than to the red light, which could be a challenge with the greenhouse vegetable crop production due to possible shading, or due to the reduced plant vegetative space. **Green and yellow sections of the spectrum** that reach the plants are reflected. Most of the absorbed sunlight belongs to the blue and red part of the spectrum. However, the recent studies have shown that plants do also absorb some green and yellow light, using it in the process of photosynthesis [8]. For the time being, in the greenhouse vegetable crop growing practice, the high-pressure sodium (HPS) lamps are used, but also the LED lightning, which gains an increasing significance in the plastic and glass greenhouses as well as in the special chambers for vegetable production without daylight. Also, in The Netherlands, the latest studies at the research centers of Wageningen and Maastricht universities have their guidelines for greenhouse lighting with little or no natural daylight for special feature vegetable crops growing—increased vitamin C content, reduced nitrates content, increased sugar content, higher yield (**Figure 18**).

8.8 Supplemental carbon dioxide in the greenhouse

The carbon dioxide (CO_2) gas is the essential component for the process of photosynthesis. Plants uptake it through their stomata on the leaves. Photosynthesis is the biochemical processes where the light energy is used to convert carbon dioxide and water into complex sugar compounds (carbohydrates) and oxygen (O_2) gas. The sugars in plants, formatted in the process of photosynthesis, are then used for the plant development and growth. In the air (outside the greenhouse), there is about 400 ppm of carbon dioxide. The CO_2 concentration in greenhouses could be increased specific installation in the greenhouses, for example, from boiler room, when energy sources are burnt. However, inside the greenhouse, the amount of CO_2 may be significantly depleted as plants use it intensively in the process of photosynthesis, or through the greenhouse windows, and it may lead to a decreased crop productivity and yield. For that reason, "CO_2 fertilization" or "CO_2 enrichment" is a standard practice in modern greenhouses and should be controlled in order not to form too high concentration (over 1500 ppm), which could be toxic for the vegetable crops (**Figure 19**).

Figure 19. Photosynthetic process equation.

Since there is about 500 times more oxygen in the air than carbon dioxide [17], it makes sense to increase the CO_2 concentrations in the greenhouse (particularly in highly equipped glasshouses). It has a positive effect on the oxygen-carbon dioxide ratio. The photosynthesis is higher by 30–50% at CO_2 concentrations of about 1000 ppm, regardless of the amount of light.

Photosynthesis depends on light, temperature, air humidity, and carbon dioxide contents in the greenhouse. There is often a question of what is the optimal concentration, but it is hard to give a correct answer to it as the process of photosyn- thesis does not depend solely on CO_2. Also, a point should be made that climatic factors affect the stomatal opening mechanism (through which the plants uptake CO_2). Generally, a small increase in the plant photosynthesis process may be achieved at 1000 ppm to 1200 ppm, but then, there is also an increased possibility of damage to the crops. One experiment done on eggplant crops showed that the first damage to the plants occurred at a constant CO_2 level of 800 ppm [17]. Quite often, the intensity of the photosynthesis may be higher at lower doses of carbon dioxide and higher intensity of light, and the other way around. Supplementing the greenhouse air with carbon dioxide may not be necessary at all as long as the processes of the crop development and growth are quite satisfactory for the vegetable grower. At the same time, in a case of intensive greenhouse ventilation, the carbon dioxide concentration may drop below a level that is necessary for the normal photosynthesis process, so increasing the CO_2 concentration may not be an economical measure. If the crop quality and production are below the satisfactory level, carbon dioxide supplementing should be the next measure. The vegetable production in cloudy period of the year, or cloudy days, increases the potential need for CO_2 supplementing the greenhouse air, which actually corresponds to a lower ventilation rate due to mainly low outdoor temperatures. According to Kamp and Timmerman [9] normal ventilation provides an amount of carbon dioxide that is similar to its levels in the outdoor air (350 ppm–400 ppm). But then, frequent ventilation in the greenhouse may not be economically desirable for the enrichment indoor air with CO_2. The necessary greenhouse carbon dioxide concentration is determined upon the type of the crops grown in the greenhouse, the greenhouse total volume and ventilation, lighting, temperature, air humidity, and stomatal opening. Since carbon dioxide is one of the products of burning (e.g., fuel for greenhouse heating system), this segment of the heating process can be used for supplementing the greenhouse air. There are various ways of extracting carbon dioxide from other products of burning (fuel), so that the CO_2 from the boiler room can be dosed and at certain times directed and distributed into the greenhouse.

Also, pure carbon dioxide can be used, which is delivered to growers in special tanks, in liquid form and then can be converted into gas and distributed in the greenhouse. That way of supplementing the CO_2 has become increasingly popular as it eliminates any potential damage to the crops, allows control of other greenhouse climatic conditions that regulate the process of photosynthesis and crop productivity, provides easy control of the carbon dioxide levels, and is more flexible for supplementing the CO_2 when necessary. Also, it would be advisable to install a proper system that registers the CO_2 concentration and then distributes it in the greenhouse. Such a system, like in other greenhouse installation operations, has corresponding sensors that are linked to special computer software that registers, monitors, and controls all the greenhouse environment parameters. In this way, it is possible to detect a cause of each change and correct it in a short period of time, and potential damages of the vegetable crops could be easily prevented (**Figure 20**).

Figure 20. Liquid carbon dioxide tank for supplementing it in the greenhouse (Serbia).

Figure 21. Plastic perforated pipes distributing CO2 (The Netherlands).

The distribution of CO_2 depends mainly on-air movement within the green- house, as CO_2 does not travel very far through diffusion systems. One of the pure CO_2 distribution ways is by a central pump that pushes it into a system of flexible perforated plastic pipes (made of polyethylene or other plastic material). The pipes for CO_2 distribution are placed below the substrate special gutters with plants (if crops are grown in such gutters), or in the lower sections of the crops (if the plants are not grown in gutters). Then, through the pipe perforation, the carbon dioxide is distributed in the air around the plants (**Figure 21**).

8.9 Greenhouse screens

Greenhouse screens provide enough light and may reduce oscillations in the indoor climatic conditions. Today, there is a vast variety of greenhouse screens made of different materials. For shading, polyethylene, polyester, acryl, and aluminium knitted cloth is used. They may be of various types of knit, physical characteristics, and colors. Besides regulating the greenhouse lighting, they may be great in controlling the indoor temperature and relative air humidity, thus saving on heating and electricity. Depending on a particular use and a usage period, they are made of cloth, plastic, or aluminium. The choice of material, fiber thickness, and the type of knitting/weaving should correspond to the screen purpose. There is also a possibility of automatic, semi-automatic, and manual screen operating system. The choice of the screen or curtain system depends on the purpose, production period, the necessary crop growing conditions, the amount of the necessary lighting, certain climatic parameters in a given geographical area, the greenhouse total volume and ultimately on the quality of the shade screens, as well. The shading screen is mostly used to prevent a negative effect of sunlight radiation during daytime and excessive cooling during night or cold periods. In regions of moderate climate and a great number of overcast days, the screen system actually saves energy, transmitting light and controlling the air temperature and humidity. The screen system can be installed horizontally (under the roof) or vertically (along the side walls). Horizontal systems have an important role in controlling the air relative humidity and can be installed at different heights in the greenhouse. Also, vertical screens or curtains can be installed as a mobile boundary and/or a partition wall, separating individual climatic units within the greenhouse. This is very useful when crops of different climatic requirements are grown. The screen systems are classified and designed according to their use in controlling the greenhouse climatic conditions, transmission of light, air temperature, and humidity, so that the environment parameter optimal values for growing crops can be reached. This can be achieved with either open or closed screen systems. If, say, only temperature has to be controlled (preserving heat), transparent energy-saving screens reduce energy consumption and allow nearly complete light transmission. If the greenhouse vegetable crop production goes on during the best part of the year, including both sunny and cloudy, both warm and cold parts of the year, then all the greenhouse parameters should be taken into account for the crop production cycle. If this is the case, a double screen system should be installed: An upper-level screens, closer to the roof, and a lower-level screen, installed below the former. The upper-level screen material should be of various woven plastic or aluminiums fibres, while the lower-level positioned ones should be transparent to transmit light, but also to maintain balanced humidity and temperature, and to save the indoor heat, too. Certain screen system is used for various purposes: protection against ultraviolet radiation or just for shading (light is not transmitted), or for transmitting and maintaining a section of the sunlight spectrum (wavelengths from 400 nm to 700 nm), which is crucial for the process of organic matter synthesis, thus producing good-quality vegetable crops (Figure 22).

Figure 22. Greenhouse screens (The Netherlands).

Also, it is very useful to install the screens on the greenhouse windows/vents. When vents are open, the screens protect against insects, birds, dust, or litter that winds may blow into the indoor area. At the same time, the screens maintain the indoor climatic conditions. The basic screen material component is polyethylene or polyester, so in a case of a greenhouse fire, the screens pose a great threat. For that reason, it is important to provide enough spacing between the screens or curtains (follow the manufacturer's recommendations) or if that is not possible, the screens should be selected from fire-resistant materials, so that the fire could be localized to just one section of the greenhouse. Such fire-resistant screens have fibers that are coated with fire-resistant material. In the modern greenhouse that covers a vast area (several hectares or tens of hectares), the screen and curtain system is moved around with special motors that automatically respond to the indoor climatic changes. Screen are necessary in greenhouses, particularly in the glass ones, as they are a great help in providing good or even optimal climatic conditions necessary for the crop's development and growth (**Figure 23**).

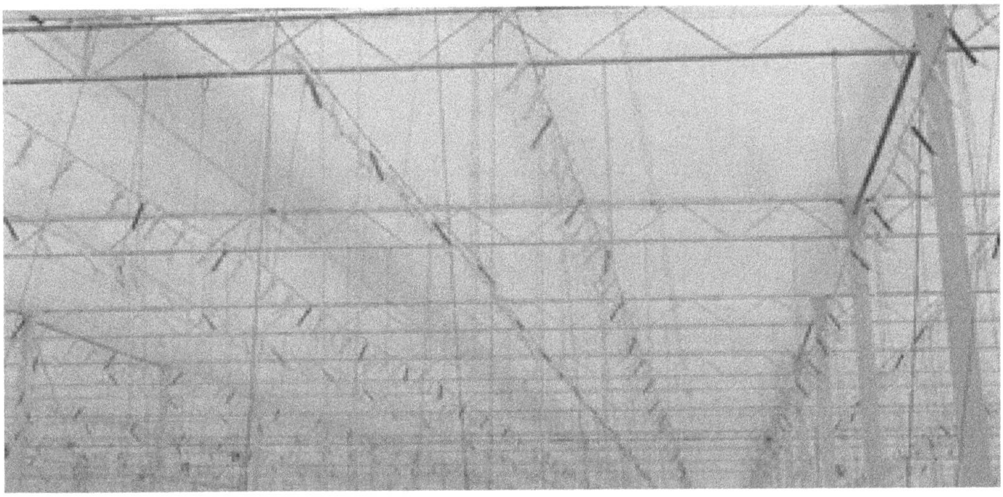

Figure 23. Curtains, Screens, guide wires, and plant-supporting strings (The Netherlands).

The screens are usually installed under the roof, and inside the greenhouse, they prevent heat from escaping (during night or cold periods) or excess heat (during summer). For this purpose, knitted polyethylene, polyester, acrylic, and aluminum curtains are used, but also curtain strips, plastic and aluminum film, or roof are painted with lime or special paints.

8.10 Crop support system

The crop support system may be permanent, temporary, or a combination of the two. The system consists of support posts, guide wires, and strings that are tied to individual plants to support their growth. The design of these structures has to correspond to the needs of the crops, so it can vary. The plant support system is made of support posts usually made of galvanized metal and horizontal guide wires to which the strings (plastic and hemp string) that support the plants are tied. The metal guide wiring is positioned at various heights or levels relative to the green- house floor. So, for instance, the guide wiring for cucumbers is positioned at a height of about 2 m and for tomatoes at 4 m – 4.7 m. In order to fix the crop to a string or guide wire to support the plant stem (cucumbers) or flower clusters (tomato), various in type, size, and purpose plastic clips and hooks are used (**Figures 24 and 25**).

Figure 24. Tomato crops tied to the supporting strings and guide wires (the Netherlands).

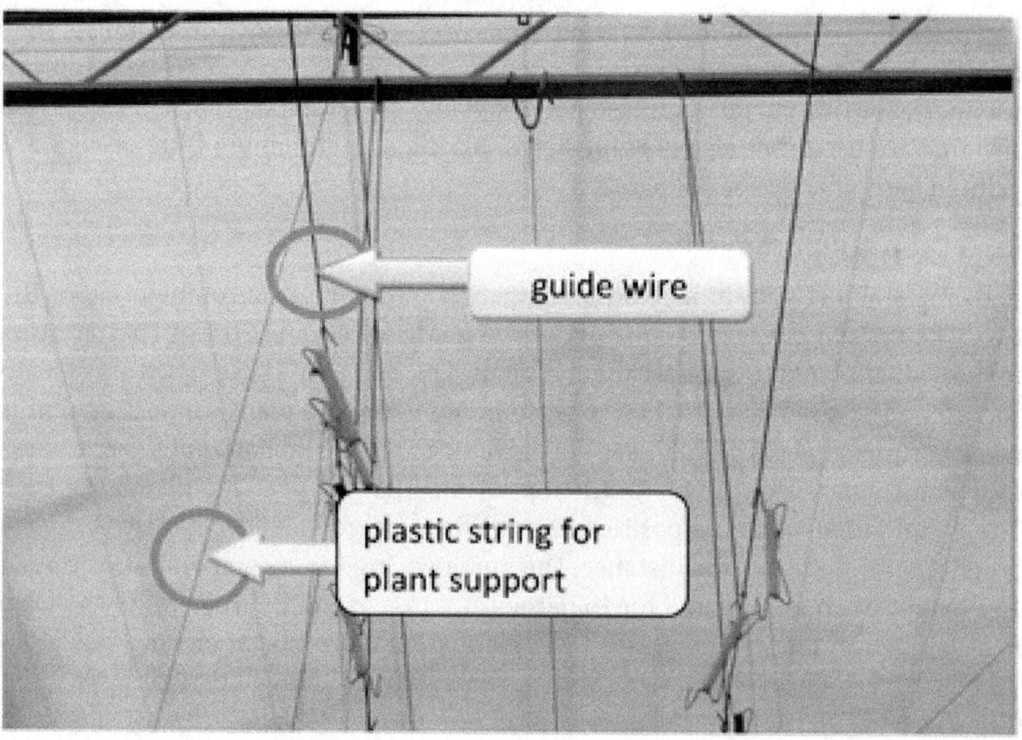

Figure 25. Crops support structure and guide wires (The Netherlands).

8.11 Greenhouse climate control system: humidity, temperature, and other factors

Here, we have various types of sensors for measuring and registering various climatic parameters in the greenhouse. The sensors are linked to a computer and computer software developed just for that purpose. In this way, the registered data can be viewed and saved and acted upon or responded to if some of the data values vary from the set parameters. Computerized monitoring of all or most of the parameters is highly significant for the development and growth of the crops as it provides that the production of the given crops is satisfactory and allows for timely response in a case of any undesirable changes to the already set production parameters (**Figures 26** and **27**).

Figure 26. Plastic film placed under the roof inside the greenhouse for catching water droplets from condensation, protecting the crops (Serbia).

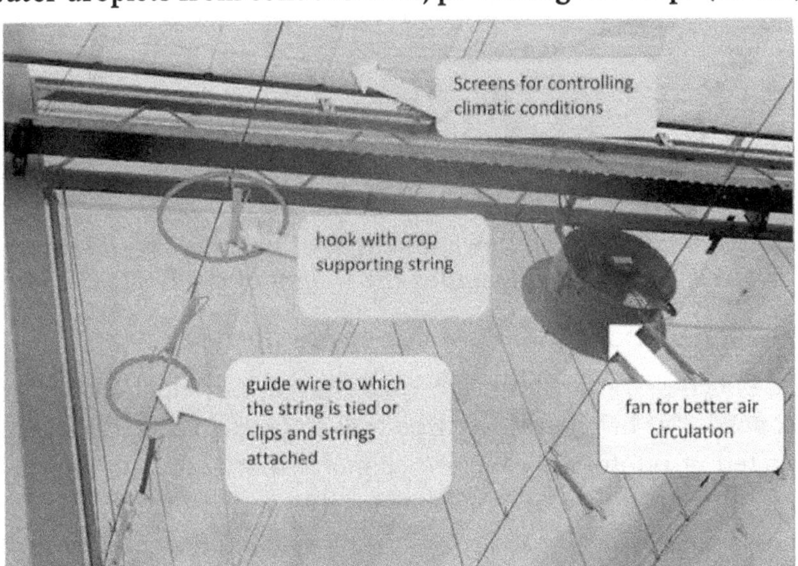

Figure 27. Clips for supporting flower clusters or branches with tomato fruits to prevent it from breaking under the weight of the fruits (Serbia).

8.12 Shelves, benches, and plant substrate gutters

Shelves and benches are now used for contemporary production of seedlings (in containers, pots, peat briquettes, etc.), as well as potted vegetables. Usually, they are constructed of reinforced concrete or metal that reacts poorly or not at all with acids from the leachate, and may be movable or immovable. Movable shelves and benches are very good for an efficient production of seedlings or potted vegetables, as well as for hydroponic production of certain leafy vegetables. The gutters should be resistant to acids that the plants have not used and are drained from the nutrient solution (for instance, due to a fast flow of the nutrient solution through the hydroponic substrate). Such compounds can damage the material the gutters are made of (but also the shelves and benches), which can then affect the nutrient solution (as they may react with the solution due to gutter corrosion, thus changing the chemical composition of the solution) and generally the whole crop production process (**Figures 28 and 29**).

Figure 28. Benches for the seedling or potted plant production (The Netherlands).

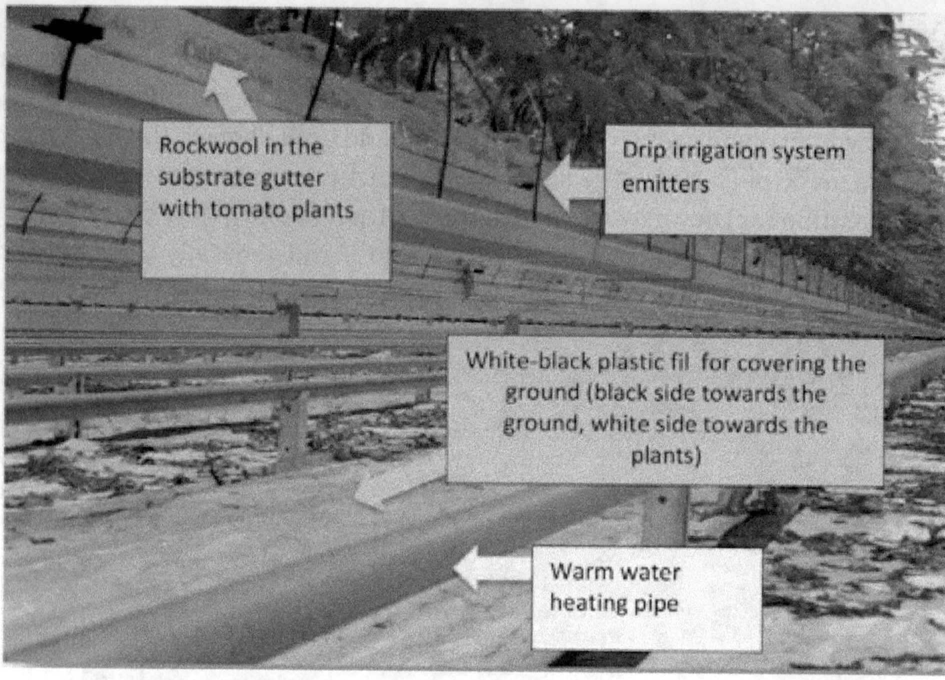

Figure 29. Gutters for plant growing substrate; below are the greenhouse heating pipes and the plastic black and white film for covering the ground (The Netherlands)

8.13 Significance of computerization and automation (robotics) in the twenty-first century greenhouse production process

The modern greenhouse vegetable crop production requires a proper environment climate control. The most important factors are efficient and proper use of energy, water, and fertilizers. In order to meet those requirements, the vegetable crop production process has to be computerized, using special computer software that can, in a simple way, monitor and control the environment climatic conditions, how the boilers are running and timely and have precise water availability. Special computers and software allow environment control in several sections in the greenhouse and reporting on the climatic conditions in those sections—so, in different greenhouse sections, it is possible to grow different vegetable crops that require different conditions for their development and growth. Such a system allows a simplified production process management where the information is available at any given time. Such computer systems are easy to use. There is a greenhouse map placed in the center of all the operations, so the climatic situation in entire area can be viewed. Both hardware and software can be adjusted to any of the production special features, but can also be upgraded easily in the case of the production improvements or enhancements. The plant ontogenesis is followed by multiple quantitative and qualitative processes occurring within the given crop, which result in its productivity. Numerous ecological factors have an impact on the crop productivity, so different development and growth stages require suitable and optimal conditions for achieving its productivity maximum. The environmental factors have an impact on the crop development, growth, and biological yield in the greenhouse because they directly influence its physiological processes [6, 20]. The environment is actually a complex and dynamic system resulting in an effect of combined climatic, soil, and biological factors [6, 21]. So, according to Krug [20], each of these factors consists of numerous elements that differ from each other in the way and intensity they influence on the plants, such as the climatic factors, sun radiation in particular, temperature, air humidity, wind, CO_2 concentration (**Figure 30**).

Figure 30. Computer with software for monitoring crop production in the glasshouse (The Netherlands).

The system comprises a control station and digital nodes positioned at various locations within the greenhouse. It also conserves cable usage in the event that the greenhouse requires expansion, as more digital nodes can simply be added.

The implementation of automation and robots in contemporary greenhouse vegetable crop production is progressively gaining traction, enabling timely and accurate harvesting, elimination of superfluous foliage, and optimal crop spacing, while also conserving labor, which is becoming increasingly limited. The utilization of robots in vegetable production is on the rise within the agricultural sector, particularly in greenhouses, as all actions and procedures are conducted according to established standards and precise timing. It also diminishes the necessity for labor in the agricultural sector. (**Figure 31**).

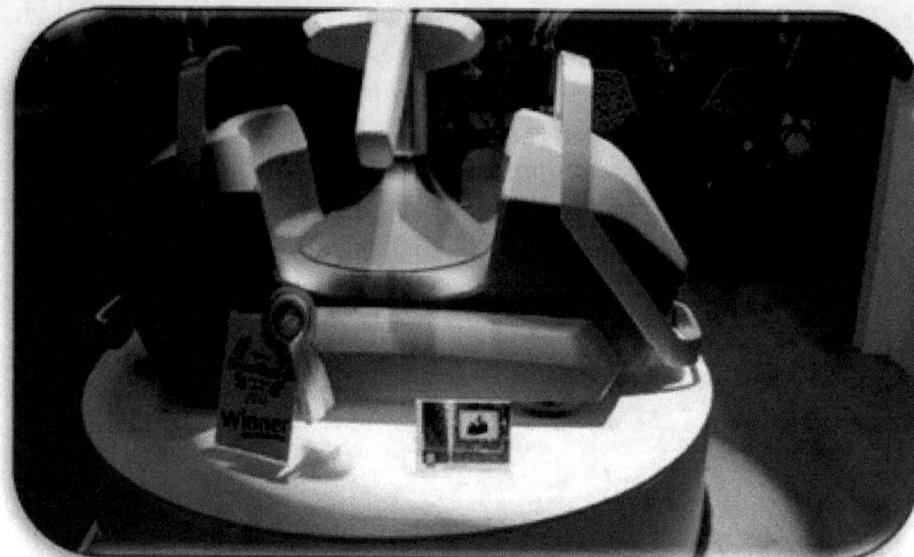

Figure 31. Tomato de-leafing robot (The Netherlands).

REFERENCES

1. UN DESA. 2015. Available from: https://www.un.org/en/development/desa/news/2015.html
2. Popović M. Povrtarstvo. Beograd: Nolit; 1989
3. Ilin MŽ, Adamović ĐB, Ilin ZS. Current status and prospects of indoor vegetable growing in the Republic of Serbia. In: Proceedings. VI Balkan Symposium on Vegetables and Potatoes. Belgium: ISHS; 2014. pp. 467-473
4. Hemming S, Speetjens SL, Wang D, Tsay JR. Greenhouse design for vegetable production in subtropical climate in Taiwan. Acta Horticulture. 2014;**1037**:65-74
5. Ishigami Y, Goto E, Watanabe M, Takahashi T, Okushima L. Development of a simulation model to evaluate environmental controls in a tomato greenhouse. Acta Horticulture. 2014; **1037**:93-98
6. Savić D. Monografija. Produktivnost praziluka (Allium porrum L.). Beograd: Izdavač: Poljoprivredni fakultet Univerziteta u Beogradu (2010); 2010
7. Maksimović SP. Proizvodnja povrća u zaštićenom prostoru. Beograd: Partenon; 2009
8. Goldammer T. Greenhouse Management. A Guide to Operations and Technology. USA: Apex Publishers; 2017

9. Kamp PGH, Timmerman GJ. Computerized Environmental Control in Greenhouses. The Netherlands: PTC +; 2003
10. Mišković A. Priručnik za proizvodnju povrćau zaštićrnom prostoru. Novi Sad: Tampograf; 2012
11. Booij R, Kreuzer ADH, Smit AL, Van der Werf A. Effect of nitrogen availability on dry matter production, nitrogen uptake and light interception of Brussels sprouts and leeks. Netherlands Journal of Agricultural Science. 1996;**44**:3-19
12. Savić D, Moravčević Đ. Praktikum. Metode za određivanje produktivnosti i parametara rastenja i razvića biljaka. Beograd: Izdavač: Poljoprivredni fakultet Univerziteta u Beogradu; 2013
13. The National Greenhouse Manufactures Association SAD, NGMA. Recommends Three Types of Greenhouse Glazing Material. 2017. Available from: https://www.ngma.com
14. USDA. Virtual Grower. 2017. Available from: https://ag.umass.edu/ greenhouse-floriculture/fact-sheets/ virtual-grower-link-to-usda-software
15. Welbaum EG. Vegetable Production and Practices. UK: CABI. CPI Group; 2015
16. WHO. Environmental Health Criteria. 2017. www.who.int/en/
17. Heuvelink E, Kierkels T. Horti-Text BV. The Netherlands; 2017
18. Ilin MŽ, Adamović ĐB, Ilin ZS, Žnidarčič DFAO. Early potato, Good Agricultural Practices for Greenhouse Vegetable Production in the South East European Countries. Principles for Sustainable Intensification of Smallholder Farms. Rome: Food and Agriculture Organization of the United Nations; 2015. pp. 389-403
19. Ilin MŽ, Adamović ĐB, Ilin ZS, Žnidarčič D. GAP Recommendations for Permanent and Temporary Protection System – Early Potato Production. PIK-REPNIK, STORITVE d.o.o. ISHS and University of Maribor, Faculty of Agriculture and Life Sciences; 2017. p. 51
20. Krug H. Environmental influences on development, growth and yield. In: Wien HC, editor. The Physiology of Vegetable Crops. Wallingford, UK: CAB International; 1998. pp. 101-180
21. Savić D, Stikić R, Jovanović Z, Prokić L. Uticaj usvajanja azota na indeks lisne površine i intercepciju svetlosti praziluka (Allium porrum L.). Savremena poljoprivreda. 53. 2001;**1-2**: 145-151

12 Chapter

AMENDING THE NATIONAL FRAMEWORK FOR DROUGHT MANAGEMENT (NFMD) TO IMPROVE VEGETABLE FARMERS' RESILIENCE TO DROUGHT IN THE NORTHERN CAPE PROVINCE OF SOUTH AFRICA

1. INTRODUCTION: AN OVERVIEW

Drought is characterized by a sustained period of precipitation deficiency in a specific area, leading to an extended scarcity of water resources, including atmospheric, surface, or groundwater. Batisani [1] asserted that drought is a naturally occurring phenomenon characterized by precipitation levels significantly below normal, resulting in severe hydrological imbalances that negatively impact land resource production systems, with effects potentially persisting long after a wet season. Predictions indicate that drought will be a significant challenge for numerous African communities in the future as a result of climate change.

Drought vulnerability pertains to the societal risk and resilience to drought conditions, and as indicated by the study conducted by Elsamian et al., [3], the drought vulnerability of a region is influenced by various elements[2]. Their analysis identified that drought sensitivity components are both natural and anthropogenic, encompassing meteorological, hydrological, and biological elements as natural factors, while anthropogenic factors comprise socio-economic and land use parameters. The study determined that the severity of drought effects fluctuates based on the interplay and spatiotemporal fluctuation of diverse natural and manmade causes.

According to a report by the National Department of Agriculture, Fisheries and Forestry [4], a national policy on the management of drought exists in South Africa and provinces are guided by the national policy framework [5]. The policy framework stipulates that provinces risk assessment basis for the development of disaster management planning must be in line with the national policy framework. In developing such a risk assessment, provinces must include contingency plans approved and accepted by all role players in the province. The contingency plan should be activated immediately after a drought disaster is declared and the long process of assessment and declaration impacts negatively on farmers.

Despite the existence of a national policy framework on drought in South Africa, Northern Cape Provincial small-scale farmers are still facing severe vulnerability with regard to prevention, adaptation, prediction, and early warning. Since 2015, 12 early warning reports have been generated for the Northern Cape Province and disseminated to farmers through extension services and communication as an integral component of the contingency plan. Awareness campaigns have been intensified with the objective of promoting a culture of risk avoidance among small farmers through integrated education, training, and public awareness programmes informed by scientific research. Drought aid schemes have been launched since the beginning of drought conditions in 2015 and the province declared a drought disaster zone on April 12, 2016 by the National Disaster Management Centre (NDMC).

Since the declaration, the drought condition has been deteriorating, and DAFF has allocated R25 million to the Northern Cape Department of Agriculture for drought relief. The Department of Agriculture, Forestry and Fisheries procured 73,000 bags of drought pellets at a cost of R20.44 million for distribution to farmers in severely affected regions of the province in 2017 [6]. As the drought condition intensifies, the management strategies in the Northern Cape appear to be inadequate and the current study intended to examine factors enhancing small-scale farmers' vulnerability to drought in the province so as to uncover new innovative management strategies to mitigate the condition. According to WSS NC [7], vulnerability embodies various factors which include environmental, physical, economic, and social factors. In order to address this problem, it is important to identify factors that cause the increased vulnerability of these farmers to drought despite the effort made by the government to mitigate the condition. Jordaan [8] argues that the situation is exacerbated by over-stocking in the Province which further complicate government efforts.

Wisner *et al.*, [9] argue that when testing the vulnerability of a community to drought, drought should be conceptualised as a trigger event that causes damage by exploiting underlying social vulnerabilities within populations exposed to its effects. It is on this basis that the United States National Drought Policy Commission Report [10] defined drought as persistent and abnormal moisture deficiency having adverse impacts on vegetation, livestock, and people. The report also states that drought is a normal feature of climate and its recurrence cannot be avoided in almost every country. However, the seriousness of its economic, environmental, and social impacts affect more people when compared to other forms of natural and anthropogenic hazards particularly on the poor and vulnerable [11]. Ngaka [12] stated that drought has become an increasingly common disaster type and major hazard in South Africa in terms of people affected and economic loss.

In the Northern Cape Province of South Africa, there is an imperative to mitigate the susceptibility of small-scale farmers to climate variability and the risks associated with climate change [13]. The occurrence of drought and the extent of agricultural losses in the Province highlight the persistent susceptibility of small-scale farmers to agriculture, despite existing mitigation methods. Farmers in the province experience differing levels of susceptibility to drought conditions in the agricultural sector, contingent upon the scale and type of their farming operations. The drought has severely impacted agricultural production in the Northern Cape, with a profound effect on small-scale farmers.

2. NATURE OF DROUGHT IN THE NORTHERN CAPE, SOUTH AFRICA

Despite the differences in the definition of drought by different researchers and scholars, they all agree that drought is a prolonged absence of moisture in a specific environment that affects livestock and plants [3, 11, 14–16]. According to Miyan [17], droughts vary from region to region and the least developed countries have become the worst affected. According to Agri SA 'status report on the current *drought crisis* [18], the severe drought has fastened its grip on the agricultural sector resulting in.

- Natural grazing becoming depleted leading to the forced slaughtering of livestock, livestock deaths due to fodder unavailability.
- Reduced plantings of summer cash crops.
- Extremely high temperatures, which have a negative impact on pollination and thus lower yields.

The description of each type is short, simple, and brief for a better understanding of the nature of drought in the Northern Cape as shown in **Figure 1**.

The current study is modelled after the study by Wisner *et al.*, [9] who conceptualised drought as a trigger event that causes damage by exploiting underlying social vulnerabilities within populations exposed to its effects. There are four main types of droughts experienced in the Northern Cape namely: a) meteorological drought, b) agricultural drought, c) hydrological drought, and d) socio-economic drought. According to Bang *et al.*, [19], these four types of droughts correspond to the levels of severity and the level of severity is determined by climatic conditions prevailing during the corresponding period. In the Northern Cape Province, annual rainfall varies from 200 to 400 mm per annum and its distribution across the province is uneven. The Western region of the Province receives winter rainfall and currently is experiencing droughts while the eastern regions receive summer rainfall.

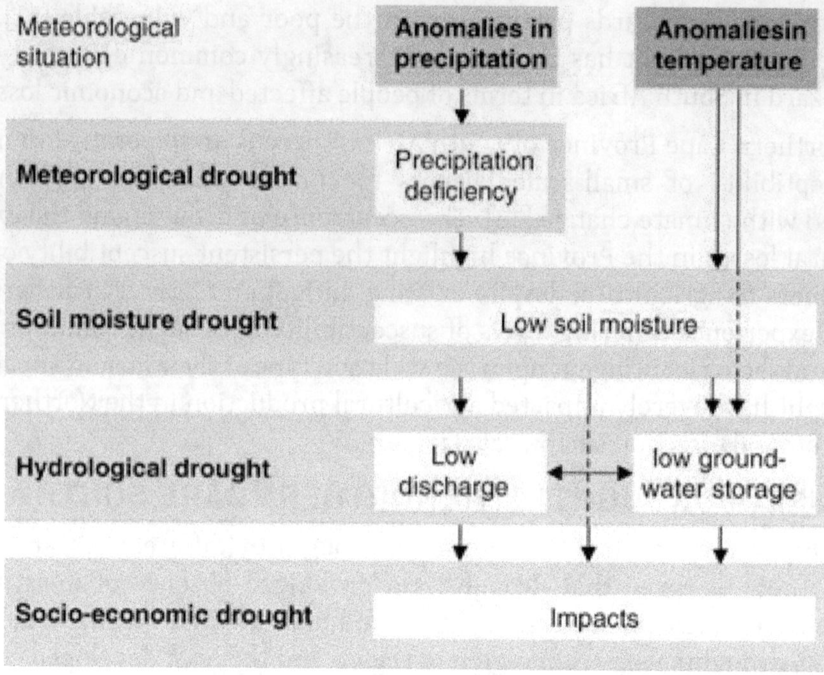

Figure 1. Different categories of droughts and their development. Source: (derived from Peter, 53, Van loon, 54 Stahl 55).

3. SOCIO-ECONOMIC IMPACT OF DROUGHT ON SOCIETY

Northern Cape economy relies mostly on farming and more people are employed in the agricultural sector, which depends entirely on precipitation for production. Drought has negative impacts on the welfare and employment of farmers, farmworkers, and their families. Farm income is dependent on the quality and quantity of livestock and crop produced at any given time and continued drought is hazardous to yields. Jordaan et al. [20] noted that socio-economic drought affects people's lives in terms of their behaviour and options such as water rationing and increased water pricing. According to Wilhite socio-economic drought is dependent on the impact of meteorological, hydrological, and agricultural droughts on the supply and demand of agricultural goods and services.

Socio-economic drought occurs when demand for economic goods exceeds the supply caused by related shortfalls in the water supply. Socio-economic droughts differ markedly from the other types of droughts because it reflects the relationship between supply and demand commodities such as livestock forage, water, and hydroelectricity that is dependent on precipitation and supply varies as a function of precipitation or water availability [21]. Small-scale farmers are mostly affected by socio-economic drought because they are highly dependent on farming for their social being and economic livelihood.

Employment on farms depends on the profitability of farm businesses and farm labours both seasonal and permanent are influenced by farm margins. During drought there is a remarkable loss of expected return from all farming enterprises and this argument is supported by Eslamian et al., [3] who defined socio-economic drought as loss of average or expected return which is measured by economic and social indicators. Socio-economic drought is different from other forms of drought since its occurrence depends on spatiotemporal supply and demand of agricultural products.

According to Van Zyl [16], there are alternatives and practical definitions for drought types usually experienced by farmers in the Province due to its unique geographical location. These include false drought where rainfall is below the long-term average, but due to overgrazing of the veld, fodder supply becomes prematurely depleted and giving the impression of prevailing drought; premature drought where a chronic drought situation is aggravated by overgrazing resulting in a premature declaration. In many instances, the adjoining farms may differ widely in intensity as a result of veldt management practices. There is also prolonged drought where for months, high intensity of livestock is maintained, and the result is more or less chronic food shortage even after good rains have fallen as plants become severely damaged. It is also possible that areas that have been declared as drought-stricken do not recover after good rains and a few months later the drought could even get worse.

Green drought occurs when excessive grazing pressures are maintained in semiarid periods and this causes food shortages even though the vegetation is appearing green and soil moisture reserves are favourable. It can also occur where natural causes like rain showers during drought promote a short spell of green growth but not enough to break the drought. A green drought can also occur where the insect (locusts, Karoo, and commando caterpillars) severely attacks the plant and deplete the fodder to such a degree that it takes the appearance of a drought situation. There is thus a shortage of fodder in spite of favourable circumstances. Finally, financial drought in which farmers exert much pressure on the government to declare drought disaster to obtain financial assistance in order to improve their cash flow.

Therefore, a region can be declared drought-stricken even though drought does not prevail as shown in **Figure 2.** The figure shows different droughts, causal factors, and the usual sequence of occurrence. Climatic factors like temperature, rainfall, evapotranspiration, and soil water deficiency are attributes that increase the vulnerability of small-scale farmers to drought in the Northern Cape Province.

The impacts of drought in relation to the different drought categories have also been determined. The table below indicates the impacts of droughts experienced by small-scale farmers in the Province who are economically, socially, and environmentally affected by drought. The table reveals various form of droughts has different impacts on small-scale farmers and **Table 1** shows the drought category as well as an impact category in the Province.

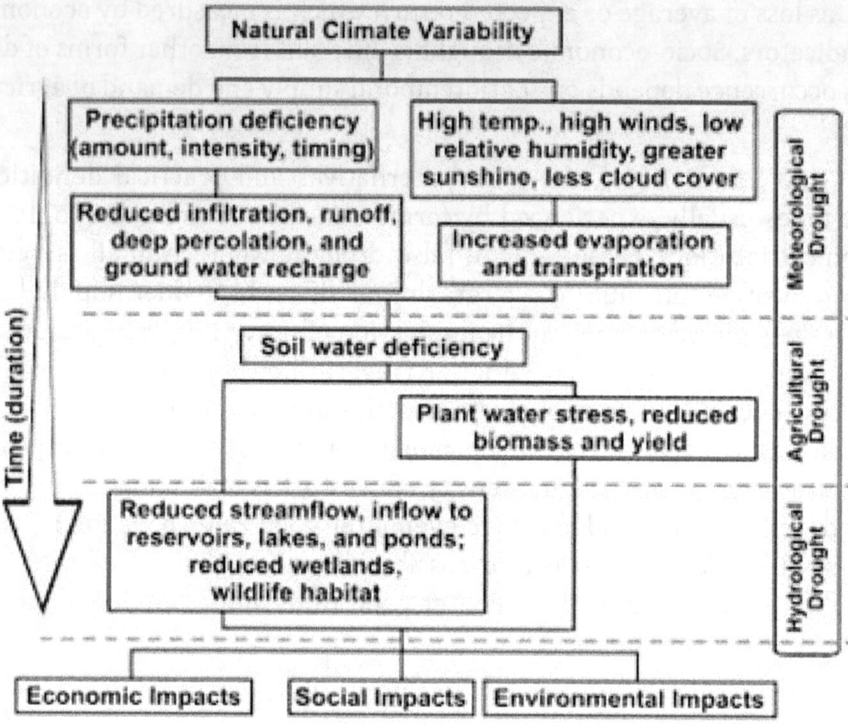

Figure 2. Drought types, causal factors, and their usual sequence of occurrence. Source: National Drought Mitigation Centre. (2005).

Table 1. Drought categories and related impact in the northern cape.

			Drought category	
Impact category		Meteorological drought	Soil moisture drought	Hydrological drought
Agriculture	Rainfed	X	x	
	Irrigated		x	X
Ecosystems	Terrestrial	X	x	
	Aquatic			X
Energy and industry	Hydropower			X
	Cooling water			X
Navigation				X
Drinking water				X
Recreation				

4. FACTORS ENHANCING SMALL-SCALE FARMERS' VULNERABILITY

The preceding section presented the description of drought, forms, and type as well as the concept of drought. This section focuses on addressing the central questions and objectives which is based on the existing management framework to analyse the social, economic, and environmental factors enhancing small-scale farmers' vulnerability to drought in the province. In this study, vulnerability is seen as a set of conditions and processes resulting from the physical, social, and economic challenges caused by persistent drought in the Northern Cape province of South Africa.

Varying degrees of vulnerability to drought conditions occur within the agricultural sector based on the size and nature of one's farming operation. In the case of South Africa, small-scale and communal farmers have proven to be more vulnerable to droughts given their concentration in less favourable climatic zones, their lack of resources and reliance on their own production for house food security [22].

Khoshnodiffer, *et al.* [23] in identifying social factors responsible for small-scale farmers' vulnerability to drought in the Northern Cape found; level of literacy, education, peace and security, access to human rights, social equity, traditional values, beliefs, and organisational systems while those causing economic vulnerability are poverty, gender, level of debt and access to credit. According to Creswell [24] opined that vulnerability to disasters like drought depends on the social, cultural, economic, and political environment. Poverty and its complex dimensions, discrimination, lack of opportunities for acquiring and developing skills and capabilities, lack of access and control over basic necessities including production resources, decent living conditions, livelihoods, and adequate incomes are the causes of vulnerabilities for millions of East Asians.

Madzwamuse [25] argues that climate change also enhances small-scale farmers' vulnerability to drought in the province and any vulnerability analysis must consider the social, economic, and institutional factors that favour large-scale commercial farmers to the neglect of small-scale farmers even though they operate under the same climatic conditions. Khoshnodiffer *et al* [23] are of the view that the identification of drought vulnerability factors is an essential step for planning drought mitigation management strategies. The current study intends not only to identify drought vulnerability factors but to test these factors (economic, social, technical, and environmental) and how they impact small-scale farmers in Northern Cape.

Eslamian et al., [3] stated that factors governing drought vulnerability may be natural (meteorological, ecological, and hydrological) while anthropogenic factors include socio-economic and land use parameters. According to United Nations International Strategy for Disaster Reduction, there exist four groups of vulnerability factors that cause small-scale farmers' vulnerability to drought. These are; physical factors that describe the exposure of vulnerable elements within a region; economic factors which describe the economic resources of individuals, populations groups, and communities; social factors, which describe non-economic factors that determine the well-being of individuals, population groups, and communities, such as the level of education, security, access to basic services, human rights, good governance, and environmental factors which describe the state of the environment within a region the small-scale farmer operate.

Jordaan *et al.* [20] are of the view that small scale farmers in Northern Cape are constantly affected by dry periods and had requested government support during dry periods. The impact of drought and vulnerability by small-scale farmers varies in degrees from place to place due to variation in climatic conditions, land use pattern/land cover, agricultural practices, social structures, and financial support. Eslamain [3] supported that drought vulnerability is closely associated with the risk and resilience of society to drought and therefore, the assessment of drought vulnerability is an important component for proper water resources management and sustainable development. Furthermore, vulnerability to drought by small-scale farmers is a function of the character, magnitude, and rate of climate variation to which the small-scale farmers are exposed to their adaptive capacity [26].

5.COPINGCAPACITYOFSMALLSCALETODROUGHTVULNERABILITY

According to Yameogo [27], coping capacity can be defined as the means by which people or organisations use available resources and the ability to face adverse consequences that could lead to a disaster. In general, this involves managing resources, both in normal times as well as during crises or adverse conditions. The strength of coping capacities usually builds resilience to withstand the effects of natural and human-induced hazards. Farmers will sometimes resort to old age traditional and indigenous knowledge of coping with drought. Rainfall harvesting with cement tanks and catchment areas develop for water storage.

Farmers sometimes will move to areas where there is water and vegetation. Nomadic farmers are common in Northern Cape. They move along the Orange river where grazing and water are available. They sleep next to their flock and will move once grazing is depleted. During good rainy days, reservoirs are allowed to be filled with water and are closed with dexterity and art to be used during the drought period. Nobody is allowed to draw water from the well until the onset of drought. Maize, sorghum, and other drought-resistant crops are planted, harvested, stored, and preserved using old skills. During drought, this storage or silo (sefalana in Setswana) is opened and good food is obtained from this storage. This small silo is built by a specialist in making mud houses and the design is such that stored foodstuff does not get decomposed.

According to Roos *et al.* [28], there are reciprocal interactions between older people and drought challenged environments. Through these reciprocal interactions and in this particular context, traditional bodies of knowledge and skills have been developed to assist older persons to cope with drought for instance older people made appropriate adjustments or implemented decisions based on their interactions with the environment and resorted to conservative, stockpiling and animal grazing patterns, vegetation, water conservation, use of plants for medical purposes, and animal feeds.

Small scale farmers are vulnerable to economic, social, and environmental effects of drought and their adaptability or coping capacity differ according to their experience of drought and mitigation ability. There is a correlation between strong financial position and coping ability among small-scale farmers. Farmers who have more money in the bank can buy more feeds or production inputs and lessen the burden of relying on state assistance. The other factor is drought preparedness and pre-disaster planning. If the farmer did his homework way in advance he can cope well when drought has occurred. Farmers normally stockpile feeds during years of bumper production, fill the storage with unused fodder and ensure that fodder banks are up and running. Their products fetch lower prices at markets due to low quality and quantity during droughts. Prices are high for production inputs like fertilisers, feeds, medications, fuel, and replacement stock. Farmers must be available for training and capacity building on drought and its vulnerability.

In order to cope well with drought vulnerability, farmers must be candid enough to do self-introspection, own lifestyle audits; reduction in the expenditure of nonessentials like clothes and eating out in big restaurants, hotels, guesthouses. Coping capacity for drought is determined by analysing the capacity of farmers, farmworkers, and rural towns to cope with droughts. Factors considered for equating coping capacity include (i) land ownership, ii) on-farm diversification to provide own feed and fodder during drought, (iii) government support during drought, (iv) institutions support during drought, (v) alternative source of income, and non- agricultural entrepreneurship opportunities [8]. According to Shiferaw *et al.*, [29], ex-post risk coping strategies are designed to prevent a shortfall in consumption when income drops below the required level as a result of climatic conditions. This includes a reduction in the selling of food, reduced consumption, and increased borrowing to higher rates of seasonal out-migration default on loans.

Drought also affects farming towns and different municipalities in Northern Cape very severely; integrated development plans (IDP) which include drought management plans to large extent increase the coping ability of farmers against drought. During drought, farmers are assisted with fodder supply and their intergovernmental cooperation. For example, the department of social service will help with clothes, food, counselling, while the department of health with coping capacity by the provision of medicine and the department of water affairs will provide water with tanks [30]. Typically, main water sensitive urban design (WSUD) activities and strategies include the following, stormwater/drainage management, re-use of water, demand reduction techniques, and greenroom installation, with each main activity having its own set of sub-activities. Water demand management (WDM) or portable water demand reduction techniques ensures that water is reduced through leaks, reduced wastewater flows, and better awareness of consumers of the environmental and financial value of water [31].

The Disaster Management Act and National Disaster Management Framework provide the legislative and policy frameworks for national and provincial drought management, mitigation practices, and strategies. Other legislative mandates are the strategic plan, Conservation of Agricultural Resources Act (CARA), and National Disaster Management Act Framework (NDMF). Drought management in Northern Cape is everyone's responsibility. Drought can only be effectively control the implementation of the mitigation strategies is approached from an integrated strategic position taking cognisance of location specifics. Research is required in view of the vulnerability of small-scale vegetable farmers to drought in the Northern Cape as the existing policies and risk management plans adapted from the Department of Agriculture, forestry and fisheries [4] frameworks appear inadequate.

The preceding subheadings discuss the vulnerability of small-scale farmers to drought disasters in the Northern Cape. Vulnerability factors like unemployment, low income, poverty, population migration, change of food and eating pattern and loss of dignity were discussed. Surely, socio-economic drought vulnerability will be with farmers until precipitation fall and other mitigation strategies and policies are fully implemented. Northern Cape is affected by climate change which is characterised by global warming resulting in high atmospheric moisture losses. Evaporation and transpiration result in extreme water losses and this affects fauna and flora on a big scale. Climate change refers to a change of climate that is attributed directly or indirectly to human activity that alters the global atmosphere and that is in addition to natural climate variability observed over comparable periods of time [33].

Climate change according to the National Climate Change Response, White Paper is the trend in change of the earth's general weather conditions as a result of an average rise in the temperature of the earth's surface often referred to as global warming. (The Government of the Republic of South Africa, 2011). Significant changes in extreme weather events such as heatwaves, very hot days, high fire danger days, and dry spells are likely to increase. Climate change is expected to increase the frequency and magnitude of many types of extreme events including floods, droughts, tropical cyclones, and wildfires [34]. The consequences of increased temperature change in rainfall patterns, extreme weather events, sea-level rise, and changes in biodiversity will have a significant influence on national economies, rural livelihoods, and development in general. According to Madzwamuse [25], Africa is said to be the most vulnerable continent to the impact of climate change as a result of national economies depending on natural resources.

- Agricultural production is projected to fall by 50% due to a reduction in precipitation and an increase in temperature while most African states will be faced with water scarcity and stress by 2050. Madzwamuse [25] and Hassan [35] stated that the impact of climate change in the South African agricultural sector will differ in different farming systems. Dry land smallholder producers will be affected mostly compared to irrigation and large-scale/commercial producers.

Northern Cape is one of the driest Province in the Republic of South Africa and it is also vulnerable to other disasters like veldt fires during the winter season when vegetation is extremely dry. Unsustainable use of land and other resources increase the vulnerability of farmers of the Northern Cape. Land degradation often stems from the nexus between poverty and lack of capacity to invest in more sustainable agricultural practices and improve land use patterns. The vulnerability to agricultural drought means that the extent of agriculture potential sensibility to the drought threat. Agricultural research on the vulnerability to agricultural drought is of great significance to mitigate drought losses and guarantee Provincial food security [31].

The economic, social, and environmental impacts of drought are huge in Northern Cape and the national costs and losses incurred threaten to undermine the wider economic and development gains made in the last few decades in the region. There is an urgent need to reduce the vulnerability of countries to climate variability and the threats posed by climate change [13]. Drought events and the magnitude of agricultural drought losses indicate the continuing vulnerability of the country to agricultural drought. Drought is among the most multifaceted and least understood of all natural hazards. Climatic variability adversely affects food production in two ways—it leads to low production, which translates into limited access, both physical and economic food. The exposure of agriculture to drought and heavy precipitation threatens food security and exacerbates poverty among smallholder farmers [36].

The impact of the drought on livestock production (beef and sheep) in the Northern Cape can be felt by the small scale at markers. Feedlots are under pressure due to shortages of maize products and high input prices and more costs will push up the price of A-grade beef prices. With good rainfall in the near future, Northern Cape producers can begin to rebuild their herds. An increase in milk prices can also cause the slaughter of cattle from this sector to decrease, putting further pressure on the factory meat prices (Agri growing greatness.co.za). In the livestock industry, below normal rainfall has almost depleted natural grazing veld, placing feed supplies for the upcoming winter season in a precarious position [18].

Southern African countries experience a dry season in winter, limiting forage. This means that any delays in the rains obviously prolong the winter period, which thus increases food shortages for livestock. Such a scenario results in high mortality of livestock, especially cattle. Losing livestock could increase poverty and negatively affect rural people's livelihoods and food security [37]. According to Nkomo [38], the impact of drought on small-scale farmers is devastating as water allocation for irrigation to farmers has been cut back in the region. The cutbacks have had a direct impact on the incomes of those farmers. Small-scale farmers are more vulnerable compared to large-scale commercial farmers because they have no insurance and savings which may serve as a mitigation strategy.

The following vulnerabilities are used to measure or assess the degree of vulnerabilities to drought disasters of small-scale farmers in Northern Cape, South Africa. Economic, social, and environmental drought vulnerabilities assessment for Northern Cape small-scale farmers. This model was proposed by Jordaan *et al.* [20] to assess drought disaster risk for Northern Cape. Social factors are related to social issues such as levels of literacy, education, the existence of peace and security, access to human rights, social equity, traditional values, beliefs, and organisational systems and economic factors are related to issues of poverty, gender, level of debt, and access to credits. Vulnerability assessment provides a framework for identifying the social, economic, technical, and environmental causes of drought impacts [23].

Identifying drought sensitivity indicators for small-scale farmers in the Northern Cape is a crucial step in planning drought mitigation management. The disaster risk assessment methodology outlined in the Disaster Management Act (Act 57 of 2002) served as the platform for drought risk assessment in the Northern Cape. Stage one encompassed the framework for primary risk variables, the constituent components, and illustrative indicators evaluated for drought risk assessment in this study. Multiple frameworks, models, methodologies, and equations are employed to evaluate diverse drought risks; developing farmers in the Northern Cape are particularly affected.

REFERENCES

1. Batisani N. The Spatiotemporal severity dynamics of drought in Botswana. Journal of Environmental Protection. 2011;2(6):803-816

2. International Strategy for Disaster Reduction (UNISDR), 2017, Drought, Living with Risk: An Integrated Approach to Reducing Societal Vulnerability to Drought.ISDR Ad Hoc Discussion Group on Drought, viewed 09 March 2017
3. Eslamian S, Eslamain, Faezah A. Handbook of Drought and Water Scarcity: Environmental Impacts and Analysis and Analysis of Drought and Water Scarcity. New York: CRC Press Taylor & Francis Group, 6000 Brooken Sound Parkerway; 2017
4. DAFF. (2018). Department of Agriculture Forestry and Fisheries National Strategic Plan 2015/2016 to 2019, retrieved from http://www.daff. gov.za/doaDev/ [Accessed: 04 July 2020]
5. Moseley WG. Agriculture on the Brink: Climate Change Lober and Smallholder, Farming in Botswana. Faculty publication; 2016 http://digita/commons.macalester.edu/geogfacpub/1
6. Department of Forestry and Fisheries DAFF. (2015). Department of Agriculture Forestry and Fisheries National Strategic Plan 2015/2016 to 2019, retrieved from http://www.daff.gov.za/doaDev/ [Accessed: 06 May 2020]
7. WSS NC, (2015). Northern Cape Drought Relief Business Plan. University of the Free State. Faculty of Natural and Agricultural Sciences. Disaster Risk Management: Study Material for DIM 605. Bloemfontein
8. Jordaan AJ, Sakulski D, Jordaan AD. *Drought Risk Assessment for Northern Cape Province*. Kimberley: Department of Agriculture, Land Reform and Rural Development; 2011
9. Wisner B, Cannon P, Davis I. At Risk: Natural Hazards, People's Vulnerability and Disasters. 2nd ed. London: Routledge; 2004
10. US Natural Drought Commission Policy Report. 2000
11. Edossa DC, Woyessa YE, Welderufel WA. Comparison Between Two Meteorological Drought Indices in the Central Region of South Africa: Agricultural Water Management. Agricultural Water Council. 2006. pp. 61-72
12. Ngaka MJ. Drought preparedness, impact and response: A case of the Eastern Cape and Free State provinces of South Africa Jamba. Journal of Disaster Risk Studies. 2012;**4**(1):1-10. DOI: 10.4102/jamba.v4.il.47
13. Kassie M et al. Managing Vulnerability to Drought and Enhancing Livelihood Resilience in Sub Saharan Africa: Technological, Institutional and Policy Options. Available from: www. elsevier.co/locate/wace2014. p. 1
14. Naumann MU, Manatam VM, Mahuni NK. Environmental Research Letters. 2015:10124013
15. Wilhite DA. Drought and Water Crises: Science, Technology and Management Issues. Boca Raton, FL: CRC Press; 2005
16. Van Zyl K. A study on the Disaster Risk Management Plan for the South African Agricultural Sector. Pretoria: Business Print Centre; 2006

17. Miyan MA. Droughts in Asia least developed countries: Vulnerabilities and sustainability. Weather and Climate Extremes. 2014
18. Agri SA. A rain drop in the drought. In: Report to the MultiStakeholder Task Team of the Drought, Agri Sa' Status Report on the Current Drought Crisis, viewed. 2016
19. Bang, S K. & Sitango, K 2003. Indigenous Drought coping strategies and risk management against El Nino in Papua New Guinea: CGPRT Centre Working Paper No 74. http:// ageconsearchumn.edu/handle/32687 [Accessed: 16 March 2016]
20. Jordaan AJ, Sakulski D, Jordaan AD. Interdisciplinary drought assessment for agriculture: the case of communal farmers in the Northern Cape Province, South Africa. South African Journal of Agricultural Extension. 2013;**41**(1): 44-58
21. Wilhite DA, Svoboda MD, Hayes MJ. Understanding the complex impact of drought: a key to enhancing drought mitigation and preparedness. Water Resources Management. 2006;**21**: 763-774
22. Rabumbulu M, Badenhorst M. Land degradation in the Wes Central Free State: Human Induced on climate variability, the perception of Abrahamskraal- Boshof district farmers. South African Geographical Journal. 2017;**99**(3):217-234. DOI: 10.1080/ 03736245.2016
23. Khoshnodifar Z, Sookhtanlo M, Gholami H. Identification and measurement of indicators of drought vulnerability among wheat farmers in Mashhad County, Iran. Annals of Biological Research. 2012;**3**(9):4593-4600
24. Creswell JW. Research Design: Qualitative, Quantitative and Mixed Methods Approach. 2nd ed. California: Sage; 2009
25. Madzwamuse M. 2010. Climate Change Vulnerability and Adaptation Preparedness in South Africa. Water conference South Africa 2010
26. Jamshidi A, Asadi K, Azadi KH, Scheffran J. *Vulnerability to climate change of smallholder farmers in the Hamadan province*. Iran: Climate Risk Management; 2018. DOI: 10.1016/j. crm,2018.6.002
27. Yameogo TB, Fonta WM, Wunscher T. Can social capital influence smallholder farmer's climate change Adaptation Decisions? Social Sciences mdpi.com.socialscience. 2018, 2018;**7**(3):33
28. Roos V et al. Coping with drought: Indigenous knowledge application in rural South Africa. Indilinga- African Journal of Indigenous Knowledge Systems. 2010;9(1):2010
29. Shiferaw B et al. Managing Vulnerability to Drought and Enhancing Livelihood Resilience in Sub Saharan Africa: Technological, Institutional and Policy Options. Available from: www.elsevier.co/locate/wace2014. p. 1
30. Sing, N.P., Bantilan, C. & Byjesh, K. 2014. Vulnerability and Policy Relevance to Drought in the Semi-Arid Tropics of Asia. A Retrospective Analysis

31. Mavhura E, Manaka D, Mushore T. Adaptation to drought in arid and semi-arid environments: Case of the Zambezi Valley, Zimbabwe. Jamba: Journal of Disaster Risks Studies. 2015;7(1):121-135
32. NDMF. National Disaster Management Framework. Pretoria: NDMC, Department of Local Government and Housing; 2005
33. UNFCCC. 2006. Climate Change: Impacts, Vulnerabilities and Adaptation in developing Countries. International Conference on Water Management
34. Bryan E, Deressa TT, Gladys AG, Claudia R. Adaptation to climate change in Ethiopia and South Africa: Options and Constraints. Environmental Science and Policy. 2009;**12, 2009**(4):413-426
35. Hassan, (2006) Impacts of Climate Change on Crop Farming in South Africa: Policy Note.21. [Online] Centre for Environmental Economics and Policy Analysis. Available from: http:// www.ceepa.co.za/docs/POLICY%2021. pdf [Accessed: 12 May 2018]
36. Derbile EK, File DJM, Dongzagla A. The double tragedy of Agriculture vulnerability to climate variability in Africa, How vulnerable is a smallholder agriculture to rainfall variability in Ghana? Jamba Journal of Disaster Risk Studies. 2016;**8**(3):9249. DOI: 10.4.102// Jamba.V813.249
37. Calvoso L, Chuluunbaartar D, Fara K. Livestock and climate change. In: Livestock Thematic papers. 2010. www.ifad.org//rkm/index.htm. [Accessed: November 2017]
38. Nkomo J.C., Nyong A.O. & Kulunda K. 2006. The Impacts of Climate Change I Africa: A Final Report Submitted at the Storm Review on the Economics of Climate Change, viewed on 21 July 2017

13 Chapter

FUNDAMENTALS OF IRRIGATION MANAGEMENT FOR VEGETABLES

1. INTRODUCTION: AN OVERVIEW

Water is a crucial inorganic resource for all living organisms to sustain their essential processes. Consequently, it significantly influences nutritional consumption in the human body, both directly and indirectly. The World Health Organization states that an individual requires 100 liters of water daily to adequately fulfill basic needs [1]. Nevertheless, 4 billion individuals globally reside in regions with water scarcity. By 2050, worldwide water consumption is projected to rise by 20–30%, and water shortage, intensified by climate change, may result in economic losses of up to 6% of gross domestic product in certain places[2]. In addition to human use, water is crucial as its scarcity severely limits plant productivity. Water in agriculture is essential for sustaining global food supply, supporting livelihoods, and enhancing resilience to climatic shocks and extremes.

Irrigation, the predominant use of water, encompasses the activities associated with the cultivation of crops. A crucial requirement for the consistent growth of plants is the availability of adequate water in the root zone throughout the growing season. Irrigation is the provision of water to the plant root zone when precipitation is insufficient to meet the necessary water requirements. The history of irrigation commences with the history of humankind. Primitive irrigation techniques were utilized for crop cultivation even prior to the emergence of civilizations. Numerous civilizations have emerged in regions where water is accessible and irrigation may be utilized. Currently, only 20% of the world's agricultural fields are irrigated, although they yield 40% of the worldwide harvest, indicating that irrigation more than doubles land productivity [3].

Most cultivated crops require irrigation during their growth period. Certain field crops, including maize, wheat, and rye, can endure rainfed agriculture. However, regarding veggies, the necessity for irrigation becomes imperative. Vegetables contain substantial amounts of water. Certain vegetables, including cucumber, tomato, lettuce, zucchini, and celery, comprise over ninety-five percent water. The elevated water content in the cells renders these plants particularly susceptible to water stress and drought situations. Their output and quality deteriorate swiftly under drought conditions. Consequently, irrigation is vital for the cultivation of most vegetables to achieve optimal output and superior quality. Nonetheless, excessive irrigation can impede germination and root growth, while diminishing vegetable quality and post-harvest longevity of the crop. Vegetable crops may undergo water stress in two distinct manners: firstly, due to inadequate water supply (drought stress), or secondly, as a result of excessive water (waterlogging or soil saturation) [4].

Given these circumstances, the quantity of irrigation water, the timing of irrigation, and the method of irrigation must be meticulously evaluated. Choosing appropriate irrigation management is essential to optimize vegetable quality, productivity, and water use efficiency, while reducing environmental implications. Determining the appropriate and efficient irrigation system for vegetables is tough because to numerous variables, including water quantity and quality, soil type, the specific crops cultivated, and cost constraints. Upon optimizing all parameters, the maximum yield can be achieved using the most advantageous quantities of water and energy.

2. SOIL–PLANT–WATER RELATIONS

Plants terminate new developments in the above ground parts by minimizing the use of water and carbohydrates in the stem and help the root develop more when they cannot get enough water from the soil, and if the situation lasts or the water in the soil is insufficient, plant activities stop completely. Nevertheless, there must be sufficient air in the soil for satisfactory root development. In the conditions where there is more water in the soil, the amount of air decreases as the spaces between the soil particles are filled with water. Therefore, it is very important to balance the amount of water and air in the root zone of the plant in order to provide the best plant growth. Thus, increasing plant production depends on knowing the relations between soil, plant, and water.

2.1 Soil properties

Knowing the physical and chemical properties of the soil is fundamental to successful irrigated agriculture. Values such as field capacity and wilting point of a soil are affected by parameters such as soil texture and structure, and organic matter content [5]. Soil moisture availability varies with the amount of water in the soil and the type of the soil. Knowing soil texture and other specific characteristics is essential for planning and using an irrigation system. Since sandy soils do not have much specific surface area, they cannot hold much water. Therefore, the field capacity value of a sandy soil can be as low as 10%. Clay particles stacked on top of each other in the form of plates have a large surface area, so the field capacity of clay soils can be over 40%. As a result, clay soil does not have to be irrigated as frequently as sandy soil. Thus, for scheduling purposes it is essential to know the texture of the soil [6]. In a soil at field capacity, a mineral soil with a high organic matter content has more water holding capacity than a mineral soil with a lower organic matter content [7].

2.2 Water quality

Irrigated agriculture depends on a sufficient supply of high-quality water [8]. Agricultural water quality is also determined based on the effect of water on plant quality, yield and soil properties [9]. Properties that determine water quality for transplant irrigation are: alkalinity, electrical conductivity (EC), sodium absorption ratio (SAR), and elemental toxicities. The direct effects of irrigation water on plant growth arise either due to the creation of high osmotic conditions from plant sap or the presence of phytotoxic compounds in the water [10].

Under the effect of high osmotic pressure, the water usage of plants decrease, which is lethal for the plants. Therefore, the total salinity content of irrigation water is extremely important [11]. Plants suffer more damage from salts in the early stages of their development compared to their ripening periods. This situation leads to either a decrease in yield or no yield at all. Negative effects of dissolved salts in water could be cessation of vegetative growth in the plant. It also appears in the form of reduced plant and seed development. In a study with broccoli plant [12], both irrigation water salinities and irrigation water amounts were effective on plant yield, while only salt levels were effective on dry matter values. There has been a significant decrease in the yield from 6 dS/m, and the increase in the amount of irrigation water has decreased the yield [12]. Not all vegetables respond to salinity in a similar way; some vegetables can give acceptable yields compared to other plants at much higher soil salinity.

pH has chemical and biological importance in water as an increase or a decrease in pH affects the toxicity of some compounds [13]. The quality of irrigation water in the root medium has a direct impact on the pH of the growing media and the nutrient availability. When the proportional amount of Na ion is high, the physical properties of the soil change negatively. Excess Na in irrigation water causes soil dispersion and structural dispersal. For this reason, the Na + content of water should be calculated using % Na SAR values.

The effects of irrigation water quality on soil and plants vary depending on the physical and chemical properties of the soil, the salt resistance of the plant grown, the climate of the region, the irrigation method applied, the irrigation interval and the amount of irrigation water [14].

Due to the depletion and insufficiency of existing water resources, there is a need to seek alternative water resources. In many countries, treated wastewater is considered as an alternative irrigation water. However, heavy metals, salts and harmful chemicals can be lethal especially in raw vegetables when not managed carefully in irrigation. To determine whether the treated wastewater is suitable for irrigation, the total concentration and electrical conductivity of the dissolved substances in the water, the sodium ion concentration, the ratio of sodium ion concentration to other cations, the concentration of boron, heavy metals, other potentially toxic substances, the total concentration of Ca^{++} and Mg^{++} ions under some conditions, total solids, organic matter load, and amount of floating matter such as oil-grease and pathogenic organisms should be examined. On the other hand, wastewater irrigation for vegetables has serious concerns since most of the vegetables are consumed raw. Moreover, in most countries using wastewater as an irrigation water for the raw eatable crops is forbidden by law.

2.3 Important soil moisture levels for irrigation

In irrigation applications, it is necessary to know the amount of moisture held in the soil at certain tensions. These reference soil moisture amounts are called soil moisture constants. Major soil moisture constants in terms of irrigation are saturation point, field capacity, and wilting point. During and immediately after irriga-tion, all the pore space in the soil is filled with water and the soil becomes saturated. As a result of water molecules filled the pores in the soil, there is a little air in the soil, and for most crops if the soil stays saturated the crop will be damaged due to this lack of air for the roots to breathe. If there are no drainage problems, the water in the soil will drain away under gravity following irrigation, leaving space for air in the soil's pore space [6]. In many types soil, after a rain or irrigation, the water immediately starts draining deeper into the soil. After 1 or 2 days, the water content in the soil will reach, Zaten after 1-2 day diyerek zaman vermissin. To larger areas underneath the surface, a nearly constant value for a particular depth in question. This somewhat arbitrary value of water content, expressed as a percentage, is called the field capacity [15]. Wilting point also called as the permanent wilting point, can be defined as the amount of water per unit weight or per unit bulk volume in the soil, expressed in percentage, that is held so tightly by the soil matrix that roots cannot absorb this water and a plant will wilt. Therefore, in practice, plants only benefit from moisture between field capacity and wilting point. The interval between the field capacity and the wilting point is called the usable/available water holding capacity. As it is known, the most important factor limiting plant growth in arid and semi-arid climates is the lack of available water in the root zone [16]. For this reason, irrigated agriculture is an inevitable necessity in arid and semiarid areas.

The objective of irrigation is to allow the soil moisture to reduce to a safe limit and then to irrigate the soil to bring it back to field capacity. The interval between irrigation will thus depend on the available moisture in the soil and the rate at which the soil water is abstracted by the crop [6].

3. MEASURING SOIL MOISTURE

Accurately measuring soil moisture and evaluating the moisture change in the soil in the vegetable root zone are very crucial in irrigation applications.

The monitoring of soil moisture is a standard way to determine when vegetable crops need to be irrigated. An effective irrigation in arid and semi-arid regions is achieved by monitoring the soil moisture and determining the soil moisture content correctly [17].

There are several methods due to the development of science and technology through the years. Gravimetric method, tensiometers, granular matrix sensors or the gypsum blocks, time domain reflectometry (TDR), frequency domain reflectometry (FDR), drill & drop soil moisture probe and neutron probe are commonly used techniques or practices.

Gravimetric method is a basic method in soil moisture measurement. It can be also used to compare different methods with one another as a standard calibration method [18]. In this approach, the moisture content of the soil is determined by drying soil samples in an oven at 105°C to a constant weight and finding the amount of water lost. The moisture content is calculated by ratio as weight of water to the weight of dry soil. Generally, samples of 50 to 100 g of soil are enough in most field tests due to the large samples requiring longer drying times [18].

In other methods, different devices ranging from inexpensive simple moisture meters to much more expensive probe systems are used to measure soil moisture. These devices provide real-time monitoring of soil moisture. In the first group, water potential in soil are measured with tensiometers and the granular matrix sensors such as gypsum block and the watermark sensors. In the second group, soil moisture content considering the time or frequency of electronic pulse traveling between or returning to electrodes and capacitance sensors are measured with water content sensors. Resistance-type moisture sensors work by measuring the resistance between electrodes inserted into the soil [17].

Tensiometer is a device that measures moisture tension inside soil. It is widely used in the practice because it is inexpensive, needs no power supply, and provides direct and continuous readings [19]. They are inserted into the soil to different depths considering the effective root zone of the plants [20]. It has water-filled generally transparent looking tube with a porous ceramic at the bottom and a vacuum gauge at the top. The readings as a suction i.e. negative pressure or a potential are in kpa (kilopascals) or centibars. Tensiometer readings are not affected by the soil temperature or the osmotic potential of the water solution in the soil and work at low water retention tensions (0–85 kPa) which represent a small part of the entire range of available moisture [21].

Granular matrix sensors or the gypsum blocks with electrodes embedded in block of porous material are used to measure a resistance that reflects. The electrodes are connected to cables that extend to the soil surface for neyi read ediyor. Resistance? by a portable resistance meter providing small voltage. If water is present in the soil, the gypsum block gets wet, thus the resistance between electrodes decreases, while on the contrary the resistance increases as soil dries. Increased resistance shows an elevated tension in the soil. Therefore, resistance readings from devices are converted to actual water contents using a calibration curve later as the granular matrix sensors indirectly measure soil water tension using electrical resistance [22]. Being an inexpensive device can be seen as an advantage but it is stated that mistakes in measurements of moisture of wet soils occur frequently [19, 23].

Capacitance sensors, TDR and FDR are the techniques that consider dielectric property of the soil to measure moisture content [24]. These electromagnetic soil moisture sensors have improved throughout the last few decades considering size, cost and precision [22]. A capacitance sensor consists of two electrodes, which provide to be immersed in the soil and measure the dielectric constant that increases. Although they are inexpensive and user friendly, the common restriction for most of the capacitance sensors is that they provide measurements considering a very small soil volume, thus they do not reflect the situation in the soil away from the sensor. Therefore, they are more suitable for small volume container-grown vegetables in greenhouses. In addition, accuracy of capacitance sensors can be affected by many soil properties such as clay and organic matter contents, salinity level, bulk density, and temperature [22].

Soil moisture content is better estimated by determining the dielectric constant based on time domain reflection principle, frequency reflection principle and standing wave principle with the advanced devices. In this context, several types of soil moisture sensors have been developed as TDR and FDR [17, 25]. TDR device works according to the principle of determining the electrical conductivity (dielectric constant) value of a material based on the propagation speed of electromagnetic waves. Although it provides rapid and repeatable measurements with no health risks, unlike neutron probe technique, it is a complex and an expensive measurement equipment. It has also some disadvantages such as reflection loss in saline soils or wet soil increases conductivity [21]. Soil moisture sensors may also require calibration, and thus calibration equations to convert readings to volumetric water content are considered [22].

In the FDR with capacitance probes, by given voltage from two electrodes, soil moisture is determined under the assumption that dielectric constant of water is much higher than soil. Capacitance is measured from variation in frequency of a reflected radio wave or resonance. When the electrodes are given voltage, which induces the frequency oscillations with an oscillator to propagate an electromagnetic signal, at a certain point resonance occurs and soil moisture content is determined through this point. The accuracy and repeatability of the FDR are high, and the FDR probes give faster response time compared to TDR probes. Moreover, FDR is relatively inexpensive and has no health risks as well. However, calibration for the results needs to be done for each soil used. To obtain correct measurements, probes need to have good contact with the soil without air gaps and the moisture measurements in saline soils are generally not reliable [19, 21].

Soil moisture profile probes or drill & drop probes provide continuous soil moisture measurements from different depths over the entire length of the probe.

Salinity and temperature sensors can also be found in the probe in addition to the moisture sensors. In this practice, the time taken for an electromagnetic wave to travel along a given length of a transmission line in the soil is measured. Soil dielectric properties are changed with moisture content in the soil. This leads to different electromagnetic wave travels rates in wet and dry soils. Thus, the soil moisture content can be estimated with this approach. This device might be a good option for fast, easy but short-term measurement for monitoring of the vegetable cultivated soils.

In neutron probe method which detects soil moisture using radioactive element, fast neutrons are continuously emitted from the neutron source to the soil environment during measurement. The probe device has a source and a detector. When fast neutrons collide with hydrogen atoms, they lose energy and decelerate. With increasing soil moisture content, the density of slow neutron clouds increases. The neutron meter determines the moisture content in the soil by determining the functional relationship between the density of the slow neutron cloud and the water molecules. Although it allows fast, reliable and repeated measurements at any soil depths, the neutron instrument is expensive, and has radiation hazard risk to health, thus it cannot be widely used [17]. Neutron probes and drill & drop probes can be effectively used in soils cultivated with deep root vegetables. A major restriction of these devices might be their expense for small farms.

4. WATER REQUIREMENTS OF THE VEGETABLES

Vegetable crops require more and frequent irrigation than other plants as they contain 60–90% water, and thus irrigation in arid and semi-arid regions plays a vital role in vegetable growth. The availability of sufficient water in soil is essential for good crop formation, growth, yield and quality in vegetable production. The application of frequent but low volumes of water for vegetable crop production has been proven to result in more-yield compared to few application [26].

Crops can experience water stress in two different ways, which are the water shortage (drought) and excess water (flooding, saturation) [27]. The excess water causes waterlogging in soils, and the symptoms are similar to the water deficit impeding the oxygen supply and respiration of roots and water uptake. Drought stress occurs when atmospheric conditions cause permanent water loss through evaporation or transpiration. Under stress conditions the stomatal closure occurs with a reduction of net photosynthesis, and these responses depend on the severity and duration of stress and crop growth stage [28]. As a practical approach in controlling water stress level, the leaf photosynthetic activity can be monitored since measuring stomatal conductance or resistance of plant leaves indicate the severity of water stress. In drought stress; plant development is regressed, woody structure occurs, bloom early and growth of the leaf area, stem height and chlorophyll content reduce [29]. In addition, water plays a considerable role in the nutrition consumption of plants by dissolving nutrients in the soil. The encounter of a dry soil layer during the growing period of the vegetables will prevent the enlargement and development of the roots. However, there are many ways to manage drought stress such as mulching, use of plant growth regulators, anti-transpirants, use of water absorbent polymers (e.g. hydrogel), grafting technique, use of resistant varieties, irrigation method selection (e.g. drip irrigation), water harvesting and protected cultivation [27].

Growing areas of tomatoes have increased intensively, green peas moderately, beans and sweet corn slowly between 1997 to 2017 [28]. Corn, soybeans, beans and peas are the crops that are moderately water stress sensitive while tomatoes are within the extremely drought sensitive group. Although most crops are less sensitive to water shortage during the early stages of vegetative growth, changes of many physiological traits causing the disturbance of fertility and reduction of yield appear during the generative stage [28]. Therefore, irrigation scheduling and irrigation water requirements are determined by the water stress tolerance and water use potential of the plant varieties. Water use potential of vegetable crops depends on crop type, field soil properties, irrigation system type, climatic conditions, and crop growth stage.

5. IRRIGATION WATER AMOUNT

Irrigation is likely to increase the size and weight of an individual fruit and to prevent defects. On the other hand, too much moisture reduces soluble solids in muskmelons (cantaloupes) and capsaicin (what makes the peppers hot) in hot peppers when it occurs during fruit development. In order to determine the amount of water needed for irrigation of plants, it is necessary to know the amount of water they consume, the percentage of this amount met by precipitation (effective precipitation) and the irrigation efficiency, which includes losses in transmission and application of the irrigation. Effective precipitation is ignored for the vegetables grown under greenhouse conditions [4]. Total irrigation water requirement for a crop in the field conditions can be calculated using below equation:

Total irrigation water requirement = (Crop evapotranspiration – Effective precipitation)/ Transmission and application efficiency. (1)

Part of water delivered from resource is not fully stored in the crop root zone as it is lost through evaporation, runoff and deep percolation within the irrigated area. Therefore, an application efficiency value used for calculating total irrigation water required is the fraction of the available water stored in effective root depth to water conveyed to the field. Irrigation water requirement can be reduced by drip irrigation method as drip irrigation method applies water directly to crop root area (only some parts of the soil root zone watered) which saves a considerable amount of irrigation water [30]. In this case, the total irrigation requirement value should be corrected with a wetting percentage or plant cover percentage value which is lower than 1.

6. IRRIGATION SCHEDULING

Irrigation scheduling simply means application of water to crops at the required time and in the required quantity. Irrigation scheduling is one of the most effective way to increase water productivity in fertile crop production. Marketable yields for most shallow rooted vegetables can be easily damaged by short-term moisture stress of two to three days. Deficit irrigation generally results yield loss and inadequate quality in vegetables, while excess irrigation increases susceptibility to diseases, irrigation energy cost and environmental pollution risk from the nutrient leaching [31].

Different techniques of irrigation scheduling in irrigation of vegetable crops are used. They can be classified as monitoring of soil water status, water balance approach from crop water demand and observing of plant traits. Continuously monitoring soil moisture throughout crop growing period is very crucial and also requires accurate measurement with precision-based devices [32]. Soil moisture sensors can be used to regulate the interval of irrigation and, possibly, the water quantity by continuously monitoring water content or tension of the soil [31].

Technological advances in automatic soil water sensor-based irrigation systems are aimed to save an optimum soil water range in the root zone for high-quality plant growth. These algorithms are? used in automated irrigation management and scheduling agricultural activities. Furthermore, these algorithms are developed to observe water content and ensure irrigation with automated activation when necessary [33]. Smart irrigation automation systems can be less used in open field agriculture, while they are used increasingly in greenhouse production with soil or without soil to save considerable amounts of water and nutrients that are heavily applied.

Irrigation scheduling in crop water demand method consists of supplying the crop evapotranspiration (ETc) (daily, five or ten days' averages) for each crop growing stage. Thus, this method is known as crop evapotranspiration method. The ETc value of a fully-irrigated vegetable crop can be calculated either empirically by multiplying daily reference of evapotranspiration (ETo) by crop coefficient (Kc) or experimentally (2). However empirical way is commonly preferred to save time, cost and labor.

$$\text{ETc} = \text{ETo} \times \text{Kc} \tag{2}$$

ETo is calculated by well-known Penman-Monteith (FAO) equation using daily or daily average (of five or ten days) air temperature, relative humidity, wind speed and solar radiation data [22, 34]. Kc values are selected from the tables prepared for different growth stages of vegetable crops [34]. Irrigation scheduling programs such as CROPWAT can be used as the model-based scheduling. This program uses Penman-Monteith (FAO) method with collected soil, crop and climatic data in the region [22, 35]. The precision of ET-based scheduling method depends strongly on the accuracy of the ETo estimation value, a correct Kc value determined with site-specific calibration approach, correct determination of the soil's available water holding capacity, and measuring site-specific precipitation [35, 36].

Irrigation interval can. be calculated by the ratio of readily available water (RAW) to the crop daily net irrigation water requirement (Inet) which are calculated as:

$$\text{Inet}\left(\text{mm day}^{-1}\right) = \text{ETc} - \text{Effective precipitation}, \tag{3}$$

$$\text{Irrigation interval}\left(\text{day}\right) = \text{RAW} / \text{Inet}. \tag{4}$$

RAW is calculated by, below equation [35].

$$\text{RAW} = (\theta\text{fc} - \theta\text{wp}) \times \text{D} \times \text{MAD} \tag{5}$$

where RAW is the readily available water content (mm), θfc is the volumetric water content at field capacity ($m^3\ m^{-3}$), θwp is the volumetric water content at the permanent wilting point ($m^3\ m^{-3}$), D is the effective rooting zone depth or the soil layer depth considered (mm) and MAD is the fraction of the total available water that is allowed to be depleted. MAD value should be kept low in vegetable irrigations to protect plants from water stress. RAW value is also equal to the net irrigation quantity applied to the soil.

Plant-based scheduling techniques have been improved from the relationship between crop water stress and soil moisture deficit to define an optimal moisture content level for crop growth. Measuring crop water stress for irrigation scheduling has been also recommended considering the variations in plant species, tissues, and phenological stages [35, 37]. The approaches have been categorized as the measurements of tissue water potential and the measurements with plant physiology-based (sap flow, stomatal conductance, thermal sensing with infrared thermometers).

Stomatal conductance is a good indicator in determination of irrigation need in many plants sensitive to water insufficiency, thus improvement of this technique among plant-based irrigation scheduling approaches has drawn increaasing attention [37]. In recent years, the use of irrigation scheduling based on the crop water stress index (CWSI), which is calculated based on the canopy temperature measured with an infrared thermometer, has gained importance. Many researchers have reported that the CWSI value can be used for preparation of irrigation scheduling [38]. This approach argues that significant increases in canopy temperature exceeding air temperature has been a good indicative for stomatal closure and water deficit stress [37].

Furthermore, a systematic method can be applied as a practical scheduling approach. In this method, water applications are managed on a time or volume basis applying every day for the same duration or in the same quantity. Moreover, it is quite practical to base irrigation scheduling on evaporation from a Class A pan as a result of the combined effect of climatic factors. This approach requires a correction coefficient (kp) (mostly changed between 0.6–0.8) to convert potential evaporation value measured in pan to the ETo value. The Kp coefficient is also expressed as Kcp when it includes the crop coefficient. Water use from fully developed vegetation can be about 75–80% of the amount of water evaporated from the pan, in other words, Kcp = 0.75–0.80. When plants do not completely cover the soil surface, actual water consumption will be less than 75% of pan evaporation. Water use for vegetable crops can be considered as 10–15% of pan evaporation during the first 1/3 of the season, 40–50% of the pan evaporation during the mid-season, and 60–80% of the pan evaporation during the last 1/3 of the growing season [39].

7. IRRIGATION METHODS

In irrigated agriculture, when operating an area for irrigation, firstly the most suitable irrigation method under the conditions should be selected, then the system required by this method must be planned, installed and operated. In general, the irrigation method to be selected must meet some conditions such as to provide uniform water distribution, to minimize deep percolation and run-off losses, not cause soil erosion, not prevent agricultural mechanization, help leaching the salt from soil.

Due to the shallow rooting depth of most vegetables and their high response to lack of water, irrigation is frequently required in small amounts. This situation is more important in greenhouses with intensive production. Vegetable growers consider drip irrigation method as an effective way to save water and that plant needs, as well as to reduce weeds, fungi and diseases. Drip irrigation minimizes water loss from run-off and deep percolation, decreases evaporation losses. It has been determined that water savings of 50–80% are achieved when compared to conventional surface irrigation methods [27]. Drip irrigation method also provides more efficient water and fertilizer usage than the sprinkler method. It also reduces disease problems because leaves are not wetted. Drip irrigation lowers energy need because of the low pump pressures required and provides more applicable opportunity of automation. However, compared to sprinkler nozzle sizes, drip system emitters have very small openings, emitters gets clogged. Therefore, it requires water quality control and some preventive solutions such as filtration and dilute acid applications. Many researchers have declared considerable benefits of drip irrigation method over other conventional irrigation methods to improve yield and water productivity (WP) of fruits and vegetables [40]. The water application efficiency is about 80–90% in drip irrigation systems [39]. Maximizing the water productivity with decreased water loss and increased yield in drip irrigation is a practical way to manage finite water supplies. Plants use large amount of the water applied from increased water efficiency. This also minimizes leaching of agro-chemicals out of the field or vegetable growing containers into the environment. In the last few decades, water productivity of vegetable crop values has been improved with the use of efficient micro irrigation techniques such as micro sprinkler and drip irrigation [41]. Jha et al. [40] determined that drip irrigation method resulted higher water productivity with more than five-folds increase in potato and cauliflower compared to the furrow method. It was also observed that drip irrigation method conserves approximately 70–80% water compared to conventional flood irrigation method.

In drip irrigation systems, to avoid possible plant stress, irrigations are usually scheduled to start when allowable percentage of usable water in the soil has been consumed. This level ranges from 30% in drought-sensitive plants to 70% in droughtresistant plants. In drip irrigation, this value is usually taken as 30% (MAD value) [39].

Irrigation volume in drip irrigation system considering soil available water depletion approach can be calculated with equation below.

$$\text{Irrigation volume (L)} = (\theta fc - \theta wp)\times D\times MAD\times P\times A \quad (6)$$

where P is the wetting factor, A is the irrigated area (m^2), and other terms are as mentioned before. Wetting ratio are considered minimum %30 in semi-arid regions, and it is 35% and 25% in arid and humid regions, respectively [42].

Wetting factor is less than 1 because especially during irrigation of plants with wide row spacing, a dry area remains between the laterals that is not wetted. In some cases, the plant covering ratio is also considered instead of this value in order to apply water according to the plant growth rate.

Irrigation volume can be also determined using the Class A pan evaporation with following equation [43].

$$\text{Irrigation volume (L)} = Ep\times Kcp\times P\times A \quad (7)$$

where Ep is the cumulative pan evaporation measured using a standard Class A pan at considered duration (mm), Kcp is the coefficient of crop-pan evaporation, P is the wetting factor and A is the irrigated area (m^2).

Due to (Ep x Kcp) is equal to the ETc (ETc = Ep × Kcp), bunun yerine (7) de can be also used to calculate irrigation volume from the ETc values determined using other approaches empirical (e.g. Penman-Monteith) or experimental.

REFERENCES

1. Howard G, Bartram J. World Health Organization. Domestic water quantity, service level and health [Internet]. 2003. Available from: https://apps.who.int/ iris/handle/10665/67884
2. TWB. Water Resource Management: Working to Improve Water Security (English). Washington, D.C.: World Bank Group, 2021.
3. FAO. Food and Agriculture Organization, 2003. Statistical databases.
4. Kemble JK, Sanders DC. Basics of Vegetable Crop Irrigation, Alabama Extension System. ANR 1169, Alabama A & M and Auburn Universities, USA., 2000.
5. Pritchett WL. Properties and Management of Forest Soils, John Willey and Sons, N.Y., 1979.
6. Burton M. Irrigation management: Principles and practices. CABI North American Office, Cambridge, MA. 2010. 375 p.
7. Soil Quality Indicators. USDA Natural Resources Conservation Service, June 2008.
8. Islam MS, Shamshad KM. Assessment of Irrigation Water Quality of Bogra, Bangladesh J. Agril. Res., 2009; 34(4): 597-608.
9. Etteieb S, Cherif S, Tarhouni J. Hydrochemical assessment of water quality for irrigation: a case study of the Medjerda River in Tunisia, Applied Water

Science. 2015; 1-12.

10. Kurunç A. Irrigation Water Quality and Salinity, Effects of Salinity on Plants, Akdeniz University, Agriculture Faculty, Department of Agricultural Structures and Irrigation. Course Note (in Turkish). 2017.
11. Yurtseven E. Irrigation Water Quality and Salinity, Ankara University, Agriculture Faculty, Department of Agricultural Structures and Irrigation. Course Note (in Turkish) 2016.
12. Yurtseven E, Baran HY. The Effects of Irrigation Water Salinity and Irrigation Water Amounts on Yield and Total Mineral Residues of Broccoli (*Brassica oleracea* botrytis). Turk J Agric For. 2000; 24: 185-190.
13. Tokatlı C, Köse E, Arslan N, Emiroğlu Ö, Çiçek A, Dayıoğlu H. Water Quality of Emet Stream Basin. Uludağ University Journal of The Faculty of Engineering. 2016; 21 (2): 9-24.
14. Rhoades JD. Quality of water for irrigation. Soil Sci. 1972; 113: 227-284.
15. Kirkham B. Chapter 10 - Field Capacity, Wilting Point, Available Water, and the Nonlimiting Water Range, Editor(s): M.B. Kirkham Principles of Soil and Plant Water Relations (Second Edition), Academic Press, 2014; 153-170p.
16. Falkenmark M, Rockstrom J. Curbing rural exodus from tropical drylands. AMBIO0122 no 71993, (United Kingdom). 1993.
17. Yu L, Gao W, Shamshiri RR, Tao S, Ren Y, Zhang Y, Su G. Review of research progress on soil moisture sensor technology. Int J Agric & Biol Eng. 2021; 14(4): 32-42.
18. Rasti A, Pineda M, Razavi M. Assessment of soil moisture content measurement methods: conventional laboratory oven versus halogen moisture analyzer. Journal of Soil and Water Science. 2020; 4(1): 151-160.
19. Kaur K, Mahajan R, Deepak B. A Review of Various Soil Moisture Measurement Techniques. International Journal of Innovative research in Science, Engineering and Technology. 2016; 5(4): 5774-5778.
20. Cregg B. Soil moisture measurements in nurseries and landscapes. Crop Advisory Team Alert. 2003; Vol. 18, No. 12. Michigan State University Extension, East Lansing, MI.
21. Nwogwu NA, Okereke NAA, Ohanyere SO, Chikwue MI. A concise review of various soil moisture measurement techniques. Proceedings of the 3rd South East Regional Conference, 27th-30th August 2018., University of Nigeria, Nsukka.
22. Cahn MD, Johnson LE. New approaches to irrigation scheduling of vegetables. Horticulturae. 2017; 3: 28.
23. Werner H. Measuring soil moisture for irrigation water management. FS876. South Dakota State University Cooperative Extension Service. Brookings, SD. 2002.

24. Brahma M, Goswami B. Electrical methods of soil moisture measurement: A Review. ADBU Journal of Electrical and Electronics Engineering (AJEEE). 2017; 1(2): 14-17.
25. Zhang Y, Ma Y H, Jiang Z H, Tan C J, Wang C S. Progress on research and application of rapid measurement of soil moisture sensor. Chinese Agricultural Science Bulletin. 2014; 30(5): 170-174.
26. Locascio SJ. Management of irrigation for vegetables: Past, present, and future. HortTechnology. 2005; 15: 482-485.
27. Patel AJ, Rathod KD, Satodiya BN. Effect of water stress and its management in vegetable crops: A review article. International Journal of Chemical Studies. 2021; 9(1): 1382-1385.
28. Nemeskéri E, Helyes L. Physiological responses of selected vegetable crop species to water stress. Agronomy. 2019; 9: 447.
29. Waśkiewicz A, Gładysz O, Beszterda M, Goliński P. Water stress and vegetable crops. In: Ahmad, P. (ed.), Water Stress and Crop Plants: A Sustainable Approach, John Wiley & Sons, Ltd., 2016; pp. 393-411.
30. Brouwer C, Prins K, Kay M, Heibloem M. Irrigation Water Management: Irrigation methods. Training Manual No. 5: FAO - Food and Agriculture Organization of the United Nation, Rome, 1985.
31. Pardossi A, Incrocci L. Traditional and new approaches to irrigation scheduling in vegetable crops. HortTechnology. 2011; 21(3): 309-313.
32. Chaudhary SK, Srivastava PK. In: Srivastava, P.K., Gupta, M., Tsakiris, G., Quinn, N.W. (Eds.), Future challenges in agricultural water management. Agricultural Water Management, Theories and Practices. 2021; pp. 445-456.
33. Saad A, Benyamina AEH, Gamatié A. Water management in agriculture: a survey on current challenges and technological solutions. IEEE Access. 2020; 8: 38082-38097.
34. Allen RG, Pereira LS, Raes D, Smith M. Crop evapotranspiration: Guidelines for computing crop water requirements. In FAO Irrigation and Drainage Paper; Food and Agriculture Organization (FAO): Rome, Italy, 1998.
35. Gu Z, Qi Z, Burghate R, Yuan S, Jiao X, Xu J. Irrigation scheduling approaches and applications: a review. J. Irrig. Drain Eng. 2020; 146(6): 04020007.
36. Davis S, Dukes M. Irrigation scheduling performance by evapotranspiration-based controllers. Agric. Water Manage. 2010; 98 (1): 19-28.
37. Jones HG. Irrigation scheduling: Advantages and pitfalls of plant-based methods. J. Exp. Bot. 2004; 55(407): 2427-2436.
38. Tekelioğlu B, Büyüktaş D, Baştuğ R, Aydınşakir K, Karaca C, Dinç N. Comparing The Methods Used to Calculate Crop Water Stress Index (CWSI) of Pomegranate and Soybean. 1st International, 14th National Congress on Agricultural Structures and Irrigation, 26-28 September 2018, Antalya, pp. 395-409.

39. Baştuğ R. Principles of drip irrigation management in vegetable production. Derim. 2014; 31 (1): 61-72.
40. Jha BK, Mali SS, Naik SK, Sengupta T. Yield, water productivity and economics of vegetable production under drip and furrow irrigation in Eastern Plateau and Hill Region of India. International Journal of Agricultural Science and Research. 2017; 7(3): 43-50.
41. Leskovar DI, Xu C, Agehara S, Sharma SP, Crosby K. Irrigation Strategies for Vegetable Crops in Water-Limited Environments. Journal of Arid Land Studies. 2014; 24-1, 133-136.
42. Yıldırım O. Field Irrigation Systems. Ankara University, Agricultural Faculty Publication No: 1449, Ankara, 1996.
43. Yavuz D, Yavuz N, Suheri S. Design and management of a drip irrigation system for an optimum potato yield. J. Agr. Sci. Tech. 2016; 18: 817-830.

14 Chapter

IMPROVEMENT OF AGRICULTURAL CROPS: A CRISPR/CAS9 METHODOLOGY

1. INTRODUCTION: AN OVERVIEW

Vegetable cultivation is a crucial element of the economic system, aimed at producing horticulture crops including vegetables, fruits, spices, tubers, and medicinal plants. These crops significantly contribute to food and nutritional security [1]. With the expansion of the global population, it is imperative to enhance agricultural production to ensure a sustainable food supply. The advancement of next-generation crops is crucial due to the large and time-consuming nature of conventional breeding techniques[2, 3]. Transgenesis represents a potential alternative method of plant breeding, whereas public acceptance poses a substantial concern for commercialization. Recently, recombinant DNA technology employing nucleases such as ZFNs[4], TALENs [5, 6], and CRISPR/Cas9 [7] has demonstrated efficacy in modifying specific genomic locations and is extensively applied in various agricultural crops. The technology is anticipated to be extensively utilized in the near future.

2. GENOME EDITING

It is a collection of sophisticated molecular biology techniques that enable the accurate, efficient, and targeted modification of particular nucleotide sequences [8–10]. Researchers employ this technology to better understand the genome's function and develop crops resistant to insects, have higher nutritional value, and thrive in dry regions. The application of genome editing methods based on site-specific nucleases (SSNs) has proven extensive gene editing across flora and fauna species during the last generation. SSNs work by assembling endonucleases capable of cleaving DNA inside a specific region in the genome. The active domain of SSN is connected to it via a DNA-binding region or an RNA sequence [11–13]. These SSNs are responsible for causing double-stranded breaks (DSBs) in the target DNA. Nonhomologous end joining (NHEJ) or homology-directed recombination (HDR) methods are used to repair DSBs, which result in insertion/deletion (INDELS) and replacement mutations in the host locus [14, 15]. Following SSN breakage of the target sequence, cellular DNA repair processes result in gene expression changes at the target sites.

1. Engineered meganuclease (MegaN),
2. Zinc finger nucleases (ZFNs),
3. Transcription activator-like effector nucleases (TALENs),
4. Clustered regularly interspaced short palindromic repeat/CRISPR/Cas9nuclease systems are the four kinds of engineered nucleases being employed in genome editing [16, 17]. All of these approaches made it possible to alter the genome in a straightforward, fast, and cost-effective way.

2.1 Engineered meganuclease (MegaN)

Meganucleases (MegaN) are endonucleases found in nature and were discovered most often in the late 1980s. Endonucleases are capable of detecting and cleaving large nucleotide sequences (ranging from 12 to 40 base pairs), which are considerably different across many genomes [13, 18]. I-SceI, a yeast mitochondrial enzyme, and I-CreI, an algal photosynthetic enzyme, are both good meganucleases. Meganucleases have been modified to recognize previous target sequences even though meganuclease receptors are still infrequent in relevant genomes. Because of the slightly longer template strand, there is more discrimination and much more minor off-target trimming. On the other hand, engineered meganucleases had a much lower utilization than some other sequential nucleases, in addition to the issue of changing meganucleases to accept novel specificities [19, 20].

2.2 Zinc finger nuclease-based engineering

Artificial sequential nucleases known as zinc finger nucleases (ZFNs) have transformed the area of programmable nucleases. ZFNs were created by attaching numerous zinc finger DNA-binding domains to the restriction endonuclease FokI's nonspecific cutting pattern [21]. The protein molecules can enlighten the difference between two DNA sequences separated by only a few nucleotides. This allows the two endonucleases to form a dimer, breaking double-stranded DNA [22]. Furthermore, because each motif in the zinc finger array reads a distinct 3-nucleotide complementary strand, the domain composition variable sequence can always be chosen to fit the particular destination. ZFNs were originally used for sequence-specific mutagenesis in tobacco in the early 2000s, which was most likely the first time a designed endonuclease identified and fractured chromosomal DNA [23, 24]. With these remarkable achievements, ZFN usage in agriculture has indeed been limited due to factors such as the technical complexity of manufacturing and the scarcity of aiming places in genomes in comparison to more recently established functional genomics approaches.

2.3 Transcription activator-like effector nucleases (TALENs)

The TALEN (transcription activator-like effector nucleases) technology was developed in 2011 to optimize the effectiveness, reliability, and availability of genome editing. Transcription activator-like effectors (TALES) were discovered [25, 26]. The TALEN, like ZFNs, creates proteins artificially with a flexible array of DNA-binding regions joined to FokI's nonspecific fragmentation site. Each repeat consists of 33–35 amino acids and identifies just one nucleotide. The last repetition is considered as a "half-repeat" since it frequently contains 20 amino acids. The varied amino acids at positions 12 and 13 provide DNA identifying distinctiveness (for example, NI accepts adenine, HD detects cytosine, NG detects thymine, and NN recognizes both guanine and adenine) [27, 28]. TAL effectors have natural segmentation grace to facilitate genome editing in TALENs, where these repetitions are organized to find individual regions of expression. Additional TALENs and gene-specific stimulators and regulatory proteins were employed as gene targeting reagents in conjunction with TAL effector assemblies [29–31]. TALENs are more adaptable compared to meganucleases and ZFNs and are used extensively in plant genome editing. However, a large number of experiments renders TALEN production as well as transport throughout plant tissue problematic.

2.4 Clustered regularly interspaced short palindromic repeat/CRISPR/Cas9 nuclease system

The CRISPR-Cas system, like TALENS, draws its inspiration from biology. The CRISPR-Cas nucleases had first been found in the adaptive immune system function of archaea and bacteria. CRISPR codes for "spacer" RNA molecules that create associations between CRISPR-associated (Cas) nucleases and instruct them to break down external nucleic acids. The spacer segments within those biological systems were obtained from bacteriophage components caused by a bacterium the prokaryote progenitor. The system's aiming specialization is based upon basic genetic platform principles.

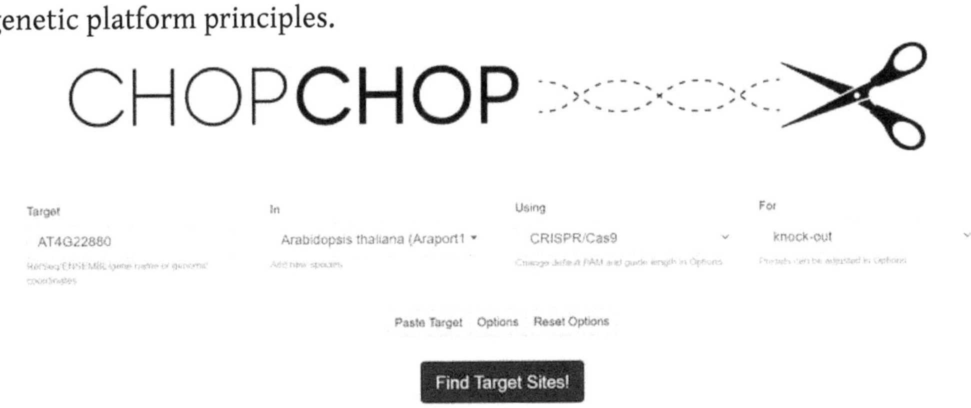

Figure 1. Home page of CHOPCHOP bioinformatics tool for designing of sgRNA.

A lateral or segmental short sequencespecific element is necessary for a target sequence site to be accepted and bisected; the above offers assurance that the prokaryote utilizing the CRISPR/ Cas system will not focus its genome; these representations are just not available among exogenously linked patterns. The Streptococcus pyogenes CRISPR-Cas9 system (CRISPR-SpCas9) was the first to be discovered for genetic manipulation, and also the term "CRISPR-Cas9" is generally used to describe this technology [32–34]. To avoid misunderstanding, CRISPR-Cas9 can only be used to relate to properties that CRISPR-SpCas9 and its orthologs share.

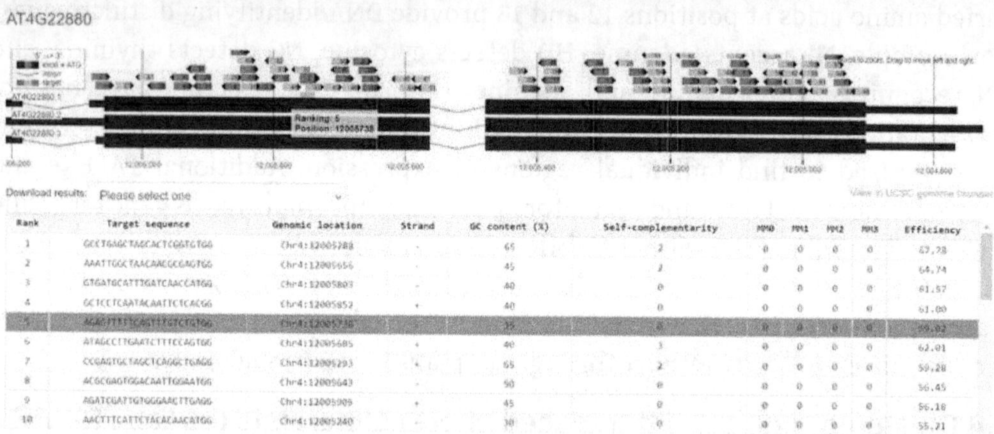

Figure 2. Preliminary stage for the identification of off target regions of our desired genomic sequence.

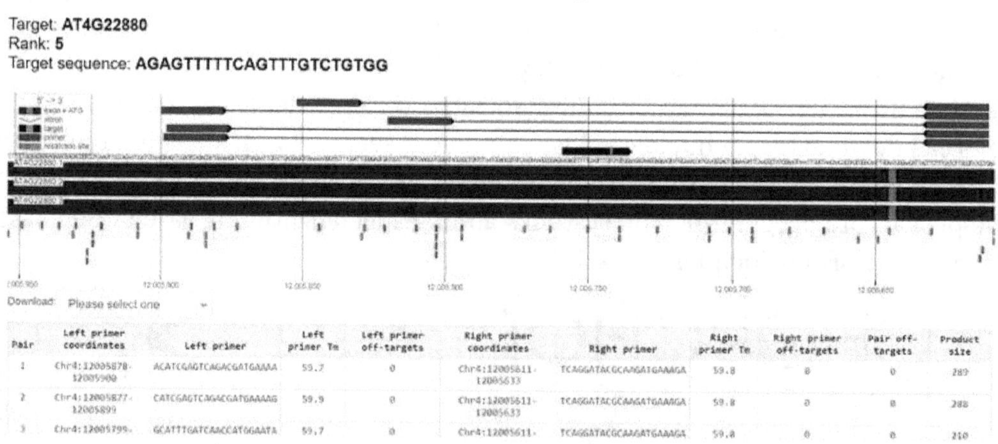

Figure 3. Intermediate stage for identification of off target regions of our desired genomic sequence. In this view, we can identify the %GC, levels of off targets and primer sequences.

The designed CRISPR-Cas9 system is made up of two parts—(1) the Cas9 nuclease and (2) a single guide RNA (sgRNA), which is made up of two RNA molecules—the spacer-containing CRISPR RNA (crRNA), and they serve to enhance crRNA (tracrRNA), which itself is needed again for retired moiety's growth and development. The sgRNA leads the nuclease combination to a specific DNA location, causing the identical nucleotide sequence to be cleaved [20, 21]. Cas9 has a composite morphology, having dual nuclease domains, RuvC and HNH [14, 20, 35]. The CRISPRCas9 system has gone through several modifications in the area of plant genome engineering methods [36], including simplicity of design, cloning, and transport into plant cells, which also has resulted in a significant success rate in genome editing technology. The design of guide RNAs using the CHOPCHOP software tool is summarized in **Figures 1–4.**

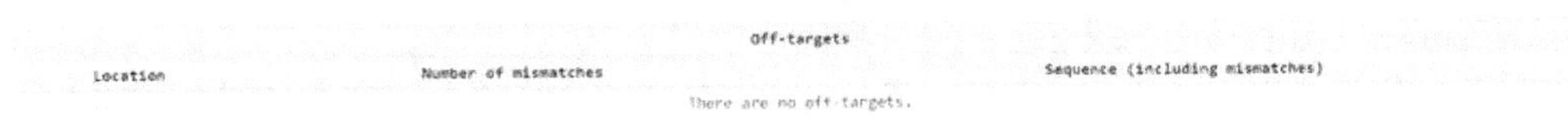

Figure 4. Final stage of off target analysis or our desired genomic sequence.

3. CRISPR/CAS9 EXOTIC VARIANTS AND CHALLENGES

CRISPR/Cas9 genome editing utilizing the SpCas9 enzyme from Streptococcus pyogenes continuously transforming the area of genome editing by offering very accurate, simple, and highly efficient gene alterations by creating nicks on the double-stranded genome of the targeted organism. Since 2013 scientists have extensively used and still exploring its vast possibilities in genome editing. Even though the efficiency of Cas9 is still high, there are some setbacks regarding their use in gene editing. One of the limitations of using them in plant biotechnology is the indistinct regulations using CRISPR/Cas9 edited plants. Another drawback is the unavailability of a standardized transformation protocol to deliver the CRISPR/Cas9 construct to some plants. These problems are now solved by the availability of novel exotic variants of genome editing enzymes that have been tested as equally efficient or perhaps more efficient than SpCas9 [37]. Scientists have identified and characterized many other kinds of microbial communities; CRISPR-RNA-guided adaptive immune systems are found. Two primary classes are distinguished, with five types and 16 subclasses [38]. These enzymes require multisubunit proteins to bind to crRNA and cleave the target genome, which is all found in class 1. "Class 2" consists mostly of two types of effectors—Type II and Type VI. Each of these types of effectors binds and cleaves crRNA and target nucleic acids. Cpf1 [38, 39], C2c1 or C2c3 [40], and C2c2 with two HEPN RNase (higher eukaryotes and prokaryotes nucleotide-binding) domains are used in Class 2 Type V and Class 2 Type VI, respectively. In contrast, Class 2 Type II is characterized by Cas9, RuvC, and HNH nuclease domains, while Class 2 Type V uses a single Ruv [41]. Some of these effectors have experimented with some plant species.

The discovery of RNA-dependent RNase enzyme systems from Class 2 Type II (FnCas9) and Class 2 Type VI (C2c2) cleared the path for novel approaches to genome editing. The bacterium Leptotrichiashahii Class 2 Type II C2c2 is directed by a single crRNA and may be trained to cleave any ssRNA with corresponding protospacers. These effectors, which are composed of two HEPN domains containing catalytic residues, preferentially cleave ssRNAs at varying distances from the crRNA binding site rather than adenine targets. C2c2 binding is controlled by a crRNA secondary structure with at least one 24-nt stem-loop structure and a 22–28-nt complementary sequence to these RNA protospacers. The latter must be flanked at the 3'end by a mononucleotide protospacer-flanking site (PFS) comprised of adenine, uracil, or cysteine [41, 42]. Another RNase- based system was identified in 2013 [43] from microbe *Francisellanovicida* (*Fn*Cas9), which could target bacterial mRNA and lead to altered gene expression and is PAM independent. This enzyme successfully inhibited the hepatitis C virus (HCV) in Huh-7.5 cells by RNA inhibition method. This enzyme targets both positive and negative strands of the virus, thus paralyzing RNA translation and replication. It was shown that mismatches up to three to six base pairs at 3'or 5′ end were tolerated by *Fn*Cas9 whereas more than six mismatches led to complete loss of activity. This enzyme is also capable of targeting DNA [44]. The above studies suggest the feasibility of developing viral infections resistant crops. The regulatory policies related to the usage of transgenic plants are still going very strong in many countries. To overcome this problem, smaller versions of genome editing enzymes are developed that can be used along with viral vectors to transform plants with desired traits. Virus vectors allow high and transient expression of heterologous genes for editing. This is proved in the case of targeted mutagenesis of *Nicotiana benthamiana* and *Petunia hybrida* using tobacco rattle virus (TRV) [45].

As *Sp*Cas9 is having a larger size (4.2 kb), the tobacco rattle virus cannot be used to express *Sp*Cas9 in plants. To resolve this problem, small genome editing enzymes were identified from different microbes such as *Staphylococcus aureus* (*Sa*Cas9, 3.2 kb), *Streptococcus thermophilus* (*St*1Cas9, 3.4 kb), and *Neisseria meningitidis* (*Nm*Cas9, 3.2 kb). These enzymes belong to the Class 2 Type II immune system and cleave double-stranded DNA using RuvC and HNH domains. Moreover, this group of enzymes cuts DNA at a specific target region, usually 21- to 24-nt long near 5′-NNGRRT-3′or 5′-NNNRRT-3′ 5′-NNAGAAW-3′ and 5′-NNNNGMTT-3′ PAM motifs, respectively. Here, in the PAM sequence, N signifies any nucleotide, R signifies A or G, M signifies A or C, and W signifies A or T [46–49]. In addition, research suggest that while using *Sa*Cas9, a greater rate of mutation (80%) was obtained by targeting the 5′-NNNGGT-3′ PAM sequence and induced homologous recombination in the selected lines. The above enzymes target a much longer PAM sequence for genome editing purposes. As an alternative, a new set of single crRNA-guided DNase enzymes with shorter PAM motifs have been recently identified again from the microbial community. This also belongs to Class 2 Type V CRISPR effectors Cpf1 from *Francisellanovicida* U112 (*Fn*Cpf1), *Acidaminococcus* sp. (*As*Cpf1), and *Lachnospiraceae* bacterium (*Lb*Cpf1) and have been successfully tried in rice and tobacco. *Fn*Cpf1 uses a single short RNA guide molecule, 42- to 44-ntcrRNA, which begins with 19 nt of the direct repeat followed by 23–25 nt of the spacer sequence. *Fn*Cpf1 identifies5'- TTN-3′region, a short T- rich PAM upstream of the 5′ end. Further, it cuts the double-stranded DNA in a staggered way after the 18th base on the nontargeted (+) strand and after the 23rd base on the targeted (-) strand [50]. Targeted mutations were observed in *NtPDS* and *NtSTF1* of *N. benthamiana* and *Os*DL, *Os*ALS, *Os*NCED1–3, and *Os*AO1–5 loci of *Oryzasativa* when codon-optimized *Fn*Cpf1 and crRNA were expressed in rice and tobacco. Interestingly, deletions were observed in both the transgenic plants as well as in transformed progenies, and mutation efficiency in rice and tobacco was around 47.2 and 28.2%, respectively [50].

Many new versions of *Sp*Cas9 have been developed with the core objective to enhance their specificity. One of the limitations that have come across was off-targeting, which will cause undesired mutations in the target. The next drawback is that some plant species have larger genomes with many duplicate genes, making genome editing technology less precise. The first *Sp*Cas9 variant that was obtained by mutating one of its domains (HNH or RuvC) was single-stranded DNA cleavage *Sp*Cas9-nickases [35, 51]. SpCas9-nickases are employed in pairs to carry out nonhomologous repair of double-stranded breaks (DSBs) using properly offset (>100 bp long) guide RNAs [47]. This strategy decreases off-target mutagenesis by extending the recognized DNA target area from 23 to 2 9 23 bp while maintaining an on-target cleavage rate comparable to that of wild-type SpCas9 [47]. In *Arabidopsis*, a single SpCas9 D10A nickase was equally efficient at initiating homologous recombination as a nuclease or homing endonuclease I, SceI, [52]. On the other hand, coupled SpCas9 nickases generated alterations comparable to those induced by SpCas9 nuclease. Furthermore, deletions were detected, not insertions, which occurred at a lower frequency [53]. Slaymaker et al. enhanced the specificity of SpCas9 by decreasing its helicase activity and created an improved form of SpCas9 (eSpCas9) [54]. Wright et al. created a split-SpCas9 system, a binary SpCas9 system, to enhance SpCas9 specificity. This was accomplished by overexpressing the nuclease and a-helical lobes in *Escherichia coli* as distinct polypeptides [55]. Komor et al. used a different approach to enhance the specificity of SpCas9, combining SpCas9-nickase with cytidine deaminase to create SpCas9-CD. While all other SpCas9 variations cause deletions or insertions in the DNA sequence, this variant enables the direct conversion of cytidine to uridine, which has the same base-pairing properties as thymine [56].

Another variant of *Sp*Cas9 is termed dead Cas9 (d*Sp*Cas9), which is developed by mutating both cleavage domains of *Sp*Cas9, and this enzyme is an RNA-guided DNA binding protein without cleavage activity [14]. In addition, it is fused with fluorescent or other types of markers and can be used in several biotechnological applications. This kind of fusion creates catalytically inactive and dead *Sp*Cas9 having the FokI nuclease domain at the N-terminus [57]. Compared to monomeric *Sp*Cas9, homodimer FokI enzymes are more precise in cleaving the target genome and can induce lesser off-targets. Piatek et al. demonstrated the fusion of synthetic transcriptional activators with the C terminus of d*Sp*Cas9 to the EDLL domain or the TAL activation domain. They developed d*Sp*Cas9—EDLL and d*Sp*Cas9—TAD synthetic transcriptional activators. This effector, guide RNA, and target molecules were transformed to *tobacco* through the agroinfiltration method. Though there were no stably transformed lines, the strong transcriptional activity of EDLL and TAD was proved in transgenic plants [58]. Fusion of d*Sp*Cas9 with methylated or demethylated promoters can lead to activation or inactivation of a gene. Some of the examples of this type of fused protein are d*Sp*Cas9-Tet1 and d*Sp*Cas9-Dnmt3a [59]. The deletion and insertion of methylases using CRISPR/Cas9 technology will allow modifications at the genetic level in living organisms [1].

4. APPLICATIONS

CRISPR/Cas9-based genome editing has contributed a lot in improving various traits of crops for the past years. Still, new challenges are being faced by the scientific community for the enhancement of the quality of various edible plants. There are numerous applications of CRISPR/Cas-based gene editing, namely enhancement of yield, improvement of resistance to pathogens, diseases, resistance to herbicides, and improvement of stress tolerance. In this chapter, details of studies related to the applications mentioned above of CRISPR are given and summarized in **Table 1**. Most of the CRISPR-related work in vegetables is done in tomatoes. The earliest report based on CRISPR-based genome editing was done by editing the *ARGONAUTE7* (*SlAGO7*) gene which is involved in the development of leaves. Moreover, mutated *SlAGO7* caused a change in the morphology of leaves which turned them into wiry leaves [77]. Another interesting work is to edit *SELF PRUNING 5G (Sp5G)* and *SELF PRUNING (Sp)* genes that caused early flowering [78]. Similar editing in developmental genes was also carried out *Brassica oleracea* [79] and *Lactuca sativa* [80]. Many genes involved in the biosynthesis of carotenoids such as *Anthocyanin 1* (*ANT1*) are involved in anthocyanin biosynthesis [81], *Phytoene desaturase* (*SlPDS*), *Phytochrome interacting factor* (*SlPIF4*) [82], and *Phytoene synthase* (*PSY1*) [83]. Mutant tomato plants with the knockout of the *Slagamous-like 6* (*SlAGL6*) gene produced parthenocarpic fruits under heat stress conditions that otherwise rigorously hinder fertilizationdependent fruit set [84]. Silencing the *eIF4E* gene in tomatoes and melons has attained resistance to RNA virus [85, 86]. Granule bound starch synthase (*GBSS*) gene using CRISPR/Cas9 that produced amylopectin and lacks amylose [87].

Table 1. Applications of CRISPR/Cas9 genome editing in various crops for improved traits.

Name of the gene	Crop	Function	References
LOGL5	Rice	Increased yield	[60]
CKX	Wheat	Increased yield	[61]
Amino acid permease	Rice	Increased yield	[62]
GS3, GW2, GW2, and GW5	Rice	Increased yield	[63]
GBBS1	Maize	Decreased low amylose	[64]
SWEET	Rice	Increased resistance to bacterial blight	[65, 66]
LOB1	Sweet orange	Resistance to Xanthomonas citri	[67]
EDR1	Wheat	Resistance to Blumeria graminis	[68]
MLO1	Tomato	Resistance to Oidiumneo lycopersici	[69]
ALS	Rice	Herbicide resistance	[70, 71]
ACCase	Rice	Herbicide resistance	[72]
ACCase	Wheat	Herbicide resistance	[73]
OsMPK5	Rice	Increased abiotic and biotic stress	[74]
MAPKs	Tomato	Increased resistance to drought stress	[75]
GmF3H1, GmF3H2 and GmFNSII-1	Soybean	Isoflavone synthesis	[76]

CRISPR/Cas9 mediated mutagenesis of genes involved in starch biosynthesis in sweet potato (*Ipomoea batatas*) was done for improving the quality of starch [88]. CRISPR/Cas9 induced mutation of CaERF28 conferred resistance to anthracnose in chili pepper (*Capsicum annuum* L.) [89].

Regulating homeostasis is a reliable way to enhance the yield of cereal. Genome editing of the C terminus of *LOGL5* in rice which codes for the cytokinin- activation enzyme increases the yield in rice during all ecological conditions [90]. Production of a high-yielding wheat variety was done by editing the gene that encodes cytokinin oxidase/dehydrogenase (*CKX*) by a knock-out mechanism [91]. Similarly, knocking out another gene that codes for amino acid permease in rice which is actively involved in nutrient portioning, led to an increase in yield [60]. In addition, some genes in rice (*PIN5b, GS3, GW2, GW2,* and *GW5*) have been edited based on CRISPR/Cas9 technology, which has led to an increase in yield [61, 62]. To increase their yield, scientists have successfully edited genes in fruits such as *CLV* [63] and *ENO* [92]. The nutritional content of crops is an essential trait for adding their commercial value to local and global markets. Genome editing of granule-bound starch synthase 1

GBBS1 gene in maize produced low amylose content variety [93]. Gluten causes celiac disease in susceptible individuals and CRISPR/Cas9 editing techniques have successfully edited the conserved region in the loci of wheat, leading to 85% loss in immunoreaction [94]. Bacterial blight caused by *Xanthomonas oryzae* is a severe threat to rice production, and genome editing of *SWEET* genes using the CRISPR technique imparted high resistance to bacterial blight [64, 95]. In addition, editing of the *LOB1* gene in *Citrus sinensis* conferred resistance to *Xanthomonas citri* [65]. Powdery mildew caused by *Blumeri agraminis* fungi also leads to various growth defects in wheat plants. Targeting three wheat homologous genes (*EDR1*) by CRISPR/Cas9 editing improved resistance to these fungi [66]. Similarly, the tomato was also conferred resistance to *Oidium neolycopersici* that also causes powdery mildew, by mutating *Solanum lycopersicum MLO1* [67].

CRISPR/Cas-based genome editing technology has to produce double- stranded breaks efficiently in the genome. Hence, scientific communities have also very well utilized their ability to edit plant viruses that attack plants and cause various diseases by employing various RNA-targeting versions of Cas enzymes such as Cas13a, Cas13b, Cas13d, and *Fn*Cas9 [68, 69]. Another problem that CRISPR has solved is to produce herbicide-resistant crops for controlling weeds that adversely affect crops' growth and decrease soil fertility and directly affect the yield from various crops. Acetolactate synthase (*ALS*) is a critical enzyme in producing branched-chain amino acids that are specifically targeted by herbicides such as sulfonylurea and imidazolinone. Using cytosine base editing, we impart herbicide resistance to *Oryza sativa ALS* by creating base transitions [70, 71]. Additionally, acetyl coenzyme A carboxylase (*ACCase*) is a critical enzyme in the production of lipids and a valuable target for genome editing with herbicides. Adenine base editing of the rice ACCase gene resulted in establishing a C2186R substitution, resulting in a mutant rice strain that is tolerant to haloxyfop-R-methyl [72]. Similarly, quizalofop-resistant wheat has been developed by adding an A1992V mutation into wheat ACCase [73]. Additionally, it has been demonstrated that altering *EPS* [96], *PPO* [97], *TubA2* [98], and *SF3B1* [99] mediates resistance to glyphosate, butafenacil, trifluralin, and herboxidiene (GEX1A). In addition to their numerous agricultural applications, these herbicide-resistant alleles can be used as selective markers to enhance gene editing processes [70]. CRISPR-based technology has been very effective in tackling stress conditions in plants and many literature surveys define the higher rate of success based on CRISPR genome editing. Xie and Yang [74] demonstrated targeted mutagenesis of the *Oryza sativa* gene (*OsMPK5*), which negatively regulates both biotic and abiotic stresses in rice.

Moreover, minimum low off-target efficiency was proved using specific guide RNAs and two appropriate vectors pRGE3 and pRGE6 [74]. Mitogen-activated protein kinases (MAPKs) are important signaling molecules that respond to drought stress in tomatoes by defending the cell membrane against oxidative damage and regulating the transcription of drought-stress-related genes. Slmapk3 mutants produced by CRISPR editing are fewer droughts tolerant and display increased wilting symptoms [75]. CRISPR/Cas9 genome editing has been used to manipulate the synthesis of secondary metabolites in plants. Zhang et al. recently showed multiplex CRISPR genome editing to enhance the isoflavone content of soybean and strengthen its resistance to soybean mosaic virus by altering the *GmF3H1*, *GmF3H2*, and *GmFNSII-1* genes [76].

Target gene identification

⇩

Bioinformatics tools for designing guide RNAs
and off target regions

⇩

Designing of guide RNAs

⇩

Cloning of guide RNA into desired expression vector

⇩

Plant transformation
1. *Agrobacterium*-mediated transformation
2. Floral dip transformation
3. Gene gun-mediated transformation
4. Protoplast transformation

⇩

Plant regeneration

⇩

Screening and selection of edited lines

⇩

1. Antibiotic selection
2. PCR screening
3. T7 endonuclease assay
4. DNA sequencing

⇩

Plants with desired traits

Flowchart showing the production of plants with the desired trait.

5. SAFETY REGULATIONS IN GENOME EDITING

Plant genome editing methods are critical for developing crops that can withstand biotic, abiotic, and extreme climatic changes and resolving global policy and governance issues. Apart from remarks on principle-based biotechnology adoption and the ethical, social, and biological issues raised by the CRISPR/Cas system, the current state of agriculture is examined. As a result of the technology's limitations, moral concerns about CRISPR have developed, necessitating intra and international attention to discover solutions that benefit the broader public interest.

However, there is a public debate regarding the direction and control of new methods' utility inside the industrialized world [100]. Governments in developing nations expect to give an alternative that promotes the alleviation of famine and poverty [101]. However, the uncertainty of agricultural biotechnology safety leads to suspicion of the regulatory framework and implies that the biological safety regulations [102] to preserve biological variety, including environmental protection and health safety, must be rejected. The unpredictability of agricultural biotechnology safety results in mistrust of the regulatory framework and implies that biosafety regulation should be rejected. Global hunger and poverty have been significantly reduced because of the green revolution. Malnutrition and the occurrence of certain chronic illnesses among otherwise healthy people have been exacerbated by inadequate amounts of specific amino acids, minerals, vitamins, and lipid acids in staple crops, as well as animal diets derived from them (the so-called diseases of civilization). The green revolution has made major contributions to the decrease of world hunger and poverty. Certain amino acids, minerals, vitamins, and fatty acids are deficient in staple crops and animal diets generated from them, contributing to malnutrition and the growing incidence of certain chronic illnesses in otherwise well-nourished people (the so-called diseases of civilization). Nevertheless, a recent trend demonstrates according to research conducted by the United Nations Food and Agricultural Organization (FAO), worldwide hunger and malnutrition have increased since 2014, with an estimated 821 million people lacking access to sufficient food in 2017 [103]. The advent of transgenic plants with insect-resistant, herbicide-tolerant, and nutritional characteristics has led to an explosion in the number of genetically modified crops grown on a wide scale. According to the latest report, 18 million farmers planted GMO crops on 181.5 million hectares in 28 countries in 2014, representing a 3–4% increase over the 2013 figures [104]. Recent marketed genetically modified crops include tomato, corn, soybean, cotton, canola, rice, potato, squash, melon, and papaya, with soybean, corn, and cotton being the most important because of their widespread cultivation critical role in the agricultural economy in many nations. America, Argentina, and Canada lead the world in producing and exporting genetically modified goods [105]. A good portion of the countries battling against GMOs in Africa are the same ones that struggled with them in Europe. Only local civil society organizations and a few NGOs have stood in the way of Uganda's effort to combat banana leaf wilt with its genetically modified bananas. All

GMO-related activities in India must be regulated by the Environmental Protection Act (EPA) [106], which was passed in 1986. The Ministry of Environment, Forests and Climate Change (MoEF&CC) is responsible for this regulation [107]. Indian genetically modified crops are subject to a multitiered regulatory system controlled by the Ministry for Environment and Forest and Department of Biotechnology, which is part of the Ministry of Science and Technology. Six competent authorities comprise this system—the Recombinant DNA Advisory Committee (RDAC), the Review Committee on Genetic Manipulation (RCGM), the Genetic Engineering Appraisal Committee (GEAC), the Institutional Biosafety Committees (IBSC), the State Biotechnology Coordination Committees (SBCC), and District Level Committees (DLC). All the various committees' tasks and responsibilities are defined in the Rules 1989 [108]. Genetic Engineering Appraisal Committee (GEAC) is a board within the Ministry of Environment, Forests and Climate Change that regulates the manufacture, import, export, and storage of hazardous microorganisms and genetically engineered organisms or cells under the Environment Protection Act 1986. Indian lawmakers have charged the Government Environmental Assessment Council (GEAC) with performing environmental evaluations of operations using GMOs and their products in research, industrial manufacturing, field application, and environmental discharge. For the production, environmental release, and marketing of GM crops, the Indian Parliament passed three essential laws. Included in these laws were the Environment Protection Act of 1986, which the Ministry of Environment manages, Forests and Climate Change (MoEF&CC), the Seeds Act of 1966, and the Seeds (Control) Order, which is administered by the Ministry of Agriculture [109]. It is generally recognized that transgenic crops offer significant advantages to society in solving concerns of food and nutrition security. Adding to the benefits of improved nutrition, herbicide tolerance, viral resistance, and tolerance to different environmental stressors, many fruits will last longer on the shelf, helping farmers in the marketplace. To accomplish its goal of food and nutrition security, India must continue its research into genetically modified crops. Despite the lack of concrete proof against the safety of genetically modified foods, the argument over whether or not they are safe will continue.

Even while few public sector intuitions share worries about genetically modified foods, it is surprising that the same is true in regards to genetically modified animals. Government of India–funded intuitions should follow the same general principles as the Indian government, demonstrating their value by cooperating with the government to battle poverty and hunger. The members of the Technical

Expert Committee constituted by the Supreme Court of India for Safety and Guidelines for genetically modified agricultural research debate the argument presented here, with some believing that it is significant and others seeing it as unimportant. However, while India has the inadequate infrastructure and lacks standards for genetically modified agricultural research and risk assessment, the initiative is of utmost importance given India's dire need. To prepare for future deregulation, India must continue researching genetically modified crops and construct basic infrastructural facilities while developing strict marketing and biosafety rules. Although portals such as the GEAC, the IGMORIS, and Biosafety Clearing-House play a role in biosafety assessment and regulations on GM plants, there is an urgent need to develop a single-window system and an online portal for the assessment, control, regulation, and approval of GM plants. The government should require every firm and public institution to register with this platform before starting any transgenic event or field testing, regardless of whether or not they are seeking clearance. To commercialize transgenic products, each new genetically modified organism (GMO) being developed must have a registration number, and registration date displayed online on a site or portal particularly intended for approval in any nation. This portal should include a publishing list that documents any genetically modified crop development activity so that any person interested in any genetically modified crop development event may access the entire information in one location. A gateway of this sort will be invaluable and broadly accessible for the public good by promoting the beneficial benefits of genetically modified food research, food safety, and the current state of genetically modified foods.

6. DEMONSTRATION FOR CRISPR-CHOPCHOP FOR SGRNA DESIGNING

In just 3 years, CRISPR genome editing dramatically changed biology, but also its acceptance and utility continue to expand. New CRISPR mechanisms and criteria for choosing ideal targets are being published all the time, highlighting the need for computerized CRISPR targeting tools to combine these guidelines to make target appropriate selection quicker. Among the most popular website software for genome editing with CRISPR and TALEN is CHOPCHOP (https://chopchop. cbu.uib. no/). It provides a user-friendly online environment for target selection, primer development, and restriction site identification, all based on the most recent large-scale investigations. In both protein-coding and noncoding genes, CHOPCHOP allows accurate localization of subsections, including coding regions, UTRs, splice sites, and particular exons. For all sgRNAs, the program detects probable off-target sites, generates primers for target sites automatically, and shows all elements in a variable graphical interface that incorporates restriction relevant details for subsequent confirmation [109].

1. Go to https://chopchop.cbu.uib.no/
2. Enter gene of interest in target field (for example *Arabidopsis thaliana* Leuco- anthocyanidin dioxygenase (LDOX), NM_001341563.1)
3. Choose the genome of interest in the In field.
4. Choose the CRISPR/CAS9 in the using field.
5. Choose the knockout in the field.
6. Press find the target site [110].

6.1 Results interpretations

- In that result window, the green color represents no off-targets, the orange color represents minimal off-targets, and the red color represents more off-targets.
- NOTE: For designing of suitable sgRNA, needs no off-targets.
- Click any green color symbol, which indicates the next level of target identification like percentage of GC content, off-target levels, and primer sequences.
- The violet color indicates primer pairs. Green and red boxes indicate the respective restriction sites (Ex. HindIII).
- Note: The sequence should include the PAM sequence.
- Any possibility of off-targets are listed in the below table with the location of the genome, possible off-target mismatches are represented in red.
- Note: Since there is no off-target it's mentioned as "there are no off targets".
- Once the target is fixed copy the target sequence with their respective primers from the web page.

7. CONCISE SUMMARY

Enhancing vegetables necessitates consideration of various elements, including increased production, insect resistance, abiotic stress tolerance, extended shelf life, processing quality, and improved nutritional value, among others. Traditional breeding makes it challenging and labor-intensive to attain the desired features. The CRISPR/Cas9 technology is a revolutionary genome editing method that can modify a plant's DNA, leading to many mutations. The introduction of the Cas9 gene into sgRNA-specific viral DNA has enabled the plant to develop virus resistance via CRISPR/Cas9 mechanisms [111]. A diverse array of viral-resistant plants was developed by incorporating many sgRNAs that target genomic regions of the entire virus into the plants. CRISPR/Cas9 can thus be utilized to augment the metabolic engineering of horticultural crops by supplying health-enhancing components.

REFERENCES

1. Slavin JL, Lloyd B. Health benefits of fruits and vegetables. Advances in Nutrition. 2012;**3**:506-516
2. Ashraf M. Inducing drought tolerance in plants: Recent advances. Biotechnology Advances. 2010;**28**:169-183
3. Tester M, Langridge P. Breeding technologies to increase cropproduction in a changing world. Science. 2010;**327**: 818-822
4. Carroll D. Genome engineering with zinc-finger nucleases. Genetics. 2011;**188**:773-782
5. Mahfouz MM, Li L, Shamimuzzaman M, Wibowo A, Fang X, Zhu JK. De novo-engineered transcription activator-like effector (TALE) hybridnuclease with novel DNA binding specificity creates double-strand breaks. Proceedings of the National Academy of Sciences of the United States of America. 2011;**108**:2623-2628
6. Li T, Liu B, Spalding MH, Weeks DP, Yang B. High-efficiency TALEN-based gene editing produces disease-resistant rice. Nature Biotechnology. 2012;**30**: 390-392
7. Cong L, Ran FA, Cox D, Lin S, Barretto R, Habib N, et al. Multiplex genome engineering using CRISPR/Cas systems. Science. 2013;**339**:819-823
8. Chen K, Gao C. Targeted genome modification technologies and their applications in crop improvements. Plant Cell Reports. 2014;**33**(4):575-583
9. Gao C. Genome editing in crops: From bench to field. National Science Review. 2015;**2**(1):13-15
10. Zhang Y, Massel K, Godwin ID, Gao C. Applications and potential of genome editing in crop improvement. Genome Biology. 2018;**19**(1):1-11
11. Gaj T, Gersbach CA, Barbas CF 3rd. ZFN, TALEN, and CRISPR/Cas-based methods for genome engineering. Trends in Biotechnology. 2013;**31**(7):397-405. DOI: 10.1016/j. tibtech.2013.04.004
12. Carroll D. Genome engineering with targetable nucleases. Annual Review of Biochemistry. 2014;**83**:409-439
13. Abdallah NA, Prakash CS, McHughen AG. Genome editing for crop improvement: Challenges and opportunities. GM Crops & Food. 2015;**6**(4):183-205
14. Jinek M, Chylinski K, Fonfara I, Hauer M, Doudna JA, Charpentier E. A programmable dual-RNA–guided DNA endonuclease in adaptive bacterial immunity. Science. 2012;**337**(6096): 816-821
15. Jaganathan D, Ramasamy K, Sellamuthu G, Jayabalan S, Venkataraman G. CRISPR for crop improvement: An update review. Frontiers in Plant Science. 2018;**9**:985
16. Voytas DF. Plant genome engineering with sequence-specific nucleases. Annual Review of Plant Biology. 2013;**64**:327-350

17. Čermák T, Baltes NJ, Čegan R, Zhang Y, Voytas DF. High-frequency, precise modification of the tomato genome. Genome Biology. 2015;**16**(1): 1-15
18. Gallagher RR, Li Z, Lewis AO, Isaacs FJ. Rapid editing and evolution of bacterial genomes using libraries of synthetic DNA. Nature Protocols. 2014;**9**(10):2301-2316
19. Daboussi F, Stoddard TJ, Zhang F. Engineering meganuclease for precise plant genome modification. In: Advances in New Technology for Targeted Modification of Plant Genomes. New York, NY: Springer; 2015. pp. 21-38
20. Zess E, Begemann M. CRISPR-Cas9 and beyond: What's next in plant genome engineering. In Vitro Cellular & Developmental Biology-Plant. 2021;**57**:584-594. DOI: 10.1007/s11627-021-10185-1
21. Chen K, Wang Y, Zhang R, Zhang H, Gao C. CRISPR/Cas genome editing and precision plant breeding in agriculture. Annual Review of Plant Biology. 2019;**70**:667-697
22. Kim YG, Cha J, Chandrasegaran S. Hybrid restriction enzymes: Zinc finger fusions to Fok I cleavage domain. Proceedings of the National Academy of Sciences. 1996;**93**(3):1156-1160
23. Lloyd A, Plaisier CL, Carroll D, Drews GN. Targeted mutagenesis using zinc-finger nucleases in *Arabidopsis*. Proceedings of the National Academy of Sciences. 2005;**102**(6):2232-2237
24. Wright DA, Townsend JA, Winfrey RJ Jr, Irwin PA, Rajagopal J, Lonosky PM, et al. High-frequency homologous recombination in plants mediated by zinc-finger nucleases. The Plant Journal. 2005;**44**(4):693-705
25. Boch J, Bonas U. Xanthomonas AvrBs3 family-type III effectors: Discovery and function. Annual Review of Phytopathology. 2010;**48**:419-436
26. Urnov FD, Rebar EJ, Holmes MC, Zhang HS, Gregory PD. Nature Reviews Genetics. 2010;**11**:636-646
27. Moscou MJ, Bogdanove AJ. A simple cipher governs DNA recognition by TAL effectors. Science. 2009;**326**(5959): 1501-1501
28. Sun N, Zhao H. Transcription activator-like effector nucleases (TALENs): A highly efficient and versatile tool for genome editing. Biotechnology and Bioengineering. 2013;**110**(7):1811-1821
29. Geizzler R, Scholze H, Hahn S, Streubel J, Bonas U, Behrens SE, et al. Transcriptional activators of human genes with programmable DNA- specificity. PLoS One. 2011;**6**(5):e19509
30. Zhang F, Cong L, Lodato S, Kosuri S, Church GM, Arlotta P. Efficient construction of sequence- specific TAL effectors for modulating mammalian transcription. Nature Biotechnology. 2011;**29**:149-153
31. Mahfouz MM, Li L, Piatek M, Fang X, Mansour H, Bangarusamy DK, et al. Targeted transcriptional repression using a chimeric TALE-SRDX repressor protein. Plant Molecular Biology. 2012;**78**:311-321

32. Sorek R, Lawrence CM, Wiedenheft B. CRISPR-mediated adaptive immune systems in bacteria and archaea. Annual Review of Biochemistry. 2013;**82**:237-266

33. Koonin EV, Makarova KS, Zhang F. Diversity, classification and evolution of CRISPR-Cas systems. Current Opinion in Microbiology. 2017;**37**:67-78

34. Shmakov S, Smargon A, Scott D, Cox D, Pyzocha N, Yan W, et al. Diversity and evolution of class 2 CRISPR–Cas systems. Nature Reviews Microbiology. 2017;**15**(3):169-182

35. Gasiunas G, Barrangou R, Horvath P, Siksnys V. Cas9–crRNA ribonucleoprotein complex mediates specific DNA cleavage for adaptive immunity in bacteria. Proceedings of the National Academy of Sciences. 2012;**109**(39):E2579-E2586

36. Andolfo G, Iovieno P, Frusciante L, Ercolano MR. Genome-editing technologies for enhancing plant disease resistance. Frontiers in Plant Science. 2016;**7**:1813

37. Murovec J, Pirc Ž, Yang B. New variants of CRISPR RNA-guided genome editing enzymes. Plant Biotechnology Journal. 2017;**15**:917-926

38. Makarova KS, Wolf YI, Alkhnbashi OS, et al. An updated evolutionary classification of CRISPR– Cas systems. Nature Reviews. Microbiology. 2015;**13**:722-736

39. Zetsche B, Gootenberg JS, Abudayyeh OO, et al. Cpf1 is a single RNA-guided endonuclease of a class 2 CRISPR-Cas system. Cell. 2015;**163**: 759-771

40. Shmakov S, Abudayyeh OO, Makarova KS, et al. Discovery and functional characterization of diverse class 2 CRISPR-Cas systems. Molecular Cell. 2015;**60**:385-397

41. Abudayyeh OO, Gootenberg JS, Konermann S, et al. C2c2 is a single- component programmable RNA-guided RNA-targeting CRISPR effector. Science. 2016;**353**:aaf5573

42. Rau K, Rentmeister A. CRISPR/ Cas9: A new tool for RNA imaging in live cells. Chembiochem. 2016;**17**: 1682-1684

43. Sampson TR, Saroj SD, Llewellyn AC, et al. A CRISPR/Cas system mediates bacterial innate immune evasion and virulence. Nature. 2013;**497**:254-257

44. Price AA, Sampson TR, Ratner HK, et al. Cas9-mediated targeting of viral RNA in eukaryotic cells. Proceedings of the National Academy of Sciences. 2015;**112**:6164-6169

45. Marton I, Zuker A, Shklarman E, et al. Nontransgenic genome modification in plant cells. Plant Physiology. 2010; **154**:1079-1087

46. Hou Z, Zhang Y, Propson NE, et al. Efficient genome engineering in human pluripotent stem cells using Cas9 from *Neisseria meningitidis*. Proceedings of the National Academy of Sciences. 2013;**110**:15644-15649

47. Ran FA, Hsu PD, Lin C-Y, et al. Double nicking by RNA-guided CRISPR Cas9 for enhanced genome editing specificity. Cell. 2013;**154**:1380-1389
48. Kleinstiver BP, Prew MS, Tsai SQ, et al. Engineered CRISPR-Cas9 nucleases with altered PAM specificities. Nature. 2015;**523**:481-485
49. Kleinstiver BP, Prew MS, Tsai SQ, et al. Broadening the targeting range of *Staphylococcus aureus* CRISPR-Cas9 by modifying PAM recognition. Nature Biotechnology. 2015;**33**:1293-1298
50. Endo A, Masafumi M, Kaya H, Toki S. Efficient targeted mutagenesis of rice and tobacco genomes using Cpf1 from *Francisella novicida.* Scientific Reports. 2016;**6**:38169
51. Jinek M, East A, Cheng A, et al. RNA-programmed genome editing in human cells. Elife. 2013;**2**:e00471
52. Fauser F, Schiml S, Puchta H. Both CRISPR/Cas-based nucleases and nickases can be used efficiently for genome engineering in *Arabidopsis thaliana.* The Plant Journal. 2014;**79**:348-359
53. Schiml S, Fauser F, Puchta H. The CRISPR/Cas system can be used as nuclease for in planta gene targeting and as paired nickases for directed mutagenesis in *Arabidopsis* resulting in heritable progeny. The Plant Journal. 2014;**80**:1139-1150
54. Slaymaker IM, Gao L, Zetsche B, et al. Rationally engineered Cas9 nucleases with improved specificity. Science. 2016;**351**:84-88
55. Wright AV, Sternberg SH, Taylor DW, et al. Rational design of a split-Cas9 enzyme complex. Proceedings of the National Academy of Sciences. 2015;**112**:2984-2989
56. Komor AC, Kim YB, Packer MS, et al. Programmable editing of a target base in genomic DNA without double- stranded DNA cleavage. Nature. 2016;**533**:420-424
57. Tsai SQ, Wyvekens N, Khayter C, et al. Dimeric CRISPR RNA-guided FokI nucleases for highly specific genome editing. Nature Biotechnology. 2014;**32**: 569-576
58. Piatek A, Ali Z, Baazim H, et al. RNA-guided transcriptional regulation in planta via synthetic dCas9-based transcription factors. Plant Biotechnology Journal. 2015;**13**:578-589
59. Liu XS, Wu H, Ji X, et al. Editing DNA methylation in the mammalian genome. Cell. 2016;**167**:233-247.e17
60. Lu K, Wu B, Wang J, et al. Blocking amino acid transporter Os AAP 3 improves grain yield by promoting outgrowth buds and increasing tiller number in rice. Plant Biotechnology Journal. 2018;**16**:1710-1722
61. Liu J, Chen J, Zheng X, et al. GW5 acts in the brassinosteroid signalling pathway to regulate grain width and weight in rice. Nature Plants. 2017;**3**:17043

62. Zhou J, Xin X, He Y, et al. Multiplex QTL editing of grain-related genes improves yield in elite rice varieties. Plant Cell Reports. 2019;**38**:475-485
63. Rodríguez-Leal D, Lemmon ZH, Man J, et al. Engineering quantitative trait variation for crop improvement by genome editing. Cell. 2017;**171**: 470-480. e8
64. Xu Z, Xu X, Gong Q, et al. Engineering broad-spectrum bacterial blight resistance by simultaneously disrupting variable TALE-binding elements of multiple susceptibility genes in rice. Molecular Plant. 2019;**12**: 1434-1446
65. Peng A, Chen S, Lei T, et al. Engineering canker-resistant plants through CRISPR/Cas9-targeted editing of the susceptibility gene CsLOB1 promoter in citrus. Plant Biotechnology Journal. 2017;**15**:1509-1519
66. Zhang Y, Bai Y, Wu G, et al. Simultaneous modification of three homoeologs of TaEDR1 by genome editing enhances powdery mildew resistance in wheat. The Plant Journal. 2017;**91**:714-724
67. Nekrasov V, Wang C, Win J, et al. Rapid generation of a transgene-free powdery mildew resistant tomato by genome deletion. Scientific Reports. 2017;**7**:482
68. Zhang T, Zheng Q, Yi X, et al. Establishing RNA virus resistance in plants by harnessing CRISPR immune system. Plant Biotechnology Journal. 2018;**16**:1415-1423
69. Mahas A, Aman R, Mahfouz M. CRISPR-Cas13d mediates robust RNA virus interference in plants. Genome Biology. 2019;**20**:263
70. Shimatani Z, Kashojiya S, Takayama M, et al. Targeted base editing in rice and tomato using a CRISPR-Cas9 cytidine deaminase fusion. Nature Biotechnology. 2017;**35**:441-443
71. Kuang Y, Li S, Ren B, et al. Base- editing-mediated artificial evolution of OsALS1 in planta to develop novel herbicide-tolerant rice germplasms. Molecular Plant. 2020;**13**:565-572
72. Li R, Li R, Li X, et al. Multiplexed CRISPR/Cas9-mediated metabolic engineering of γ-aminobutyric acid levels in *Solanum lycopersicum*. Plant Biotechnology Journal. 2018;**16**:415-427
73. Zhang R, Liu J, Chai Z, et al. Generation of herbicide tolerance traits and a new selectable marker in wheat using base editing. Nature Plants. 2019;**5**:480-485
74. Xie K, Yang Y. RNA-guided genome editing in plants using a CRISPR–Cas system. Molecular Plant. 2013;**6**: 1975-1983
75. Wang L, Chen L, Li R, et al. Reduced drought tolerance by CRISPR/ Cas9-mediated SlMAPK3 mutagenesis in tomato plants. Journal of Agricultural and Food Chemistry. 2017;**65**:8674-8682

76. Zhang P, Du H, Wang J, et al. Multiplex CRISPR/Cas9 mediated metabolic engineering increases soya bean isoflavone content and resistance to soya bean mosaic virus. Plant Biotechnology Journal. 2020;**18**:1384-1395
77. Brooks C, Nekrasov V, Lippman ZB, Van Eck J. Efficient gene editing in tomato in the first generation using the clustered regularly interspaced short palindromic repeats/CRISPR-associated 9 system. Plant Physiology. 2014;**166**: 1292-1297. DOI: 10.1104/pp.114.247577
78. Soyk S, Müller NA, Park SJ, et al. Variation in the flowering gene SELF PRUNING 5G promotes day-neutrality and early yield in tomato. Nature Genetics. 2017;**49**:162-168.DOI: 10.1038/ng.3733
79. Lawrenson T, Shorinola O, Stacey N, et al. Induction of targeted, heritable mutations in barley and *Brassica oleracea* using RNA-guided Cas9 nuclease. Genome Biology. 2015;**16**:258. DOI: 10.1186/s13059-015-0826-7
80. Woo JW, Kim J, Kwon S, et al. DNA-free genome editing in plants with preassembled CRISPR-Cas9 ribonucleoproteins. Nature Biotechnology. 2015;**33**:1162-1164. DOI: 10.1038/nbt.3389
81. Čermák T, Baltes NJ, Čegan R, et al. High-frequency, precise modification of the tomato genome. Genome Biology. 2015;**16**:232. DOI: 10.1186/s13059-015-0796-9
82. Pan C, Ye L, Qin L, et al. CRISPR/ Cas9-mediated efficient and heritable targeted mutagenesis in tomato plants in the first and later generations. Scientific Reports. 2016;**6**:24765. DOI: 10.1038/ srep24765
83. Filler Hayut S, Melamed Bessudo C, Levy AA. Targeted recombination between homologous chromosomes for precise breeding in tomato. Nature Communications. 2017;**8**:15605. DOI: 10.1038/ncomms15605
84. Klap C, Yeshayahou E, Bolger AM, et al. Tomato facultative parthenocarpy results from Sl AGAMOUS-LIKE 6 loss of function. Plant Biotechnology Journal. 2017;**15**:634-647. DOI: 10.1111/ pbi.12662
85. Mazier M, Flamain F, Nicolaï M, et al. Knock-down of both eIF4E1 and eIF4E2 genes confers broad-spectrum resistance against potyviruses in tomato. PLoS One. 2011;**6**:e29595. DOI: 10.1371/ journal.pone.0029595
86. Rodríguez-hernández AM, Gosalvez B, Sempere RN, et al. Melon RNA interference (RNAi) lines silenced for Cm-eIF4E show broad virus resistance. Molecular Plant Pathology. 2012;**13**:755-763. DOI: 10.1111/j.1364-3703.2012.00785.x
87. Andersson M, Turesson H, Nicolia A, et al. Efficient targeted multiallelic mutagenesis in tetraploid potato (*Solanum tuberosum*) by transient CRISPR-Cas9 expression in protoplasts. Plant Cell Reports. 2017;**36**:117-128.] DOI: 10.1007/s00299-016-2062-3
88. Wang H, Wu Y, Zhang Y, et al. CRISPR/Cas9-based mutagenesis of starch biosynthetic genes in sweet potato (*Ipomoea batatas*) for the improvement

of starch quality. International Journal of Molecular Sciences. 2019;**20**:4702. DOI: 10.3390/ijms20194702

89. Mishra R, Mohanty JN, Mahanty B, Joshi RK. A single transcript CRISPR/Cas9 mediated mutagenesis of CaERF28 confers anthracnose resistance in chilli pepper (*Capsicum annuum* L.). Planta. 2021;**254**:5. DOI: 10.1007/s00425-021-03660-x
90. Wang C, Wang G, Gao Y, et al. A cytokinin-activation enzyme-like gene improves grain yield under various field conditions in rice. Plant Molecular Biology. 2020;**102**:373-388
91. Zhang Z, Hua L, Gupta A, et al. Development of an Agrobacterium-delivered CRISPR/Cas9 system for wheat genome editing. Plant Biotechnology Journal. 2019;**17**:1623-1635
92. Yuste-Lisbona FJ, Fernández- Lozano A, Pineda B, et al. ENO regulates tomato fruit size through the floral meristem development network. Proceedings of the National Academy of Sciences. 2020;**117**:8187-8195
93. Gao H, Gadlage MJ, Lafitte HR, et al. Superior field performance of waxy corn engineered using CRISPR– Cas9. Nature Biotechnology. 2020; **38**:579-581
94. Sánchez-León S, Gil-Humanes J, Ozuna CV, et al. Low-gluten, nontransgenic wheat engineered with CRISPR/Cas9. Plant Biotechnology Journal. 2018;**16**(4):902-910
95. Oliva R, Ji C, Atienza-Grande G, et al. Broad-spectrum resistance to bacterial blight in rice using genome editing. Nature Biotechnology. 2019;**37**: 1344-1350
96. Hummel AW, Chauhan RD, Cermak T, et al. Allele exchange at the EPSPS locus confers glyphosate tolerance in cassava. Plant Biotechnology Journal. 2018;**16**:1275-1282
97. de Pater S, Klemann BJPM, Hooykaas PJJ. True gene-targeting events by CRISPR/Cas-induced DSB repair of the PPO locus with an ectopically integrated repair template. Scientific Reports. 2018;**8**:3338
98. Liu L, Kuang Y, Yan F, et al. Developing a novel artificial rice germplasm for dinitroaniline herbicide resistance by base editing of OsTubA2. Plant Biotechnology Journal. 2021;**19**:5-7
99. Butt H, Eid A, Momin AA, et al. CRISPR directed evolution of the spliceosome for resistance to splicing inhibitors. Genome Biology. 2019;**20**:73
100. Bartkowski B, Theesfeld I, Pirscher F, Timaeus J. Snipping around for food: Economic, ethical and policy implications of CRISPR/Cas genome editing. Geoforum. 2018;**96**:172-180
101. Azadi H, Ho P. Genetically modified and organiccrops in developingcountries: A review of options for foodsecurity. Biotechnology Advances. 2010;**28**(1):160-168. DOI: 10.1016/j.biotechadv.2009.11.003

102. Ames GCW. Bananas, beef, and biotechnology: Three contentious US-EU trade disputes. Applied EconomicPerspectives and Policy. 2001;**23**(1):214-222. DOI: 10.1111/ 1058-7195.00056
103. Food and Agricultural Organization. The State of Food Security and Nutrition in the World. 2018. Available from: http://www.fao. org/state-of-food-security-nutrition/en/
104. Lucht JM. Public acceptance of plant biotechnologyand GM crops. Viruses. 2015;**7**(8):4254-4281
105. James C. Global status of commercialized biotech and gm crops: 2010. In: International Service for the Acquisition of Agribiotech Applications. Ithaca, NY: ISAAA; 2010
106. MoEF and BCIL New Delhi. Regulatory Framework for Genetically Engineered (GE) Plants in India. 2015. Available from: http://www.geacindia.gov.in/resource-documents/13_2- Regulatory_Framework_for_GE_Plants_in_India.pdf
107. Mo EF& CC. Genetic Engineering Approval Committee. 2016. Available from: www.moef.nic.in/division/ genetic-engineering-approval- committee-ge-ac
108. Choudhary B, Gheysen G, Buysse J, Meer P, Burssens S. Regulatory options for genetically modified crops in India. Plant Biotechnology Journal. 2014;**12**: 135-146
109. Labun K, Montague TG, Gagnon JA, Thyme SB, Valen E. CHOPCHOP v2: A web tool for the next generation of CRISPR genome engineering. Nucleic Acids Research. 2016;**44**(W1): W272-W276
110. Labun K, Krause M, Torres Cleuren Y, Valen E. CRISPR genome editing made easy through the CHOPCHOP website. Current Protocols. 2021;**1**(4):e46
111. Ali Z, Abulfaraj A, Idris A, Ali S, Tashkandi M, Mahfouz MM. CRISPR/Cas9-mediated viral interference in plants. Genome Biology. 2015;**16**(1):1-11

Index

www.ingramcontent.com/pod-product-compliance
Lightning Source LLC
LaVergne TN
LVHW081938200926
848484LV00009B/2466